Materials Science and Technologies

nova
science publishers
www.novapublishers.com

Materials Science and Technologies

High-Performance Calcium-Carbonate Concrete
Natt Makul
2022. ISBN: 978-1-68507-412-8 (eBook)

Analytical Models of Interstitial-Atom-Induced Stresses in Isotropic Metallic Materials
Ladislav Ceniga (Editor)
2022. ISBN: 978-1-68507-429-6 (eBook)

Analytical Models of Coherent-Interface-Induced Stresses in Composite Materials II
Ladislav Ceniga
2021. ISBN: 978-1-68507-003-8 (eBook)

Analytical Models of Coherent-Interface-Induced Stresses in Composite Materials III
Ladislav Ceniga
2021. ISBN: 978-1-53619-996-3 (eBook)

Microgels: Synthesis, Properties and Applications
Wing-Fu Lai (Editor)
2018. ISBN: 978-1-53613-522-0 (Hardcover)
2018. ISBN: 978-1-53613-523-7 (eBook)

Zirconia and Alumina Bioceramic Biocompatibility
Giulio Maccauro, Pierfrancesco Rossi Iommetti, Luca Raffaelli and Paolo Francesco Manicone (Editors)
2012. ISBN: 978-1-62081-650-9 (Online Book)

More information about this series can be found at https://novapublishers.com/product-category/series/materials-science-and-technologies/

Kalpana Awasthi, Arti Srivastava and Mridula Tripathi
Editors

Materials Science

Future Aspects

nova
science publishers

www.novapublishers.com

DOI: https://doi.org/10.52305/RRTP1482

Library of Congress Cataloging-in-Publication Data

ISBN: 978-1-68507-843-0

Published by Nova Science Publishers, Inc. † New York

Dedicated to Professor Onkar Nath Srivastava (1942-2021)

Prof. Onkar Nath Srivastava was an Indian material physicist, an Emeritus professor of Banaras Hindu University (BHU) and the vice president for India and South Asia of the International Association for Hydrogen Energy, who is known for his contributions to the disciplines of nanotechnology and hydrogen energy. He is the author of two books and over 440 scientific papers and a recipient of several honors including Shanti Swarup Bhatnagar award, the highest Indian award in the science and technology categories. The Government of India awarded him the fourth highest civilian honour of the Padma Shri, in 2016, for his contributions to science and engineering. Prof. Srivastava was born on the last day of the year 1942 in Varanasi, Uttar Pradesh, India. He secured his master's degree in Physics (M.Sc.) in 1961 from BHU and followed it up with a doctoral degree (Ph.D.) from the same institution, under the guidance of renowned physicist Ajit Ram Verma, in 1966. After doing his post doctoral research at Cornell University, USA, he returned to India to start his career as a lecturer at Banaras Hindu University where he served in different capacities as that of a reader, professor, head of the department of physics and the coordinator of the *Centre of Advance*

Studies-Hydrogen Energy Centre. After his superannuation from service, he continues his association with the university as a professor emeritus and as an associate faculty member of *Condensed Matter Experiment* research program of the university.

He has undertaken several projects for various government agencies; Nanoscience and Technology of the Department of Science and Technology (2005–2010), Support to Hydrogen Energy Centre (2007–2012), Development & Demonstration of Hydrogen Catalytic Combustion Cookers (2007–2010), Development & Demonstration of Hydrogen Fueled three wheelers (2009–2012), Mission Mode Project on Hydrogen Storage Materials (Hydride) (2009–2014), all of the Ministry of New and Renewable Energy and Synthesis Characterization and Properties of Single Walled Carbon Nanotubes (2009–2012) of the Defence Research and Development Organization (DRDO) are some of the notable ones. He has mentored 57 doctoral students. He is also an elected fellow of the Asia Pacific Academy of Materials, International Academy of Physical Sciences and the New York Academy of Sciences.

Contents

Preface

Nanomaterials are materials with individual units ranging in size from 1 to 1000 nanometers. Nanomaterials research necessitates collaboration between materials science and nanotechnology, improved materials metrology, and synthesis. This book is made up of chapters that cover the most important aspects of nanotechnology. The majority of the subjects covered in this book are related to nanomaterial properties, synthesis, procedures, and applications. This edited book provides a comprehensive overview of the current state of this vital area.

The different materials and their presentations in energy, ceramics, alloys, catalysis, membrane, pollution, and biomedical are covered in this edited book. It addresses a range of aspects because these materials' structure can be tailored at extremely small scales to achieve specific properties, thus greatly expanding the materials science toolkit. It explores several applications that could potentially be used to improve the environment and to more efficiently and cost-effectively produce energy, producing solar cells that generate electricity at a competitive so the book offers a valuable asset for a broad readership, including professionals, students, and researchers from materials science/engineering, polymer science, composite technology, nanotechnology, and biotechnology whose work involves various types of nanomaterials. Increased coverage of important background science makes this a valuable self-contained text, and extensive expanded referencing engages readers with the newest research and industrial advancements in the subject. This edited book contains twelve chapters of valuable studies from recent years.

In Chapter 1, *Sunita Kumari Pandey* and *Thakur Prasad Yadav* have given idea about hydrogen as alternative energy carrier. The chapter started with the explanation of different sources of energy then silent features of hydrogen as alternative energy resource along with its production, storage and transmission in different field.

In Chapter 2, *Kuldeep Panwar* and *Sumita Srivastava* have presents the theoretical model on correlation of all the effects produced by ball-milling parameters on the microstructure of the alloy and their impact on the hydrogenation characteristics.

In Chapter 3, *Satish Kumar Verma* and *Thakur Prasad Yadav* have discussed different properties, synthesis processes and applications of the MgH_2 as hydrogen storage material. They have given details of different methods for preparation of the MgH_2 and various techniques for making MgH_2 a viable storage material.

In Chapter 4, *Vivek Shukla* and *Thakur Prasad Yadav* have discussed the current scenario of green hydrogen energy.

Amit K Gupta, Aashish Prakash and *Rohit R Shahi* have focused on electrochemical properties of High entropy type materials such as High Entropy Alloys (HEAs) and High Entropy Oxide which are used as emergent electrode materials for Battery and Super capacitor in Chapter 5.

Subhash, Pinki and *Ashu Chaudhary* have given details of properties and applications of high entropy ceramic materials like High-entropy nitride, carbide, boride and oxide thin films in Chapter 6

The historical research and development of Heusler alloys have been summarized by *S. S. Mishra* and *T. P. Yadav* in the book Chapter 7

In Chapter 8, *C. Ravi Prakash Patel* and *Thakur Prasad Yadav* have given the details about one-dimensional carbon nanomaterial in respect of its properties, methods of synthesis and applications in diverse field.

In Chapter 9, *Chandani Singh, Arti Srivastava, Bhaskar Sharma* and *S. K. Singh* have given details of outstanding performances of Graphene as catalyst in various fields by explaining the catalysis process of many known reactions like Michael addition, condensation, cyclocondensation, multicomponent reaction, and trans esterification with good yield and excellent recyclability in an eco-friendly manner.

The Chapter 10 focuses on the recent development of electrochemical sensors using graphene-based materials for heavy metal ions detection.

In the Chapter 11, the author *Goutam Kumar Patra* and *et al.* have reported crystal engineering pathways for developing binary pyridyl and ternary pyridyl and triphenylphosphine based supramolecular polymeric complexes of Cu (I).

Isha Gupta presented a review on preparation methods and applications of porous EVAL based asymmetric membranes in Chapter 12.

Acknowledgments

We would like to dedicate this edited book in the memory of respected teacher and PhD supervisor (Prof. O. N. Srivastava) who passed away due to COVID-19 ailment. KA thanks the Higher Education Department, Uttar Pradesh, India for providing financial support under the research development scheme.

Chapter 1

Hydrogen: Ultimate Alternative Energy Carrier

Sunita Kumari Pandey and Thakur Prasad Yadav*

Hydrogen Energy Centre, Department of Physics, Institute of Science,
Banaras Hindu University, Varanasi, India

Abstract

Hydrocarbon fuel is increasing severe environmental problems such as global warming and climate change therefore significant awareness is paid to a new energy supply for a sustainable and cleaner environment. To restrict the climate change effect, efforts are being made through reducing CO_2 emission by searching carbon-free energy systems globally. Different renewable energies such as solar, wind, tidal, biomass, sea waves and hydrogen energy, etc. eco-friendly energy carriers are being explored in this perspective. The hydrogen is the only zero-carbon energy carrier that can be produced from water. However, hydrogen storage is considered to be the crucial component of the hydrogen economy, cutting across production, distribution, and applications. It is generally believed that storage in the form of hydrides is the optimum storage mode with the highest hydrogen density. This storage mode operates at low pressure and temperature, leading to safe operation for energy carriers. This book chapter discusses the available scenario of alternative energy sources and hydrogen energy in detail.

Keywords: hydrogen energy, hydrogen economy, hydrogen storage, activation energy

---

*Corresponding Author's Email: yadavtp@gmail.com.

In: Materials Science: Future Aspects
Editors: Kalpana Awasthi, Arti Srivastava and Mridula Tripathi
ISBN: 978-1-68507-843-0

1. Introduction

Energy is an essential requirement globally, and the more developed a nation, the more energy it needs. It has been anticipated that energy is directly related to the currency of any country [1]. The total primary energy production has increased from 35.54 Quadrillions Btu (Btu is short term of British thermal unit, which is a traditional unit of energy and 1 Btu = 1.06kJ) in 1950 to 87.60 Quadrillions Btu in 2017. On the other hand, the consumption has far increased from 34.61 Quadrillions Btu in 1950 to 97.72 Quadrillions Btu in 2017. From this profile of production and consumption, it is estimated that this gap between demand and supply will increase further in the future years (EIA). The total primary energy overview from the year 2050 to 2017 is shown in Table 1.

This increase is the result of growth in the world population and a general rise in prosperity. Due to then ever-ending and increasing energy demands throughout the world, it has become a matter of concern that the availability of fossil fuels (mainly oil) which is depleting at an alarming rate, would last for about 40 years [2-3]. It is a fact that fossil fuels have played a vital role in the advancements of life during the industrial era, which no one can deny. Still, it is also true that fossil fuels have generated discouraging environmental concerns. The studies reveal that mainly human activities such as the combustion of fossil fuels and deforestation have led to an increase in the atmospheric carbon dioxide concentration by about 35% since the beginning of industrialization [4].

Table 1. The total primary energy overview (1950-2017)

Total Primary Energy Overview Unit: Quadrillion Btu								
Year	Production			Total	Consumption			Total
	Fossil Fuels	Nuclear Electric Power	Renewable Energy		Fossil Fuels	Nuclear Electric Power	Renewable Energy	
1950	32.563	0.000	2.978	35.540	31.632	0.000	2.978	34.616
1975	54.733	1.900	4.687	61.320	65.357	1.900	4.687	71.965
2000	57.366	7.862	6.102	71.330	84.735	7.862	6.104	98.817
2010	58.216	8.434	8.212	74.863	80.891	8.434	8.166	97.580
2015	70.213	8.337	9.650	88.200	79.328	8.337	9.634	97.526
2017	68.052	8.419	11.137	87.607	78.120	8.419	11.016	97.728

Fossil fuels are the main contributor to global warming, and therefore, they have a harmful impact not only on human health but also on the environment. The Paris Agreement related to climate change was held in Paris from 30th November to 13th Dec. and it was adopted on 12th Dec. 2015. This convention asserted that a substantial fraction of reserves of coal, oil, and gas will have to remain underground and unburned (Stranded Assets) to put a cap on CO2 emission, which is the main cause of global warming, keep the temperature rise within 2°C limit. There are so many reasons like global warming, air pollution, and the gap between demand and supply, which leads us to think and switch over to renewable sources of energy [5-10].

2. Different Sources of Energy

To meet the gap between demand and supply, there are ten different sources of energy that are currently in use, and an overview of each of the various sources of energy are explained briefly as follows:

- **Solar Energy**: Solar power harvests the energy of the sun through collector panels to create conditions that can then be turned into a kind of power. Large solar panel fields are often used in the desert to gather enough electrical power to charge small substations. Many homes use solar systems to carry out cooking, cooling and supplement their electricity.
- *Disadvantage*: The issue with solar is that while there are plentiful amounts of sun available, only specific geographical ranges of the world get enough of the direct power of the sun for long enough to generate usable energy from this source.
- **Wind Energy**: Wind power is becoming more and more common, and the innovations that allow wind farms to appear are making them a more common sight. Using large turbines to take available wind as the power to turn, the turbine can then turn a generator to produce electricity.
- *Disadvantage*: While this seemed like an ideal solution to many, the reality of the wind farms is starting to reveal an unforeseen ecological impact that may not make it a perfect choice.
- **Geothermal Energy**: Geothermal energy is the energy that is produced from beneath the Earth. It is clean, sustainable, and

environment-friendly. High temperatures are produced continuously inside the 'Earth's crust by the slow decay of radioactive particles. Hot rocks below the Earth heat the water that produces steam. The steam is then captured which helps to move turbines. The rotating turbines then power the generators.

- *Disadvantage*: The most significant disadvantage of geothermal energy is that it can only be produced at selected sites throughout the world. The largest group of geothermal power plants globally is located at The Geysers, a geothermal field in California, United States.
- **Hydrogen Energy**: Hydrogen is available in water (H_2O) and is the most common element available on Earth. Water contains two-thirds of hydrogen and can be found in combination with other elements. Once separated, it can be used as a fuel for generating electricity. Hydrogen is a tremendous source of energy and can be used as a source of fuel to power ships, vehicles, homes, industries, and rockets. It is entirely renewable, can be produced on demand, and does not leave any toxic emissions in the atmosphere.
- *Disadvantage*: Although hydrogen has higher energy content about 3 times than fossil fuel like oil, it has less energy content on a volumetric basis because of its low density. The storage of hydrogen issue is not solved. Hydrides that are being vigorously researched are considered efficient and safe storage mode. However, no hydrides having all the required properties have been found yet.
- **Tidal Energy**: Tidal energy uses the rise and fall of tides to convert the kinetic energy of incoming and outgoing tides into electrical energy. The generation of energy through tidal power is primarily prevalent in coastal areas.
- *Disadvantage*: Huge investment and limited availability of sites are a few of the drawbacks of tidal energy. When there is increased height of water levels in the ocean, tides are produced, which rush back and forth in the sea. Tidal energy is a renewable source of energy and produces large energy even when the tides are at low speed.
- **Wave Energy**: Wave energy is produced from the waves that are produced in the oceans. Wave energy is renewable, environment friendly and causes no harm to the atmosphere. It can be harnessed along coastal regions of many countries and can help a country reduce its dependence on foreign countries for fuel.

- *Disadvantage*: Producing wave energy can damage the marine ecosystem and be a source of disturbance to private and commercial vessels. It is highly dependent on wavelength and can also be a visual and noise pollution source.
- **Hydroelectric Energy**: Many people are unaware that besides coal, hydropower (power by water) can produce electricity by turning the generators which are then used.
- *Disadvantages*: The problems faced with hydropower right now have to do with the aging of the dams. Many of them need major restoration work to remain functional and safe, which costs enormous sums of money. In addition, the drain on the 'world's drinkable water supply is also causing issues as townships may wind up needing to consume the water that provides them power.
- **Biomass Energy**: It is produced from organic material and is commonly used worldwide. Chlorophyll present in plants captures the 'sun's energy by converting carbon dioxide from the air and water from the ground into carbohydrates through photosynthesis. When the plants are burned, the water and carbon dioxide are again released back into the atmosphere. It generally includes crops, plants, trees, yard clippings, wood chips, and animal wastes. Biomass energy is used for heating and cooking in homes and fuels industrial production.
- *Disadvantage*: This type of energy produces a large amount of carbon dioxide into the atmosphere.
- **Nuclear Energy**: While nuclear power remains a great subject of debate as to how safe it is to use and whether or not it is energy efficient when you take into account the waste it produces – the fact is it remains one of the significant non-renewable sources of energy available to the world. The energy is created through a specific nuclear reaction, collected, and used as power generator.
- *Disadvantage*: While almost every country has nuclear generators, there are moratoriums on their use or construction as scientists try to resolve safety and disposal issues for waste.
- **Fossil Fuels** (Coal, oil, and natural gas): Fossil fuels provide power for most of the world, primarily using coal and oil. Oil is converted into many products, the most used of which is gasoline. Natural gas is starting to become more common but is used mainly for heating

applications although more and more natural gas-powered vehicles are appearing on the streets.

- *Disadvantage*: The issue with fossil fuels is twofold. To get to the fossil fuel and convert it to use, there has to be heavy destruction and pollution of the environment. The fossil fuel reserves are also limited; expected to last only another 100 years given is a basic rate of consumption.

3. Ultimate Alternative Energy Carrier

Out of all the available energy sources described above, hydrogen is considered an alternative energy carrier to resolve the problems as it offers the long-term prospect of plentiful clean energy supplies. In terms of mobile applications, hydrogen is seen as a promising candidate [11-18]. It is clean (pollution-free), renewable, and environmentally friendly, unlike petroleum fuels. The overwhelming majority of hydrogen is chemically bound as H_2O in water, and some are bound to liquid or gaseous hydrocarbons. The clean way to produce hydrogen from water is to use sunlight with photovoltaic cells and water electrolysis [19] (Figure 1). It is a sustainable solution for reducing global fossil fuel consumption and combating global warming. Cold combustion of hydrogen in fuel cells leads to the creation of electrical power. Hot combustion in motor vehicles' internal combustion (IC) engines provides power in the same way fossil fuels do [20]. Both the cold and hot combustion processes lead to the emission of water. The main advantage of hydrogen as a fuel is the absence of CO_2 emissions [21].

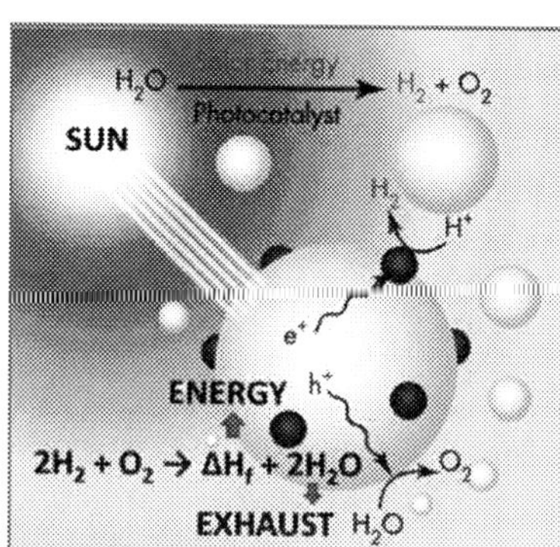

Figure 1. Cyclic process of the hydrogen energy.

4. Hydrogen

4.1. Characteristics and Properties

In hydrogen, the interaction between molecules is weak compared to other gases therefore the critical temperature is low (Tc = 33.0 K). The hydrogen atom is the lightest gas, even lighter than air, with its most common isotope consisting of only one proton and one electron. Hydrogen atoms are not found by themselves on Earth; instead, hydrogen is always combined with other elements in chemical compounds. Hydrogen atoms readily form H_2 molecules, which are smaller in size when compared to most other molecules. Hydrogen compounds are all around us in the natural world and they are found in coal, oil, natural gas, and animal and plant materials. The molecular form simply referred to as hydrogen, is colorless, odorless, and tasteless. Its density is about 14.4 times lighter than air (density of hydrogen at 1 atm is 0.0000899 g/cm^3) diffuses faster than any other gas. The physical properties of hydrogen are summarized in Table 2.

Hydrogen can be considered an ideal gas over a wide temperature range and even at high pressure [23]. At standard temperature and pressure conditions, it is a colorless, odorless, tasteless, non-toxic, non-corrosive, non-metallic diatomic gas, which is in principle physiologically not dangerous. One of its most important characteristics is its low density, which makes it difficult to store in small volumes.

It is buoyant above the temperature of 22 K, i.e., over (almost) the whole temperature range of its gaseous state. Hydrogen gas is highly diffusive; hence it rapidly mixes with the ambient air upon release. There is a wide flammability range of hydrogen (at room temperature) between 4 and 75 Vol.% of the concentration in air and up to 95 Vol.% in oxygen. The burning velocity of hydrogen in air at stoichiometric ambient conditions is 2.55 m/s, reaching a maximum of 3.20 m/s at a concentration of 40.1%, which would even increase to 11.75 m/s in pure oxygen. Compared to hydrocarbon fuels like methane and gasoline, hydrogen provides more energy upon combustion with zero carbon emission. Some of the significant properties of hydrogen are compared with methane and gasoline fuels in Table 3.

Table 2. Physical properties of hydrogen [22]

	Property	Values and Units
	Molar mass	2.016 kg $kmol^{-1}$
	Particular gas constant	4124 J Kg^{-1} K^{-1}
	Calorific value (gravimetric)	120 MJ kg^{-1} = 33.33 $kWhkg^{-1}$
At triple point	Temperature	-259.35°C (13.80 K)
	Pressure	0.07 bar
	Density gaseous	0.125 kg m^{-3}
	Density liquid	77 kg m^{-3}
	Heat of fusion	58.5 kJ kg^{-1} = 16.25 $kWhkg^{-1}$
At boiling point	Boiling temperature	-252.85°C (20.30 K)
At 1.01325 bar	Heat of vaporization	445.4 kJ kg^{-1}= 123.7 $kWhkg^{-1}$
Liquid phase	Density	70.8 kg m^{-3}
	Calorific value (volumetric)	8.5 MJ dm^{-3} = 2.36 $kWhkg^{-1}$
	Specific heat capacity C_p	9.8 kJ $kg^{-1}K^{-1}$
	Specific heat capacity C_v	5.8 kJ $kg^{-1}K^{-1}$
	Thermal conductivity	0.099 $Wm^{-1}K^{-1}$
	Dynamic viscosity	11.9×10^{-6} N s m^{-2}
	Speed of sound	1089 m s^{-1}
Gaseous phase	Density	1.34 kg m^{-3}
	Calorific value (volumetric)	0.16 MJ dm^{-3} = 0.044 $kWhdm^{-3}$
	Specific heat capacity C_p	12.2 kJ $kg^{-1}K^{-1}$
	Specific heat capacity C_v	6.6 kJ $kg^{-1}K^{-1}$
	Thermal conductivity	0.017 $Wm^{-1}K^{-1}$
	Dynamic viscosity	1.11×10^{-6} N s m^{-2}
	Speed of sound	355 m s^{-1}
At critical point	Temperature	-239.95°C (33.20 K)
	Pressure	13.1 bar
	Density	31.4 kg m^{-3}
At std. conditions (0°C and 1.01325 bar)	Density	0.09 kg m^{-3}
	Calorific value (volumetric)	0.01 MJ dm^{-3} = 2.8 $Whdm^{-3}$
	Specific heat capacity C_p	14.32 kJ $kg^{-1}K^{-1}$
	Specific heat capacity C_v	10.17 kJ $kg^{-1}K^{-1}$
	Thermal conductivity	0.184 $Wm^{-1}K^{-1}$
	Coefficient of diffusion	0.61 cm^2s^{-1}
	Dynamic viscosity	8.91×10^{-6} N s m^{-2}
	Speed of sound	1246 m s^{-1}
Mixtures with air	Lower explosion limit	4 vol% H_2 (λ=10.1)
	Lower detonation limit	18 vol% H_2 (λ=1.9)
	Stoichiometric mixture	29.6 vol% H_2 (λ=1)
	Upper detonation limit	58.9 vol% H_2 (λ=0.29)
	Upper explosion limit	75.6 vol% H_2 (λ=0.13)
	Ignition temperature	585°C (858 K)
	Minimal ignition energy	0.017 mJ
	Maximum Laminar flame speed	~ 3 m s^{-1}
	Adiabatic combustion temperature	~ 2100°C

Table 3. Comparison of some important physical properties of fuels such as hydrogen, methane and gasoline

	Unit	Hydrogen (H_2)	Methane (CH_4)	Gasoline (C_8H_{18})
Lower heating value	kWh/kg	33.33	13.90	12.40
Self-ignition temperature	°C	585	540	228-501
Flame temperature	°C	2045	1875	2200
Ignition limits in air	Vol %	4-75	5.30-150	1.00-7.60
Min. ignition energy	mWs	0.02	0.29	0.24
Flame propagation in air	m/s	2.65	0.40	0.40
Detonation limits	Vol %	13-65	6.30-13.50	1.10-3.30
Detonation velocity	Km/s	1.48-2.15	1.39-1.64	1.40-1.70
Explosion energy	Vol %	13-65	6.3-13.5	1.1-3.3

4.2. Economy

To meet the never-ending energy demand of the world, to stabilize and eventually lower harmful gases such as CO_2 emissions, to encourage more energy conservation/efficient lifestyles, and to develop more renewable energy technologies to overtake the dirty fossil sources of energy; new alternative energy schemes which mainly impart hydrogen as a future energy carrier were given much consideration. In order to resolve the above-said problems, the concept "hydrogen economy" was coined by John Bockris during a talk he gave in 1970 at General Motors (GM) Technical Center.

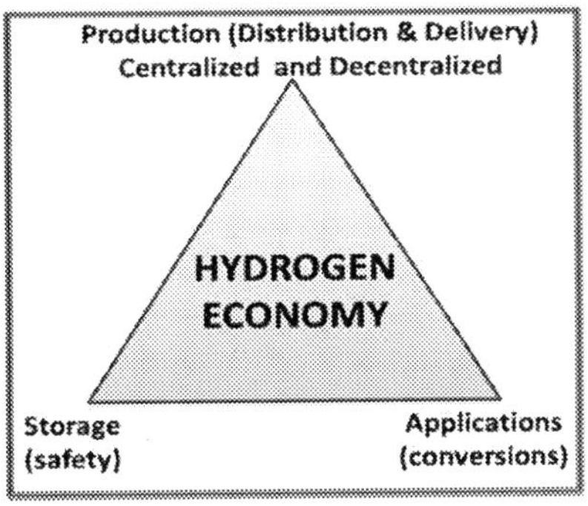

Figure 2. Hydrogen economy depends on H_2 production, H_2 storage, H_2 application and its cyclic process.

The term "Hydrogen Economy" "ic" refers to the infrastructure to support society's energy requirements, based on the use of hydrogen rather than fossil fuels [24-25]. Hydrogen economy argues that hydrogen can be an environmentally cleaner energy source for end-users, particularly in transportation applications, without releasing pollutants and carbon dioxide at the point of end-use. A technological breakthrough is required to bring hydrogen a cost-effective and easily storable fuel for the success of the hydrogen economy [26]. It consists of essential three ingredients: (i) H_2 Production, (ii) H_2 storage, and (iii) H_2 application, and its cyclic process is shown in Figure 2.

4.3. Production

Unfortunately, pure hydrogen is not widely available on our planet, and to obtain hydrogen from natural compounds, energy expenditure is required. Most of it is locked in water or hydrocarbon fuels. It can be produced using other high-energy fuels, i.e., fossil fuels, but such methods require fossil fuels and generate CO_2 to a greater extent than conventional engines and thus contribute to global warming more than if those fossil fuels were to be used directly to power automobiles for example. It can also be produced using vast amounts of energy and water. Nuclear power can provide energy, but it has well-known disadvantages. Some 'green' energy sources are capable of generating energy in a cost-effective way if the externalities of conventional energy sources are factored in, but the policies of the world's major governments do not factor them in. However, most 'green' energy sources produce low-intensity energy rather than the significant amounts of energy required for extracting significant amounts of hydrogen using thermochemical electrolysis, for example [9-10]. This is called the production problem. The clean way to produce hydrogen from water is to use sunlight with photovoltaic cells and water electrolysis. Temperatures above 2,000°C are required for direct thermal dissociation of H2O (>900°C with a Pt/Ru catalyst). Water electrolysis is a well-established technology that is used today to produce high purity hydrogen. At room temperature and pressure, electrolysis necessitates a minimum voltage of 1.481V and, as a result, a minimum energy of 39.4 $kWhkg^{-1}$ hydrogen. Today, electrolyzer systems consume approximately 47 $kWhkg^{-1}$ hydrogen, with an efficiency of approximately 82%. To meet the world's demand for fossil fuels, more than 31012 kg of hydrogen must be produced per year, which is 100 times the current hydrogen production.

4.4. Storage

After hydrogen production, it can be stored as a compressed gas, as a liquid, in a chemical compound (*e.g.,* chemical hydrides or metal hydrides), or physically held within nanoporous structures [27-30]. A major element of the cost of most of these storage modes (and a major consideration in terms of their energy efficiency) is the energy required to get the hydrogen in and out of storage. Table 4 shows the cost of a number of means of storage, including liquefaction, gas compression above ground and underground, and chemical and metal hydrides. In each case, the hydrogen storage method's cost depends on various factors like the cost of the requisite energy to get the hydrogen into the required form for storage, the scale (and throughput). It sometimes depends on the storage medium that is envisaged.

Table 4. Costs for various hydrogen storage technologies [32]

Technology	Cost range, US$(2000) per kg H_2	Comments
Liquefaction (>45 kg/h)	1–1.5	Cost highly dependent on scale, efficiency, cost of electricity
Compressed gas (<1 week)	0.15–0.6	For stand-alone (i.e., not on-board) storage only. Strong economies of scale
Bulk underground	0.12–0.3	Costs rise with increased storage time/reduced throughput
Chemical hydrides	1.5–2.5c	Large economies of scale, figures apply to 3.6kt–9 Mt H_2;on-board storage
Metal hydrides	0.4–4.5	For storage times of 1–14 days; on-board storage
Methanol	n/a	Cost not calculated; methanol can be produced from other sources than H_2

Source: Hawkins, 2006.

4.5. Transmission, Distribution (T&D) and Application

In addition to being stored with the associated cost, hydrogen must often be transmitted or distributed to its place of use. Clearly, the means of T&D is closely related to the standards of storage. Pipelines distribute compressed gas, and an extensive pipeline network stores a sizeable quantity of gas (the amount stored can be increased by increasing the pipeline pressure). At higher flow rates and distances, pipelines are the cheapest. At higher distances and lower flow-rates, liquid transport is most affordable, with tube trucks being the cheapest at common distances and flow rates.

Hydrogen production, large-scale storage facilities, and T&D equipment are all part of what is sometimes termed hydrogen 'infrastructure' 'infrastructure.' It can be seen that they can all be essential components of the cost of delivering hydrogen to end-users. Another critical component of such infrastructure is the network of refueling stations that will be required if hydrogen is to be widely used as a motor fuel [31]. Application: Hydrogen is a wonderful source of energy and can be used as a source of fuel to power ships, vehicles, homes, industries, and rockets. Following is the hydrogen timeline in brief, which led to 'today's application:

- 1766: British scientist Henry Cavendish identified hydrogen as a distinct element.
- 1783: The first hydrogen balloon is launched in France.
- 1800: British scientists William Nicholson and Sir Anthony Carlisle learn that applying an electric current to water produces oxygen and hydrogen gases.
- 1838: Swiss-German chemist Christian Friedrich Schönbein discovers that hydrogen and oxygen gases can be combined in a way to produce water and electricity.
- 1843: British scientist Sir William Grove—later called the "Father of the Fuel Cell"- creates a "gas battery," which is basically a fuel cell.
- 1920s: German engineer Rudolph Erren converts a standard internal-combustion engine to use hydrogen or hydrogen mixtures.
- 1959: Professor Francis T. Bacon of Cambridge University in England builds the first hydrogen-air fuel cell, which powers a welding machine.
- 1959: Harry Karl Ihrig, an engineer at the Allis Chalmers Manufacturing Company in the United States, demonstrates the first vehicle powered by a fuel cell of a 20-horsepower farm tractor.
- 1972: Ford Gremlin's internal-combustion engine was modified by students at the University of California at Los Angeles to run on hydrogen. The modified Gremlin wins first prize in a competition for the vehicle with the lowest tailpipe emissions.
- 1988: Scientists in the Soviet Union successfully converted a 164-passenger commercial jet to fly with one of the 'jet's three engines powered by liquid hydrogen. Even though the flight lasted only 21 minutes, it suggests the use of liquid hydrogen in jet engines.

- 1900: The 'world's first solar-powered production plant to produce hydrogen opened in southern Germany.
- 1994: The Daimler Benz Motor Company in Germany demonstrates its first fuel cell vehicle
- 2004: Germany tests the 'world's first fuel cell-powered submarine in deep water
- 2017: Hydrogen-powered 3-wheeler vehicle developed by Hydrogen Energy Centre-Banaras Hindu University, Varanasi, India was demonstrated [(Figure 3 (a)].
- 2018: TATA Motors in association with Indian Oil Corporation flagged the trials of India's First ever Hydrogen Fuel Cell powered Bus [Figure 3 (b)].

Figure 3. (a) Hydrogen-powered 3-wheeler vehicle developed by Hydrogen energy centre-Banaras Hindu University, Varanasi, (b) Hydrogen powered Tata Starbus fuel cell bus.

Conclusion

Hydrogen is the simplest and most abundant element on Earth and it can store and deliver usable energy. Therefore, hydrogen is an energy carrier, an energy source, and can store and deliver a tremendous amount of energy. Hydrogen can be used in fuel cells to generate electricity and power and heat. For the hydrogen economy to succeed, hydrogen storage has become a bottleneck out of all the ingredients present. So, there is a need for a safe and efficient H_2 storage system. But for the effective realization of hydrogen economy requires an optimized hydrogen storage system capable of reversibly storing hydrogen at moderate temperature (~100-150°C) and pressure (1-10 atm), gravimetric storage capacity ~5.50wt.%, volumetric storage capacity ~40g/L H_2. Hydrogen transportation and utilities are emerging markets.

Acknowledgments

TPY is grateful to the University Grants Commission (UGC) for the Raman Fellowship, which allowed a portion of the work to be done during the fellowship. The authors would like to dedicate this article to the memory of their beloved mentor and distinguished researcher in the area of hydrogen storage and nanomaterials, Prof. O. N. Srivastava, who departed prematurely from this world on 24th April 2021 due to COVID-19.

References

[1] Hickey W. *Energy as the 'Non-Devaluing' Currency: A Store of Wealth in 'Today's World: Energy and Human Resource Development in Developing Countries*. Palgrave Macmillan, New York 2017; 71-100.

[2] Rusman, N.A.A. and Dahari M. A. (2016).Review on the current progress of metal hydrides material for solid-state hydrogen storage applications. *Int. J. Hyd. Energy* 41: 12108-12126.

[3] Zhou W., Tang Z., Zhu D., Ma Z., Wu C., Huang L., Chen Y. (2017) Low-temperature and instantaneous high-rate output performance of AB_5-type hydrogen storage alloy with duplex surface hot-alkali treatment. *J. Alloys&Comp.*692: 364-374.

[4] Srivastava O.N., Yadav T.P., Shahi R.R., Pandey S.K., Shaz M.A., Bhatnagar A. (2015). Hydrogen energy in India: storage to application. *Proceedings of the Indian National Science Academy*81(4):915-937.

[5] Shukla V., Bhatnagar A., Verma S.K., Pandey A.P., Vishwakarma A.K., Srivastava P., Yadav T.P., Srivastava O.N. (2021). Simultaneous improvement of kinetics and thermodynamics based on SrF_2 and SrF_2@Gr additives on hydrogen sorption in MgH_2. *Mat. Adv.* 2: 4277-4290.

[6] Verma S.K., Bhatnagar A., Shukla V.,Soni P.K., Pandey A.P., Yadav T.P., Srivastava O.N.(2020). Multiple improvements of hydrogen sorption and their mechanism for MgH_2 catalyzed through TiH_2@ Gr. *Int. J. Hyd. Energy* 45(38): 19516-19530.

[7] Lai B., Singh S.C., Bindra J.K., Saraj C.S., Shukla A., Yadav T.P., Wu W., McGill S.A., Dalal N.S., Srivastava A., Guo C. (2019). Hydrogen evolution reaction from bare and surface-functionalized few-layered MoS_2nanosheets in acidic and alkaline electrolytes. *Mat. Today Chem.* 14: 100207.

[8] Shukla V., Bhatnagar A., Singh S., Soni P.K., Verma S.K., Yadav T.P., Shaz M.A.,Srivastava O.N. (2019).A dual borohydride (Li and Na borohydride) catalyst/additive together with intermetallic FeTi for the optimization of the hydrogen sorption characteristics of $Mg(NH_2)_2/2LiH$. *Dalton Trans.* 48(30): 11391-11403.

[9] Yadav T.P., Woellner C.F., Sinha S.K., Sharifi T., Apte A., Mukhopadhyay N.K., Srivastava O.N., Vajtai R., Galvao D.S., Tiwary C.S., Ajayan P.M. (2018).Liquid exfoliation of icosahedral quasicrystals. *Adv. Funct. Mat.* 28: 1801181.

[10] Patel C.R.P., Tripathi P., Vishwakarma A.K., Talat M., Soni P.K., Yadav T.P., Srivastava O.N. (2018). Enhanced hydrogen generation by water electrolysis employing carbon nano-structure composites. *Int. J. Hyd. Energy* 43(6): 3180-3189.

[11] Pandey S.K., Bhatnagar A., Mishra S.S., Yadav T.P., Shaz M.A., Srivastava O.N.(2017). Curious Catalytic Characteristics of Al–Cu–Fe Quasicrystal for De/Rehydrogenation of MgH_2. *The J. Phys.Chem.C*121(45): 24936-24944.

[12] Shahi R.R., Yadav T.P., Shaz M.A., Srivastava O.N. Synthesis characterization and hydrogenation behaviour of as quenched $Ti_{41.5}+XZr_{41.5-x}Ni_{17(x= 0, 3.5, 11.5 \text{ and } 13.5)}$nanoquasicrystalline ribbons. *Journal of Physics: Conference Series 809 (1): 012011* 2017.

[13] Shukla V., Bhatnagar A., Soni P.K., Vishwakarma A.K., Shaz M.A., Yadav T.P., Srivastava O.N. (2017). Enhanced hydrogen sorption in a Li–Mg–N–H system by the synergistic role of $Li_4 (NH_2)_3 BH_4$ and $ZrFe_2$. *Phys. Chem. Chem. Phys.*19: 9444-9456.

[14] Dixit V., Bhatnagar A., Pandey S., Yadav T. P., Shaz M.A., Sinha A.S.K., Srivastava O.N. (2017). Carbon Derived from Solid Kernel of Coconut: Its Structural, Microstructural Characteristics and Use as New Hydrogen Storage Material. *Mats. Focus* 6(5): 502-507.

[15] Shahi R.R., Bhatanagar A., Pandey S.K., Shukla V., Yadav T.P., Shaz M.A., Srivastava O.N.(2015). MgH_2–$ZrFe_2H_x$ nanocomposites for improved hydrogen storage characteristics of MgH_2. *Intl. J. of Hyd. Energy* 40(35): 11506-11513.

[16] Shukla V., Bhatnagar A., Pandey S.K., Shahi R.R., Yadav T.P., Shaz M.A., Srivastava O.N.(2015). On the synthesis, characterization and hydrogen storage behavior of $ZrFe_2$ catalyzed Li-Mg-N-H hydrogen storage material. *Intl J. Hyd. Energy* 40(36): 12294-12302.

[17] Pandey S.K., Bhatnagar A., Shahi R.R., Yadav T.P., Srivastava O.N.(2014). Co-Catalytic Effect of Carbon Based Nanostructures and TiO_2 on Sorption Behavior of Nanocrystalline MgH_2. *Adv. Sci. Lett.* 20(5-6): 1120-1123.

[18] Pandey S.K., Singh J., Yadav T.P., Srivastava O.N.(2014). Material Tailoring of Ti–V–Fe–Zr for Improving Sorption Behaviour Employing Off Stoichiometric Concentration of Native Element Ti. *Mats. Focus* 3(1): 28-35.

[19] Pukazhselvan D., Saravanan R.S.S., Yadav T.P.(2012). Towards Sustainable Green Energy Development and Insights on Few Scientific Problems Leading to Less Carbon Econom. *Rev. Adv. Sci. & Engg.*1: 302-318.

[20] Schlapbach L., Züttel (2001). Hydrogen-storage materials for mobile applications, *Nature* 414: 353-358.

[21] Preuster P., Alekseev A., Wasserscheid P., Hydrogen Storage Technologies for Future Energy Systems. *Ann. Rev. Chem. & Biomole. Engg.*8: 445–471.

[22] Hirscher M. *Handbook of Hydrogen Storage: New Materials for Future Energy Storage*. ISBN: 978-3-527-32273-2, 2010.

[23] Simon H.A., Liu C.S., Harnett J.P. Properties of hydrogen: nitrogen, hydrogen: carbon-dioxide, and carbon-dioxide: nitrogen mixtures. *NASA contractor report* 1966.

[24] Berry G.D., Pasternak A.D., Rambach G.D., Smith J.R., Schock R.N(1996). Hydrogen as a future transportation fuel. *Energy.* 21(4): 289-303.

[25] Penner S.S. (2006). Steps toward the hydrogen economy. *Energy* 31(1): 33-43.

[26] Ogden J.M. Modeling Infrastructure for a Fossil Hydrogen Energy System with CO_2 Sequestration, Greenhouse Gas Control Technologies. *6th International Conference* 2003;1069-1074.

[27] Yadav T.P., Shahi R.R., Srivastava O.N. (2012). Synthesis, characterization and hydrogen storage behaviour of AB_2 ($ZrFe_2$, $Zr(Fe_{0.75}V_{0.25})_2$, $Zr(Fe_{0.5}V_{0.5})_2$ type materials. *Int. J. Hyd. Energy* 37: 3689-3696.

[28] Shahi R.R., Yadav T.P., Shaz M.A., Srivastva O.N. (2011). Effect of processing parameter on hydrogen storage characteristics of as quenched $Ti_{45}Zr_{38}Ni_{17}$ quasi crystalline alloys. *Int. J. Hyd. Energy;*36: 592-599.

[29] Shahi R.R., Yadav T.P., Shaz M.A., Srivastava O.N. (2010). Studies on dehydrogenation characteristic of $Mg(NH_2)_2$/LiH mixture admixed with vanadium and vanadium based catalysts (V, V_2O_5 and VCl_3). *Int. J. Hyd. Energy*35(1):238-246.

[30] Shahi R.R., Yadav T.P., Shaz M.A., Srivastava O.N.(2008). Effects of mechanical milling on desorption kinetics and phase transformation of $LiNH_2/MgH_2$ mixture. *Int. J. Hyd. Energy*33(21): 6188-6194.

[31] Agnolucci P. (2007). Hydrogen infrastructure for the transport sector. *Int. J. Hyd. Energy* 32: 3526-3544.

[32] Hawkins S. (2006). *Technological Characterisation of Hydrogen Storage and Distribution Technologies*, UKSHEC Social Science Working Paper No. 21:1-36.

Chapter 2

Theoretical Modelling on Hydrogenation Characteristics of Ball-Milled AB_5-Type Hydrogen Storage Nanomaterial

Kuldeep Panwar[1] and Sumita Srivastava[2,*]
[1]Department of Physics, Pt. L.M.S. Government Post Graduate College, Rishikesh, India
[2]Government Degree College, Pokhri Quili, Tehri Garhwal, India

Abstract

The ball-milling process is a method to break large particles into smaller ones. It has versatile application in the synthesis of new material. A nanostructured version of the given bulk alloy can be obtained through ball-milling. Ball-milling can be utilized in the synthesis of new material through mechanical alloying initiated from the raw ingredients. It can be applied for reactive milling as well, by proper addition of ingredients in a specific environment. Ball-milling is further applicable in producing composite nanostructured material. The materials obtained after ball-milling have nanostructure, which introduces many microstructural changes within the material. Few noticeable changes are reduced particle and grain size, higher density of grain boundaries, large surface area, homogeneous mixing of added ingredients into major phases etc. All these properties have a direct effect on the hydrogenation behaviour of hydrogen storage alloys. Such properties have been studied earlier, however, their correlation with various hydrogenation characteristics still need to be defined more deeply. The present study is focused on the theoretical perspectives and modelling of the hydrogenation behaviour of ball-milled AB_5-type alloys. The model developed in the present study includes the properties like activation, kinetics, p-c isotherms, hydrogen

* Corresponding Author's Email: sumita_uki1@rediffmail.com.

In: Materials Science: Future Aspects
Editors: Kalpana Awasthi, Arti Srivastava and Mridula Tripathi
ISBN: 978-1-68507-843-0

storage capacity and their correlation with structural changes. Hydrogen storage capacity has been further correlated with thermodynamic properties, structural parameters and electronic properties. Based on the proposed theoretical model, the ball-milling index (I) has been formulated. I has been correlated with ball-milling parameters and hydrogenation characteristics. I corresponding to the optimized hydrogenation properties is described as optimized index I_0. On applying the formulation of the optimized index to various studies done earlier by several researchers, it was found that I_0 depends on the type of ball-mill used in the process. It was concluded that through the help of the theoretical model and optimized index, many hydrogenation characteristics can be predicted in the process of the synthesis of new material. It may help to choose the ball-milling parameters for desired hydrogenation properties.

Keywords: hydrogen storage alloy, ball-milling, nanomaterial, AB_5-type, theoretical modelling, optimized index

1. Introduction

Metal hydrides are important hydrogen storage alloys due to their versatile applications in various fields. Some of the applications are: in vehicles, heat pumps, heat exchangers, compressors, stationary hydrogen storage, Ni-MH cell etc. [1-7]. With the advancement in technology, new applications are coming forward. These applications require specific characteristic hydrogenation properties of the hydrogen storage material. Although hydrogen storage alloys cover a variety of materials naming; metal hydrides, alanates, borohydride, fullerenes, carbon nanotubules, active carbon and many more, the hydrogenation characteristics of a particular class of parent alloy representing the whole family are almost fixed [8-14]. Hence, their uses in applications are also getting limited. To overcome this problem, many investigations have been done in terms of improving the hydrogenation characteristics. The hydrogenation behaviour of a particular material is characterized by many properties. Some of them belong to activation treatment, hydrogen storage capacity, plateau pressure, operating temperature, kinetics corresponding to hydrogenation/dehydrogenation, cyclic stability, the heat of formation etc. [15-17]. Among various hydrogen storage materials, AB_5-type metal hydride offers excellent property being operated at ambient temperature and pressure and is easy to activate at room temperature [18-21].

There are various methods of improving the hydrogenation properties of different types of hydrogen storage materials.

The methods to improve the hydrogenation properties can be broadly categorized into two types. Before discussing such methods it may be mentioned here, that for metal hydrides representative parent alloys are AB_5 ($LaNi_5$, $MmNi_5$), AB_2 ($TiMn_2$), A_2B (Mg_2Ni) and AB (FeTi). The first method is to substitute other elements at A or/and B [22-26]. This leads to multi-element composition. Some properties of all the elements differ, giving new hydrogenation characteristics to the substituted version of the alloy. Another method corresponds to the variation in the synthesis route. Conventionally metal hydrides are synthesized through arc/induction melting.

Along with the conventional route, other synthesis processes may also lead to different and improved hydrogenation properties. Famous of them are melt-spinning [27, 28], formation of composite materials [29, 30], synthesis through mechanical alloying [31, 32], reactive milling [33, 34], formation of nanostructured material [35-38] etc.

One more significant process of synthesis is ball-milling, which is the subject matter of the present investigation. The ball-milling process is a method to break large particles into smaller ones. It has versatile application in the synthesis of new materials. The synthesis process via ball-mill involves various milling parameters; like the ball to powder ratio, milling time, milling speed (rpm), environment within the mill and added material. The optimizations of all these parameters are needed to get the best results. The materials obtained after ball-milling has nanostructure, which introduces many microstructural changes within the material. Few noticeable changes are reduced particle and grain size, higher density of grain boundaries, large surface area, homogeneous mixing of added ingredients into major phases etc. All these properties have a direct effect on the hydrogenation behaviour of the hydrogen storage alloys.

A large number of experimental studies have been performed on ball-milled Mg-based hydrogen storage alloys. Comparatively sparse studies are available on the AB_5-type alloys. Keeping this in mind experimental studies on ball-milled AB_5-type alloys were undertaken by us, which are published elsewhere [38].

The studies made by us and by earlier researchers in this context have revealed the microstructural changes induced in the alloy during ball-milling. The various hydrogenation properties, like thermodynamic, kinetics, p-c isotherm and hydrogen storage capacity of the ball-milled alloy have also been addressed separately earlier. In search of new materials with potential use in applications, it will be beneficial for researchers to develop some models to predict the hydrogenation properties of new materials before their actual synthesis. At present no such model is available for the ball-milled alloy. The present investigation is focused on the correlation of the several effects introduced in the alloy during ball-milling and their impact on the hydrogenation properties of the ball-milled alloy based on the studies of AB_5-type hydrogen storage alloy. It may be mentioned here that, although the effect of ball-milling and its parameters on hydrogenation characteristics have been studied by earlier researchers, their overall correlation to predict the hydrogenation properties has not been reported. Therefore, the present study has undertaken two objectives. The first is focused on developing a theoretical model to predict the hydrogenation properties. The model developed in the present study includes the properties like activation, kinetics, p-c isotherms, hydrogen storage capacity and their correlation with structural changes. Hydrogen storage capacity has been further correlated with thermodynamic properties, structural parameters and electronic properties. The second objective is the formulation of the optimized index based on the theoretical model. Hence, a theoretical model in form of a flow chart has been proposed to obtain the desired hydrogenation properties after ball-milling. The proposed model may also be applied to adjust the ball-mill parameters for getting a set of specific hydrogenation characteristics after ball-milling. In another way, one can say that the proposed theoretical model may help to predict the various properties after ball-milling. Although the present model has been developed based on the studies done on the AB_5-type hydrogen storage alloy, the same may be extended to the other alloys also; as the model does not involve any specific characteristics of the AB_5-type alloy.

2. Methodology

The methodology of the present investigation is the study of the theoretical background of the impact of various ball-mill parameters on the hydrogenation properties of the AB_5-type alloy. Based on the theoretical background, a model in form of a flow chart has been proposed to correlate all the effects produced in the alloy and its hydrogenation properties after ball-milling. With the help of the proposed theoretical model, the ball-milling index has been formulated. This model and the index have been applied to the experimental results of the already reported work of others. Through proper analysis of the model, a method has been proposed to predict the hydrogenation properties of the alloy.

Ball–milling is a process to synthesize the nanocrystalline form of any material. In the context of hydrogen storage in the solid metal hydride, nanocrystalline versions of hydrogen storage alloys show remarkable changes in the hydrogenation properties. Many properties like activation treatment, rate of hydrogen absorption–desorption, hydrogen storage capacity and cycle life have been found to improve after ball–milling. The degree of improvement depends on the proper choice of ball-milling parameters, like time of milling, rotation per minute (rpm) during milling, the environment of milling, ball size, the ball to powder ratio etc. All these parameters affect the microstructure of the as-milled alloy and thus the hydrogen sorption properties of the hydrogen storage alloy. The improvement in hydrogenation characteristics and the mechanism of such improvement has been described in the following.

Structural and microstructural characterization of ball-milled alloy reveals many changes in the morphology. Changes have been noticed in lattice parameters, unit cell volume, as well as in particle and crystallite size. These changes in the ball-milled alloys may introduce some new features in hydrogenation behaviour. Many reports have been published on ball-milling properties of Mg-based alloys; however, fewer studies have been reported for AB_5-type alloys. Few studies are focused on structural properties, while others on hydrogenation properties. Complete studies of all the aspects of hydrogenation behaviour for AB_5-type alloys are not available. Experimental investigations on hydrogenation behaviour of ball-milled AB_5-type alloys like activation, kinetics, p-c isotherms and hydrogen storage capacity have been already studied by us [38], however, their theoretical modelling and correlation with structural changes have not been studied. Therefore the present study is focused on the theoretical perspectives and modelling of hydrogenation behaviour of ball-milled AB_5-type alloys.

2.1. Theoretical Background of the Proposed Model

The role of various factors affecting the hydrogenation characteristics has been discussed in the following. This may help the readers to understand the correlation of all such hydrogenation properties in the proposed theoretical model.

2.1.1. Activation and Kinetics

Mg-based alloys and alanates need high temperatures for activation and desorption of hydrogen [39-46]. Ball-milling reduces the activation energy and decomposition temperature in these alloys [47-49]. AB_5-type alloys generally do not need a high temperature for activation and hydrogen desorption. These materials can be operated at room temperature. Due to the coating of the oxide layer on the surface, full activation is achieved after a few hydrogenation-dehydrogenation cycles. Ball-milling creates small particles of fresh oxide-free surface, which absorbs hydrogen more readily. Figure 1 shows a model of change in surface morphology due to ball-milling to favour activation and absorption-desorption process. The smaller particle and grain size upon ball-milling enhances the density of fresh surface and grain boundaries respectively. Eventually, it reduces the diffusion length for hydrogen atom diffusion into the bulk. Thus activation process and absorption-desorption kinetics become faster in comparison to as received bulk alloy. Earlier studies have been reported the increase in activation and kinetics of AB_5-type alloys upon ball-milling [19, 38, 50].

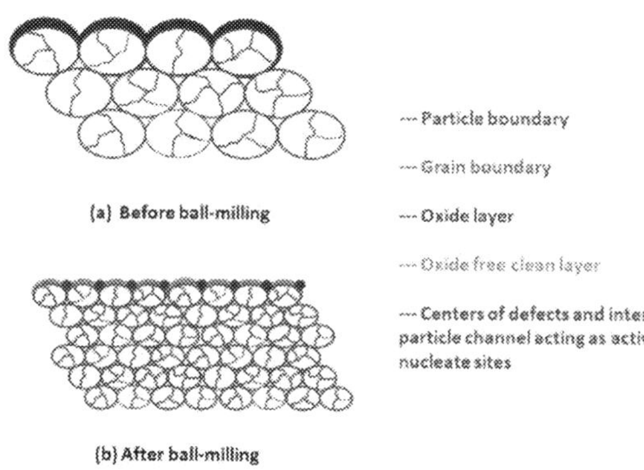

Figure 1. Structural changes in the alloy after ball-milling. Microstructural comparison has been shown (a) before ball-milling and (b) after ball-milling.

During the ball-milling process, the powder particles which are trapped between the colliding balls are subjected to compressive impact forces. The powder particles are deformed, fractured and cold-welded, leading to defects in the system [51]. Thus ball-milling process results in average particle size reduction and the creation of defects and strain. Both factors contribute to the faster activation and hydrogen absorption-desorption kinetics. Several crystal defects such as dislocations, stacking faults, vacancies and an increased number of grain boundaries enhance the diffusivity of hydrogen into the matrix [52, 53]. Such defect centres act as nucleation sites for hydrogenation. A graphical model incorporating all such features of ball-milling has been developed in the present study and is shown in Figure 1. This Figure demonstrates that the density of the fresh oxygen-free surface, grain density and nucleation sites (like defects and grain boundaries) are more in the ball-milled alloy in comparison to the bulk sample. This may be the reason for enhanced activation and kinetics in the alloy after ball-milling.

2.1.2. Pressure-Composition Isotherms

Pressure-Composition Isotherm (p-c isotherm) is the most important characteristic of any hydrogen storage alloy. One may determine many important properties from p-c isotherm as given below:

- Absorption-desorption plateau pressure
- Hydrogen storage capacity
- The slope of the p-c isotherm tells about the nature of α - β phase change during hydrogenation
- Collective p-c isotherms at different temperatures are used to calculate the heat of formation and entropy (ΔH and ΔS) through Van't Hoff equation

The earlier studies suggest that as the unit cell volume increases, the plateau pressure decreases [54]. This is because of easy hydrogen absorption-desorption due to enlarged void size and therefore less driving force is required for hydrogenation. In the case of ball-milled alloy, the properties of the sample are very much dependent on the ball-milling parameters. Therefore some differences are seen in the studies reported earlier about changes in lattice parameters and unit cell volumes upon ball-milling. In the present investigation together with many other reports, an increase in lattice parameter and unit cell volume has been noticed after 5-hour milling. All the earlier studies on the ball-milled AB_5-type alloys reveal an increase in plateau pressure together with an increase in unit cell volume [51, 52, 55-57]. This anomaly reflects that the hydrogen absorption-desorption mechanism has different nature in the ball-milled sample of the bulk alloy.

Upon ball-milling, the crystalline nature of the alloy gets disturbed due to the creation of many defects including a higher density of grain boundaries. This may result in the lower stability of the hydrides. Lower stability means the more positive value of heat of formation of hydride. The earlier studies suggest that plateau pressure decreases with the increasing stability of hydrides (more negative value of ΔH) [58]. Thus in the case of ball-milled alloy, the stability decreases giving rise to an increase in plateau pressure. This may be the probable reason for high plateau pressure in the ball-milled sample in comparison to the bulk alloy.

It is also observed that the p-c isotherm of AB_5-type ball-milled alloy has more slopes in comparison to the bulk alloy [19, 38, 51, 52, 55-57]. The larger slope is again due to less crystallinity of the ball-milled sample in comparison to the bulk version. It is also known that milling for a larger time may result in amorphous material [38]. The flatten plateau represents a clear cut phase change from α to β phase at constant pressure. Due to the poor crystalline nature, the phase change is not clear in the p-c isotherm of ball-milled alloy. It can also be seen that in a normal p-c isotherm of bulk alloy hydrogen absorption does not start before a certain hydrogen pressure. While in the case of a ball-milled sample, a small amount of hydrogen is absorbed from initial hydrogen pressure [19, 38]. After that, hydrogen concentration increases with increasing value of hydrogen pressure without any plateau region up to the maximum value of hydrogen absorption capacity of the specific sample. Hence, the slope of the p-c isotherm is resulted again due to the poor crystalline nature of ball-milled alloy in comparison to the bulk alloy.

2.1.3. Hydrogen Storage Capacity

As for as the storage capacity of the ball-milled samples is concerned, the earlier studies show an increase in storage capacity up to the certain limit of ball-milling time [19, 38, 50]. Prolonged milling results in a decrease in storage capacity due to the amorphous nature of alloy after long milling.

The earlier studies accomplished by us dealt with thermodynamic, structural and electronic properties of bulk AB_5-type metal hydrides [58, 59]. A detailed discussion made in these studies reveals that the hydrogen storage capacity of metal hydrides depends on the following factors:

- Thermodynamic properties
- Structural parameters
- Electronic properties

The earlier investigations done by us on the calculation of heat of formation [58], structural [59] and electronic (in the process of publication) parameters associated with hydrogenation properties have shown that all such studies are equally applicable to the bulk as well as on the ball-milled alloy. It is due to the fact that such studies have involved the fundamental parameters like atomic radius and electronic configuration of the substituted element in the AB_5-type alloy. Similarly, the calculation of thermodynamic property heat of formation is based on some equations including Meidema's rule of reverse stability. These equations do not differentiate between bulk and ball-milled state. Hence, it was realized that the hydrogenation properties induced after ball-milling have some other basis together with fundamental properties associated with the bulk version.

2.1.3.1. Thermodynamic Properties

The most important thermodynamic parameter for metal hydride is the heat of formation of hydride (ΔH). The stability of hydride is related to ΔH. If the ΔH value is more positive, it means hydride is not stable. So hydrogen will not be absorbed and hydrogen storage capacity will be less. On the other hand, if ΔH value is more negative, it implies a very stable hydride. In such a case again hydrogen desorption will not be easy. So such type of alloy may not be able to work as a reversible metal hydride. Hence, one can see that an appropriate value of ΔH is required for acting as a reversible metal hydride. After ball-milling the alloy hydride may become less stable due to the introduction of any defects in the structure.

2.1.3.2. Structural Parameters

As already mentioned, the earlier investigations suggest the increased value of hydrogen storage capacity after ball-milling. On applying the structural model developed by us earlier [59], one may find that the key parameter as equivalent radius r_B^* in the AB_5-type alloys remains the same for the ball-milled samples as well. Thus, it may be said that the earlier developed structural model will not work in the case of ball-milled alloy. The major role is played by morphological features developed during the ball-milling process. The changes introduced through ball-milling are not at the elemental level and within the unit cell, rather these are created at the grain level. As already said, the ball-milling introduces many defects including the higher density of grain boundaries and fresh oxide-free surfaces, which act as nucleation sites for the hydrogenation process inside the bulk. These changes are responsible for easy activation and fast kinetics as well as for increased hydrogen storage capacity. It may be thought that several types of defects are playing a major role in hydrogen absorption. In the ball-milled sample, hydrogen is not only stored at interstitial sites of the crystalline structure but also stored at defects and grain boundaries.

2.1.3.3. Electronic Properties

The studies on the effect of electronic properties on hydrogen storage capacity have been done by us and in a way of communication. Now, if one compares the electronic properties of ball-milled alloy with bulk alloy based on broad electronic properties developed, no differences can be noticed. Broad electronic properties are based on the number of electrons and the electronic configuration of elements forming the alloy. Such electronic values will not change after ball-milling. This forces one to re-visit the mechanism of hydrogen absorption in the ball-milled alloy. Defects created during ball-milling have modified electronic structures [61]. During hydrogen absorption, electrons of hydrogen are donated to the empty energy levels of metal hydride below the Fermi level. Defects may create empty energy levels to attract more electrons of hydrogen, thus promoting the bondage of hydrogen with metal hydride and acting as nucleation sites for hydrogenation.

Hence, from the above discussion, it may be realized that various types of defects are responsible for all the improved hydrogenation properties of the ball-milled alloy. It is worth mentioning here that the nature of the ball-milled sample depends very much on the selection of various ball-milling parameters. The most important one is the milling time. The larger milling time may completely damage the crystalline property of intermetallic, thus giving suppressed hydrogenation properties instead of the improved one.

2.2. Theoretical Modelling of Hydrogenation Characteristics of Ball-Milled AB_5-Type Alloy

In the earlier studies by us, key factors of theoretical modelling of hydrogenation characteristics of AB_5-type bulk hydrogen storage alloy have been investigated [58, 59]. These key factors are equivalent radius r_B^* of elements substituted at B site of AB_5 alloy and electronic configuration of elements substituted at B site. On applying this model to ball-milled samples, no change as such is possible in these key factors, as they represent the properties of substituted elements at A or/and B site in the parent AB_5-type alloy. Merely ball-milling of the parent alloy cannot make changes at the elemental level. Rather great and noticeable changes occur at the morphological level. As already mentioned, various types of defects, the creation of new and fresh oxide-free surfaces, smaller grain size resulting in a higher density of grain boundaries create structural changes. Thereafter all such changes are responsible for affecting the hydrogenation characteristics. The improvement in the hydrogenation properties depends very much on the extent of milling time and other milling parameters. The model developed in the present investigation correlates all the effects produced upon ball-milling on the hydrogenation behaviour of AB_5-type alloy, which has not been reported earlier.

The most important parameter of the ball-milling is the time of milling, which can be easily adjusted to get the optimized microstructure. Whereas milling for a moderate time creates a fine and homogeneous crystalline microstructure, larger milling may turn the alloy into an amorphous one. Because of the ball-milling, particle and grain size of the bulk alloy reduces. Due to this, the long-range order in the alloy is broken. More often, this results in increased unit cell volume. On the other hand, the ball-milling introduces the various defects and strains in the lattice. The effects of the occurrence of defects are also seen in the electronic structure of the alloy. The higher density of grain boundaries and defects gives rise to an increase in nucleation sites. All these effects together result in a smaller diffusion path during hydrogen absorption-desorption. This may lead to the poor crystalline structure of the ball-milled alloy in comparison to the bulk alloy.

The defects in the electronic structure create vacant energy levels to trap the electrons of the hydrogen. Similarly, the defects in the microstructure are responsible for easy activation and fast kinetics of hydrogen absorption-desorption. All these three factors; increased unit cell volume, hydrogen trapping on the defects and fast activation and kinetics; together may increase the hydrogen storage capacity of the ball-milled alloy. On the other hand, the changes introduced in the microstructure of the ball-milled alloy in terms of the poor crystallinity leads to the decreased stability of the hydride, resulting in the higher plateau pressure. The strain developed in the lattice during ball-milling increases the slope in the p-c isotherm. If the plateau pressure is high and ΔH has a more positive value, hydride will be less stable with difficulty in absorbing the hydrogen. In the same way, lower plateau pressure and more negative value of ΔH indicate the large stability to hydrides with difficulty in desorbing the hydrogen. Therefore for reversible hydrogen absorption-desorption, the middle value of ΔH is chosen.

All the above analysis of the proposed theoretical model, in form of the flow chart, is represented in Figure 2. Thus, by applying the proposed model, ball-milling parameters may be adjusted to get the desired p-c isotherm. In another way, by predicting the value of hydrogen storage capacity and plateau pressure, the nature of the p-c isotherm may be predicted.

2.3. Ball-Milling Index (I)

The superiority of the ball-milled alloys over bulk alloy concerning hydrogenation properties lies in the fact, that the ball-milling process introduces many microstructural changes in the alloy. These changes correspond to the reduction in particle and crystallite size and the creation of defects. To get the optimized microstructure of ball-milled alloy, the important factors are ball-milling parameters and the brittleness of the alloy. The combined effect of all these factors ultimately results in an optimized particle and crystallite size. The defects introduced during milling and increased density of grain boundaries give a unique identity to the microstructure of the alloy after milling.

It is also clearly visible in the flow chart of the theoretical model that the milling parameters play an important role in getting the optimized microstructure to impart specific hydrogenation properties. The amount of the energy transferred to the alloy particles during milling is directly proportional to the milling time and milling rpm (revolution per minute) of the ball-mill. The other parameters in this context are the ball to powder ratio and grinding media. Hence, ball-milling index I is proposed in the present investigation for the ball-milling of the AB_5-type alloy as:

$$I = \text{rpm} \times \text{time of milling (minute)}$$

The ball-milling index I for the optimized microstructure has been proposed as I_0 and termed as the optimized index. It is supposed that corresponding to I_0, the improved hydrogenation properties will be observed. On changing the rpm or time of milling, the value of I will change. Accordingly, the hydrogenation properties will also change. The value of I has a direct effect on the hydrogenation properties. The value of I also depends on the choice of other ball-milling parameters and the type of alloy. The effect of other parameters on the value of I is described in the following:

i. Ball to powder ratio– The larger be this ratio, the more energy will be transferred to the alloy during milling. Therefore for the larger ball to powder ratio, I will be comparatively less.

ii. Grinding media– Wet grinding media produce finer and homogeneous particles. So, dry milling may have a comparatively large value of I.

iii. Type of alloy– The degree of the brittleness of the alloy is also important for getting an optimized microstructure. For the brittle alloy, I may be less. In the case of the AB_5-type of the alloy, it has been observed that the substitutions at B site make the alloy more brittle in comparison to the parent alloy.

In the next section of 'results and discussions', the proposed model and formulation of the ball-milling/optimized index will be applied to the earlier published reports on the AB_5- type alloys. The validity of the formula will be checked on this basis. A detailed discussion will be made on the correlation of the optimized index with hydrogenation properties.

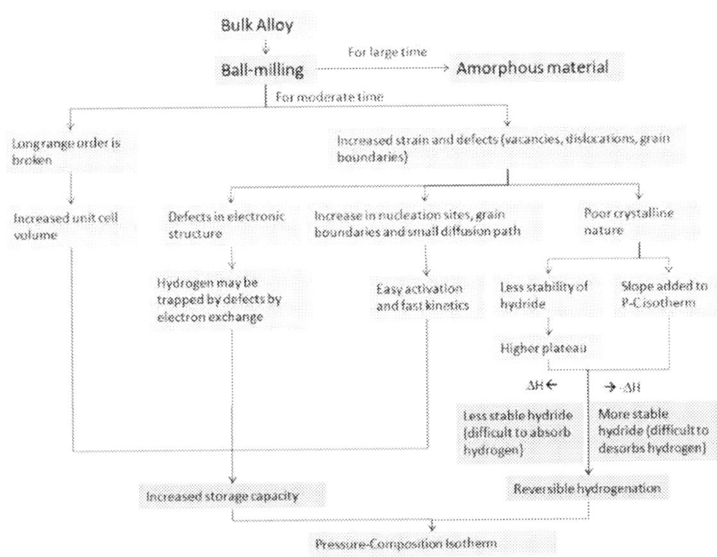

Figure 2. Theoretical modelling of hydrogenation characteristics of AB_5-type alloy upon ball-milling.

3. Results and Discussions

The theoretical model proposed in the present investigation in form of the flow chart (Figure 2) correlates several hydrogenation properties with the ball-milling parameters together with the effects produced in the microstructure of the alloy during the ball-milling process. Now, the earlier published reports will be analyzed and discussed based on the present model and formulation. In the next step, the discussion will be made that how the present theoretical model and formulation may help in predicting the hydrogenation properties before the actual synthesis of the alloy.

3.1. Application of the Present Model and Formulation

The experimental results of earlier published reports have been presented in Table 1. In this Table, a few experimental conditions have been mentioned corresponding to which maximum storage capacity was achieved for the respected alloy. By the interpretation of the results given in Table 1, it is clear that the optimized index is different in different experiments performed by researchers. As already mentioned that the optimized index depends on many factors, therefore it has a wide range of values in different experiments. It is also noteworthy that the amount of energy transferred to the alloy particle during ball-milling also depends on the choice of ball-mill used. Because of this, in some cases, a small time of milling may result in a similar microstructure instead of milling for a few hours. So now the question arises what is the role of the optimized index when it has a different value in the different experiments. The role of the optimized index is not to equate all the experimental setup; rather it is helpful for a particular experiment using the same ball-mill again and again. Through knowing the optimized index for one experiment, the ball-mill parameters can be adjusted to get desired hydrogenation properties in other sets of experiments. Now, the optimized index will be correlated with the hydrogenation properties.

i. Heat of formation

   The thermodynamic property heat of formation is related to the stability of the hydride and the plateau pressure. It is known that the more negative value of ΔH indicates larger stability of the hydride and less plateau pressure [58]. The increase in the value of I more than I_0 may result in a comparatively poor crystalline structure. So, the stability of the hydride will decrease, giving rise to the more positive value of ΔH and higher plateau pressure.

ii. Kinetics

   The increased kinetics of hydrogen absorption-desorption in the ball-milled alloy is due to the presence of a larger number of defects and a higher density of grain boundaries in the alloy. Increased value of I will naturally increase the number of defects and grain boundaries. Thus resulting in the increased rate of hydrogen absorption-desorption. The further increase in I will saturate the kinetics.

iii. Hydrogen storage capacity

The increased value of hydrogen storage capacity of the ball-milled version of the alloy is again due to an optimized microstructure. The larger number of defects and grain boundaries provide a favourable path for hydrogen absorption. Simultaneously the crystallinity of the alloy becomes poor during the ball-milling process in comparison to the bulk alloy. These two factors contradict each other in attaining a higher storage capacity. Whereas the more value of I favours the first condition, it suppresses the capacity due to the second condition. Hence one has to compromise between the two factors while choosing the optimized index.

iv. The effect of rpm

The ball-milling index I defined in the present study is the multiplication of rpm and time of milling. Hence, the different combination of rpm and time of milling will result in the same value of I. It may be mentioned here that for the same value of I, less rpm will result in a comparatively more homogeneous microstructure with better crystallinity. This will eventually reflect better hydrogenation characteristics. The less rpm will produce a more stable hydride with a higher negative value of ΔH, low plateau pressure, better kinetics and increased storage capacity.

Table 1. The comparison of the ball-milled data of earlier reports

S.N.	Alloy sample	rpm	Time of milling (min)	Particle size (μm)	Ball to powder ratio	Optimized index
1	$MmNi_5$ [38]	100	60	2	100:1	6000
2	$LaNi_5$ [57]	250	60	1.5	-	15000
3	$MmNi_{4.6}Fe_{0.4}$ [19]	200	10	2.5	-	2000
4	$LaNi_5$ [50]	150	0.25	-	-	37.5
5	$LaNi_5$ [51]	240	120	-	10:1	28800

3.2. Verification of the Optimized Index

Based on the above discussion, the results presented in Table 1 may be verified for the optimized index. All the data represented in Table 1 corresponds to the maximum storage capacity in the respected experiments. Any increase or decrease in the optimized index has resulted in a lower storage capacity. It may be noticed from Table 1 that the value of the optimized index has a large variation. It is obvious due to the different ball-mill applied for the purpose and due to the variation in the choice of ball-mill parameters. In the first sample $MmNi_5$, the ball to powder ratio was taken as 100:1. Therefore, the optimized index is comparatively less. The common value of the ball to powder ratio is 10:1. In all the cases, absorption-desorption kinetics increases with the milling time. The plateau pressure has also been found to increase with the milling time. This suggests that all the analysis related to the optimized index proposed in the present study is in good agreement with the experimental results. It is also clear that there is not a single value of the optimized index. It is found for the specific experiment using the same ball-mill each time.

3.3. Prediction of Hydrogenation Properties Using the Model Proposed in the Present Study

The proposed theoretical model shown in Figure 2 demonstrates the steps of changes/effects produced in the microstructure of the alloy during ball-milling and their impacts on the hydrogenation characteristics. The very first point is that the prediction about hydrogenation properties will be more accurate for a specific ball-mill. It may also be mentioned here that a particular group of researchers apply the same ball-mill in many experiments; therefore they may predict the properties for their future experiments. In the first step of prediction, one experiment is performed with the best ball-mill parameters decided by the literature survey. Microstructure and hydrogenation properties are monitored for the ball-milled version of the alloy. The first set of the hydrogenation properties are tabulated and the ball-milling index is calculated. Now, analyzing the first result, desired hydrogenation properties may be achieved by adjusting the ball-mill parameters. The adjustment of the ball-mill parameters may be achieved by following the discussions made in the previous sections. In this way, by altering the ball-mill parameters optimized index can be calculated. This optimized index will give superior hydrogenation characteristics as desired at the beginning of the experiments. The application of the proposed model will save time and money in synthesizing many alloys for obtaining the best results using the trial and error method.

Conclusion

The present investigation is focused on proposing the theoretical model on the correlation of all the effects produced by ball-milling parameters on the microstructure of the alloy and their impact on the hydrogenation characteristics. For this, a theoretical model has been presented in form of the flow chart and a ball-milling/optimized index has been formulated. The proposed model has been applied to the earlier published reports of other researchers and the optimized index has been calculated. During analysis of the earlier published reports based on the proposed model, a large variation in the value of the optimized index has been observed. This was attributed to the different ball-mill applied by individual groups of researchers for the purpose. The proposed model was found in good agreement with the earlier published results in correlating the optimized index with hydrogenation characteristics. The mechanism of predicting the hydrogenation characteristics of the ball-milled alloy has also been discussed based on the proposed model. This may help the researchers to synthesize new alloys with desired hydrogenation properties instead of using the trial and error method.

Acknowledgments

The authors are grateful to Late Prof. O. N. Srivastava (Bhu, India) and Prof. I. P. Jain (Rajasthan University, India) for helpful discussions.

References

[1] Kolbig, M., Burger, I. and Linder, M. (2019). Characterization of metal hydrides for thermal applications in vehicles below 0°C. *Int J Hydrogen Energy* 44:4878–4888.

[2] Lieutenant, K. and Borissova, A. (2020). A landscape of hydride compounds for off-board refilling of transport vehicles. *Int J Hydrogen Energy* 45:2954–2966.

[3] Lototskyy, M.V., Tolj, I., Klochko, Y., Davids, M.W., Swanepoel, D. and Linkov, V. (2020). Metal hydride hydrogen storage tank for fuel cell utility vehicles. *Int J Hydrogen Energy* 45:7958–7967.

[4] Raju, M. and Kumar, S. (2012). Optimization of heat exchanger designs in metal hydride based hydrogen storage systems. *Int J Hydrogen Energy* 37:2767–2778.

[5] Lototskyy, M.V., Yartys, V.A., Pollet, B.G. and Bowman, R.C. Jr. (2014). Metal hydride hydrogen compressors: A review. *Int J Hydrogen Energy* 39:5818–5851.

[6] Zhu, D., Zhou, W., Tang, Z., Heng, Y. and Chen, Y. (2019). SOC-dependent high-rate dischargeability of AB_5-type metal hydride anode: Mechanism linking phase transition to electrochemical H-desorption kinetics. *Int J Hydrogen Energy* 44:15278-15286.

[7] Kazakov, A.N., Romanov, I.A., Mitrokhin, S.V. and Kiseleva, E.A. (2020). Experimental investigations of AB_5-type alloys for hydrogen separation from biological gas streams. *Int J Hydrogen Energy* 45:4685–4692.

[8] Kim, J.B., Han, G., Kwon, Y.K., Bae, J., Cho, E.A., Cho, S.B. and Lee, B.J. (2020). Thermal design of a hydrogen storage system using $La(Ce)Ni_5$. *Int J Hydrogen Energy* 45: 8742–8749.

[9] Ali, N.A., Idris, N.H., Sazelee, N.A., Yahya, M.S., Yap, F.A.H. and Ismail, M. (2019). Catalytic effects of $MgFe_2O_4$ addition on the dehydrogenation properties of $LiAlH_4$, *Int J Hydrogen Energy* 44: 28227–28234.

[10] Beatrice, C.A.G., Moreira, B.R., Oliveira, A.D., Passador, F.R., Neto, G.R.A., Leiva, D.R. and Pessan, L.A. (2020). Development of polymer nanocomposites with sodium alanate for hydrogen storage. *Int J Hydrogen Energy* 45:5337–5346.

[11] Wang, Y., Zou, K., Zhang, D., Cao, Z., Zhang, K., Xie, Y., Zhou, G., Li, G. and Bai, S. (2020). Cobalt-copper-boron nanoparticles as catalysts for the efficient hydrolysis of alkaline sodium borohydride solution. *Int J Hydrogen Energy* 45: 9845–9853.

[12] Zhao, Q., Song, Q., Zhu, H., Zhang, Z., Jiang, Z. and Zhou, B. (2020). Origin of the structural diversity of the alkaline metal borohydride MBH_4 (M = Li, Na, K, Rb and Cs): Insights from first-principles calculations. *Int J Hydrogen Energy* 45: 9946–9958.

[13] Samantaray, S.S., Sangeetha, V., Abinaya, S. and Ramaprabhu, S. (2020). Diatom frustule-graphene based nanomaterial for room temperature hydrogen storage. *Int J Hydrogen Energy* 45:764–773.

[14] Sharma, A. (2020). Investigation on platinum loaded multi-walled carbon nanotubes for hydrogen storage applications. *Int J Hydrogen Energy* 45:2967–2974.

[15] Rusman,N.A.A. and Dahari, M. (2016). A review on the current progress of metal hydrides material for solid-state hydrogen storage applications. *Int J Hydrogen Energy* 41:12108–12126.

[16] Zhang, F., Zhao, P., Niu, M. and Maddy J. (2016). The survey of key technologies in hydrogen energy storage. *Int J Hydrogen Energy* 41:14535–14552.

[17] Chanchetti, L.F., Leiva, D.R., Faria, L.I.L. and Ishikawa, T.T. (2020). A scientometric review of research in hydrogen storage material. *Int J Hydrogen Energy* 45:5356–5366.

[18] Barbir, F., Veziroglu, T. N. and Plass,Jr. H.J. (1990). Environmental damage due to fossil fuels use. *Int J Hydrogen Energy* 15:739–749.

[19] Sarma, V.V., Raman, S.S.S., Davidson, D.J. and Srivastava O.N. (2001). On the mechanically pulverized $MmNi_{4.6}Fe_{0.4}$ as a viable hydrogen storage material. *Int J Hydrogen Energy* 26:231–236.

[20] Sharma, V.K. and Kumar, E.A. (2014). Effect of measurement parameters on thermodynamic properties of La–based metal hydrides. *Int J Hydrogen Energy* 39:5888–5898.

[21] Briki, C., Bouzid, M., Dhaou, M.H., Jemni, A. and Lamine, A.B. (2018). Experimental and theoretical study of hydrogen absorption by $LaNi_{3.6}Mn_{0.3}Al_{0.4}Co_{0.7}$ alloy using statistical physics modelling. *Int J Hydrogen Energy* 43: 9722–9732.

[22] Srivastava, S. and Upadhyaya, R.K. (2011). Investigations of AB_5-type hydrogen storage materials with enhanced hydrogen storage capacity. *Int J Hydrogen Energy* 36:7114–7121.

[23] Jain, R.K., Jain, A. and Jain, I.P. (2012). Effect of La-content on the hydrogenation properties of the $Ce_{1-x}La_xNi_3Cr_2$ (x=0.2, 0.4, 0.6, 0.8, 1) alloys. *Int J Hydrogen Energy* 37: 3683–3688.

[24] Yılmaz, F., Ergen, S., Hong, S.J. andUzun, O. (2018). Effect of Bismuth on hydrogen storage properties of melt-spun $LaNi_{4.7-x}Al_{0.3}Bi_x$ (x = 0.0, 0.1, 0.2, 0.3) ribbons. *Int J Hydrogen Energy* 43:20243–20251.

[25] Yamagishi, R., Kojima, T., Kameoka, S., Okuyama, D., Sato, T.J., Nishimura, C. and Tsai, A.P. (2017). Creating the hydrogen absorption capability of $CeNi_5$ through the addition of Al. *Int J Hydrogen Energy* 42:21832–21840.

[26] Lodziana, Z., Debski, A., Cios, G. and Budziak, A. (2019). Ternary $LaNi_{4.75}M_{0.25}$hydrogenstoragealloys: Surface segregation, hydrogen sorption and thermodynamic stability. *Int J Hydrogen Energy* 44:1760–1773.

[27] Srivastava, S. and Srivastava, O.N. (1998). Investigations on synthesis and characterization of $MmNi_{4.3}Al_{0.3}Mn_{0.4}$ and $MmNi_{4.0}Al_{0.3}Mn._{0.4}Si_{0.3}$, hydrogen storage materials through thermal and spin melting processes. *Int J Hydrogen Energy* 23(1):7–13.

[28] Srivastava, S. and Srivastava, O.N. (1998). Investigations on synthesis, characterization and hydrogenation behaviour of the spin and thermal–melted versions of $LaNi_{5-x}Si_x$ (x = 0.1, 0.3, 0.5) hydrogen storage materials. *J Alloys Compd* 267: 240–245.

[29] Srivastava, S. and Srivastava, O.N. (1999). Synthesis, characterization and hydrogenation behaviour of composite hydrogen storage alloys $LaNi_5/La_2Ni_7$, $LaNi_3$. *J Alloys Compd* 282:197–205.

[30] Srivastava, S. and Srivastava, O.N. (1999). Hydrogenation behaviour with regard to storage capacity, kinetics, stability and thermodynamic behaviour of hydrogen storage composite alloy $LaNi_5/La_2Ni_7$, $LaNi_3$. *J Alloys Compd* 290 (1-2): 250–256.

[31] Zhang, Z., Elkedim, O., Balcerzak, M. and Jurczyk, M. (2016). Structural and electrochemical hydrogen storage properties of $MgTiNi_x$ (x = 0.1, 0.5, 1, 2) alloys prepared by ball milling. *Int J Hydrogen Energy* 41: 11761–11766.

[32] Nobuki, T., Crivello, J.C., Cuevas, F. and Joubert, J.M. (2019). Fast synthesis of TiNi by mechanical alloying and its hydrogenation properties. *Int J Hydrogen Energy* 44: 10770–10776.

[33] Puszkiel, J.A., Larochette, P.A., Baruj, A., Meyer, G. and Gennari, F.C. (2016) Hydrogen cycling properties of xMg-Fe materials (x: 2, 3 and 15) produced by reactive ball milling. *Int J Hydrogen Energy* 41: 1688–1698.

[34] Farina, L., Brutti, S., Trequattrini, F., Palumbo, O., Gatto, S., Reale, P., Silvestri, L., Panero, S. and Paolone, A. (2017). An extensive study of the Mg-Fe-H material

obtained by reactive ball milling of MgH_2 and Fe in a molar ratio 3:1. *Int J Hydrogen Energy* 42: 22333–22341.

[35] Gajdics, M., Calizzi, M., Pasquini, L., Schafler, E. and Revesz, A. (2016). Characterization of a nanocrystalline Mg-Ni alloy processed by high-pressure torsion during hydrogenation and dehydrogenation. *Int J Hydrogen Energy* 41: 9803–9809.

[36] Callini, E., Aguey-Zinsou, K.F., Ahuja, R. et al. (2016). Nanostructured materials for solid-state hydrogen storage: A review of the achievement of COST Action MP1103. *Int J Hydrogen Energy* 41: 14404–14428.

[37] Liu, P., Chen, H., Yu, H., Liu, X., Jiang, R. and Li, X., Zhou, S. (2019). Oxygen vacancy in magnesium/cerium composite from ball milling for hydrogen storage improvement. *Int J Hydrogen Energy* 44:13606–13612.

[38] Srivastava, S. and Panwar, K. (2016). Investigations on Microstructures of Ball–milled $MmNi_5$ Hydrogen Storage Alloy. *Materials Research Bulletin* 73 : 284–289.

[39] El-Eskandarany, M.S., Al-Ajmi, F., Banyan, M. and Al-Duweesh, A. (2019). Synergetic effect of reactive ball milling and cold pressing on enhancing the hydrogen storage behavior of nanocomposite MgH_2/10 wt% $TiMn_2$ binary system. *Int J Hydrogen Energy* 44:26428–26443.

[40] Antiqueira, F.J., Leiva, D.R., Zepon, G. and Cunha, B.F.R.F., Figueroa, S.J.A. and Botta, W.J. (2020). Fast hydrogen absorption/desorption kinetics in reactive milled Mg-8 mol% Fe nanocomposites. *Int J Hydrogen Energy* 45:12408–12418.

[41] Fadonougbo, J.O., Jung, J.Y., Suh, J.Y., Lee, Y.S., Shim, J.H., Fleury, E. and Cho, Y.W.(2019). The role of Fe particle size and oxide distribution on the hydrogenation properties of ball-milled nano-crystalline powder mixtures of Fe and Mg. *J Alloys and Compd* 806:1039–1046.

[42] Fadonougbo, J.O.,Jung, J.Y., Suh, J.Y., Lee, Y.S., Shim, J.H. and Cho, Y.W. (2017). Low temperature formation of Mg_2FeH_6 by hydrogenation of ball-milled nano-crystalline powder mixture of Mg and Fe. *Materials and Design* 135:239–245.

[43] Xiao, X., Yu, K., Fan, X., Wu, Z., Wang, X., Chen, C., Wang, Q. and Chen, L.(2011). Synthesis and hydriding/dehydriding properties of nanosized sodium alanates prepared by reactive ball-milling. *Int J Hydrogen Energy* 36: 539–548.

[44] Ismail, M., Zhao, Y., Yu, X.B., Ranjbar, A. and Dou, S.X. (2011). Improved hydrogen desorption in lithium alanate by addition of SWCNT–metallic catalyst composite. *Int J Hydrogen Energy* 36: 3593-3599.

[45] Ortiz, R.G., Tena-Garcia, J.R.,Flores-Jacobo, A. and Suarez-Alcantara, S.(2018). From the can to the tank: $NaAlH_4$ from recycled aluminium. *Int J Hydrogen Energy* 43: 20183–20190.

[46] Ali, N.A., Idris, N.H., Sazelee, N.A., Yahya, M.S., Yap, F.A.H. and Ismail, M. (2019). Catalytic effects of $MgFe_2O_4$ addition on the dehydrogenation properties of $LiAlH_4$. *Int J Hydrogen Energy* 44: 28227–28234.

[47] Pei, Z.W., Wu, C., Bai, Y., Liu, X. and Wu, F. (2017). $NaNH_2$-$NaBH_4$ hydrogen storage composite materials synthesized via liquid phase ball-milling: Influence of Co-Ni-B catalyst on the dehydrogenation performances. *Int J Hydrogen Energy* 42:14725–14733.

[48] Manna, J., Tougas, B. and Huot, J. (2018). Mechanical activation of air exposed TiFe + 4 wt% Zr alloy for hydrogenation by cold rolling and ball milling. *Int J Hydrogen Energy* 43: 20795–20800.

[49] Gu, J., Gao, M., Wen, L., Huang, J., Liu, Y.F. and Pan, H. (2015). Improved hydrogen storage properties of combined $Ca(BH_4)_2$ and $LiBH_4$ system motivated by addition of $LaMg_3$ assisted with ball milling in H_2. *Int J Hydrogen Energy* 40:12325-12335.

[50] Aoyagi, H., Aoki, K. and Masumoto, T. (1995). Effect of ball milling on hydrogen absorption properties of FeTi, Mg_2Ni and $LaNi_5$. *J Alloys Compd* 231:804−809.

[51] Joseph, B. and Schiavo, B. (2009). Effects of ball−milling on the hydrogen sorption properties of $LaNi_5$. *J Alloys Compd* 480: 912−916.

[52] Esquivel, M.R. and Meyer, G. (2007). A comparison of the evolution during the mechanical alloying of both a $MmNi_5$−Ni and a Mm−Ni mixtures: Stages of milling and microsructural characterization. *J Alloys Compd* 446−447: 212−217.

[53] Porcu, M., Petford-Long, A.K. and Sykes, J.M. (2008). TEM studies of Nb_2O_5 catalyst in ball-milled MgH_2 for hydrogen storage. *J Alloys Compd* 453:341−346.

[54] Srivastava, S. and Upadhyaya, R.K. (2011). Investigations of AB_5-type hydrogen storage materials with enhanced hydrogen storage capacity. *Int J Hydrogen Energy* 36: 7114-7121.

[55] Liang, G., Huot, J. and Schulz, R. (2001). Hydrogen storage properties of the mechanically alloyed $LaNi_5$-based materials. *J Alloys Compd* 320:133-139.

[56] Joseph, B., Schiavo, B., Staiti, G.D. and Sekhar, B.R. (2011). An experimental investigation on the poor hydrogen sorption properties of nano-structured $LaNi_5$ prepared by ball-milling. *Int J Hydrogen Energy* 36: 7914-7919.

[57] Corre, S., Bououdina, M., Kuriyama, N., Fruchart, D. and Adachi, G. (1999). Effects of mechanical grinding on the hydrogen storage and electrochemical properties of $LaNi_5$. *J Alloys Compd* 292:166-173.

[58] Panwar, K. and Srivastava, S. (2018). Investigations on calculation of heat of formation for multi-element AB_5-type hydrogen storage alloy. *Int J of Hydrogen Energy* 43:11079−11084.

[59] Panwar, K. and Srivastava, S. (2019). On structural model of AB_5-type multi-element hydrogen storage alloy. *Int J Hydrogen Energy* 44: 30208-30217.

[60] Talaganis, B.A., Esquivel, M.R. and Meyer, G. (2011) Improvement of as-milled properties of mechanically alloyed $LaNi_5$ and application to hydrogen thermal compression. *Int J Hydrogen Energy* 36:11961-11968.

[61] Oelerich, W., Klassen, T. and Bormann, R. (2001). Comparison of the catalytic effects of V, V_2O_5, VN, and VC on the hydrogen sorption of nanocrystalline Mg. *J Alloys Compd* 322: L5-L9.

Chapter 3

Elemental Metal Hydride MgH_2 for Promising Hydrogen Storage Applications

Satish Kumar Verma and Thakur Prasad Yadav*

Hydrogen Energy Centre, Department of Physics, Institute of Science, Banaras Hindu University, Varanasi, India

Abstract

Metal hydrides are the most promising materials for hydrogen storage in solid-state mode. A variety of metal hydrides, such as AB, A_2B, AB_2, AB_5, and elemental-type, are well known. Due to very high hydrogen storage capacity, reversibility, light-weight, and low cast, elemental hydrides are the critical materials for the hydrogen storage application among these available metal hydrides. Based on the availability (3rd most abundant in sea-water and 8th largest in earth crust), storage capacity (7.6 wt %), low cost, and light-weight, MgH_2 is the vital candidate of the elemental hydrides for hydrogen storage application. However, MgH_2 has some issues related to its high thermodynamics and sluggish kinetics. These issues have been encountered through different techniques like alloying, scaffolding, nanoconfinement and additive/catalyst. This book chapter shall discuss elemental hydride, especially MgH_2 for hydrogen storage material, in detail.

Keywords: hydrogen storage, metal hydrides, elemental hydrides, magnesium hydride

---

*Corresponding Author's Email: yadavtp@gmail.com.

In: Materials Science: Future Aspects
Editors: Kalpana Awasthi, Arti Srivastava and Mridula Tripathi
ISBN: 978-1-68507-843-0

1. Introduction

Hydrogen storage is a crucial issue for the hydrogen economy, cutting across production, distribution, and applications. Hydrogen storage can rightly be said as a bottleneck to realizing the application of hydrogen as a fuel. The implementation of hydrogen energy systems relies on hydrogen fuel cell/IC engine-based vehicles and other portable and stationary hydrogen-powered devices. Specifically, on-board hydrogen storage systems need to enable hydrogen fuel cell/IC engine vehicles performance comparable to conventional motorized vehicles. However, this is not easy to achieve with existing pressurized and liquefied hydrogen storage techniques [1]. The hydrogen storage in solid-state mode is comparatively safe, efficient, and cost-effective than liquid and compressed gases mode. The solid-state materials like hydrides, unlike gaseous storage, do not require very high pressure (>200 atm), nor do they require low temperature (~ -252°C) as needed for storage as a liquid [2, 3]. The United States Department of Energy (US-DOE) targets [4] for on-board hydrogen storage systems by 2025 serve as a helpful reference, and viable hydrogen storage materials are accessed concerning these targets. It should be noted that the DOE 2025 targets for gravimetric and volumetric hydrogen densities of 5.5 wt% and 40 g-H_2/L are system-based, which include the weight and volume of the container and controlling accessories. Also, cyclability up to 1500 cycles of de/re-hydrogenation, enthalpy ~30-40 kJ/mol is the US-DOE criteria that need to be satisfied by viable storage materials. Hydrides generally have relatively high gravimetric and volumetric densities but fall short of the DOE targets regarding reversibility, kinetics, and operating temperature. By contrast, physisorbents store hydrogen reversibly with fast kinetics but suffer from low gravimetric and volumetric hydrogen densities and weak hydrogen binding under ambient conditions. In the upcoming section, we will discuss various hydrides highlighting the advantage they offer in storing hydrogen and work that needs to be done to make them viable hydrogen storage materials.

2. Properties of MgH_2 and Related Challenges

Decades of research reveal that MgH_2 is the most prominent candidate for hydrogen storage in solid-state mode. It has several properties which make it front running material for hydrogen storage. It has a high hydrogen content of 7.60 wt% (gravimetric) and 110 g/L (volumetric) [5, 6]. Our planet has plenty of Mg metal; 8th most abundant in the earth crust (Figure 1) and 3rd most abundant in sea-water (Table 1). Apart from these merits, it also has a light-weight and low cost [7, 8].

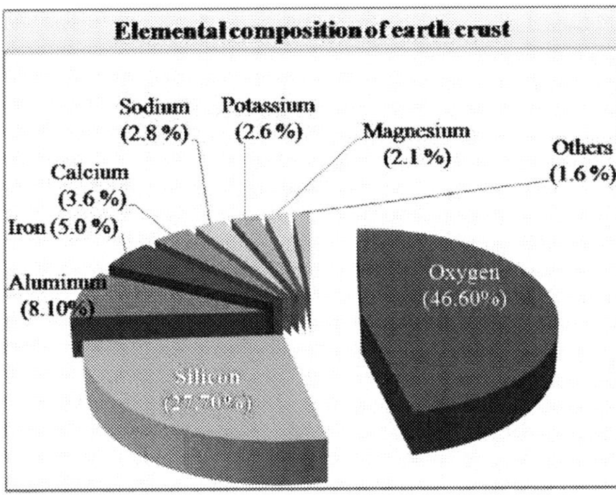

Figure 1. Pi-chart for elemental composition of the earth's crust.

Table 1. Table for chemical composition in sea-water

Chemical constituent	g/kg of seawater	moles/kg	Relative concentration
Chloride	19.162	0.5405	1.0000
Sodium	10.679	0.4645	0.8593
Magnesium	1.278	0.0526	0.0974
Sulfate	2.680	0.0279	0.0517
Calcium	0.4096	0.01022	0.0189
Potassium	0.3953	0.01011	0.0187
Carbon (inorganic)	0.0276	0.0023	0.0043
Bromide	0.0663	0.00083	0.00154
Boron	0.0044	0.00041	0.00075
Strontium	0.0079	0.00009	0.000165

Despite these merits, MgH_2 has some certain challenges associated with its high thermodynamics (high reaction enthalpy ~74 kJ/mol) and kinetics barrier (sluggish hydrogen de/re-hydrogenation kinetics < 0.4 kgH_2/min) [9, 10]. Different roots have been adopted to defeat these challenges associated with MgH_2.In the present scenario, the promising methods used for improving the hydrogen sorption properties and thermodynamics of $MgH_2$are alloying, scaffolding, nanoconfinement, composite with complex hydride, and the addition of catalyst [10, 11]. Different research groups have adopted various methods for making MgH_2 a viable hydrogen storage material [12–18]. Therefore, a detailed description of MgH_2 is given in the "MgH_2 as hydrogen storage material" section.

3. Synthesis Roots of Elemental Metal Hydride MgH_2

Decades of research reveal that MgH_2 can be synthesized in different ways like high energy milling, chemical reaction method, etc. This section describes well the various synthesis roots for MgH_2.

3.1. High-Pressure Direct Hydrogenation Synthesis Root

The synthesis of MgH_2 was first reported by Wiberg et al., [19] in 1951. They have synthesized MgH_2 with direct magnesium metal (Mg) hydrogenation under a very high hydrogen pressure of 200 atm at a high temperature of 570°C. However, they have reported that even under 160 atm, 500°C metallic Mg can not absorb the hydrogen without any catalyst. Therefore, they used MgI_2 as a catalyst for this synthesis process which works as a hydrogen carrier.

$$Mg + H_2 \xrightarrow[200\text{ atm, }570\text{ °C}]{MgI_2} MgH_2 \quad (1)$$

Zaluska et al., [20] have also reported the synthesis of MgH_2 from the nanocrystalline Mg power hydrogenated under low pressure (~ 30 atm) and high temperature (~500°C). The Mg nanopowder was produced through the high-energy ball-milling under the argon atmosphere with a variable milling time of 20 hours.

3.2. Chemical Reaction Method

The most satisfactory synthesis for MgH_2 was designed by Barbaras et al., in 1951 using the chemical reaction method. They have prepared MgH_2 using chemical treatment of diethyl magnesium with lithium aluminum hydride in diethyl ether solution [21]. A simple chemical reaction method for preparing MgH_2 was reported by Michael J. Michalczyk [22]. In this study, MgH_2 has synthesized by using phenyl silane's reaction with dibutyl magnesium. The prepared magnesium hydride can be found from the reaction mixture either as a solvated hydride or as a N,N,N',N'-tetramethyl ethylene-diamine (TMEDA) adduct. The mild conditions synthesis of highly reactive MgH_2 was reported in 1980 by Bogdanovic et al., [23]. They have synthesized MgH_2 from a three-step chemical reaction using anthracene and magnesium powder under 20-60°C with THF and CrTi as catalysts. The chemical reaction [23] for the synthesis of MgH_2 can be given as-

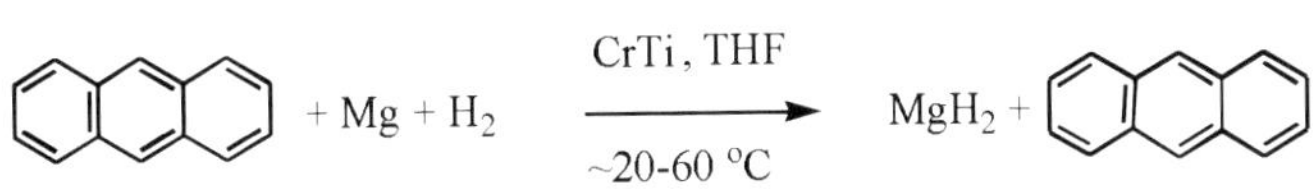

(2)

3.3. Autocatalytic Reaction Method

It is a simple and novel method for the synthesis of MgH_2. In this method, MgH_2 can be synthesized using Mg powder with MgH_2 as a self-catalyst. This method was designed and developed by Bhatnagar et al., [24]. They have synthesized MgH_2 under the various concentration of catalyst (MgH_2) ball-milled with Mg powder under different pressure (15, 30, 45 atm) and temperatures of 350°C followed by heat treatment. They found Mg catalyzed with 5 wt% of MgH_2 under 30 atm hydrogen pressure and annealed at 350°C for 10 hours, resulting in the formation of MgH_2 having X-ray diffraction similar to but obtained from Alfa-Aesar. The synthesized MgH_2 had shown 6.60 wt% of hydrogen capacity with fast kinetics (absorb 6.6 wt% in 30 minutes at 300°C under 20 atm and desorb 6.0 wt% in 30 minutes at 300°C under 1atm H_2 pressure). This synthesized MgH_2 has better kinetics and low cost (~ 4 times low) than standard MgH_2 of Alfa-Aesar.

3.4. Solvothermal Method

Solvothermal is a technique used to prepare various materials like metals, semiconductors, polymers, etc., with high yield [25]. A typical pressure range from 1 atm to 10000 atm during the synthesis of materials with a temperature range from 100°C to 1000°C is used. In this process, if the solvent used is water, the method is called hydrothermal. The hydrothermal treatment generally performs below the supercritical temperature of water (374°C). This process can prepare geometric samples like thin films, bulk powders, single crystals, nanomaterials, etc.

In 1993, Koerner et al., [26] filed a United States patent to describe the detailed solvothermal synthesis of MgH_2. They have synthesized MgH_2 using Mg with 1.2-2.0 wt% of MgH_2 inside a stainless steel autoclave under constant stirring followed by heat treatment at 350°C at ~ 5 atm hydrogen pressure. In addition, a multichannel recorder was employed for recording the variation in temperature and pressure during the hydrogenation process at 225°C under a hydrogen pressure of 5.8 atm for 7 hours. After cooling, the reactor is evacuated and then flushed with argon. Next, the magnesium hydride is filled with a free-flowing powder into a baked-out vessel filled with argon.

4. MgH_2 as Hydrogen Storage Material

Different research groups have done several investigations to examine MgH_2 as the most prominent material for hydrogen storage. In addition, they have reported various techniques for making MgH_2 a viable storage material. This section describe show MgH_2 shows the possible hydrogen storage material.

4.1. Mechanical Alloying

A widely used method of synthesizing nanocrystalline materials like metal nitrides, oxides at room temperature is mechanical alloying (MA) [18, 27]. Therefore, this method is the most suitable method for synthesizing the Mg-based materials for hydrogen storage applications [28].

Shang et al., [29] have synthesized the alloy of MgH_2 with different metals like Al, Ti, Fe, Ni, Cu, and Nb for hydrogen storage applications. They have reported that the MA of MgH_2 with various metals formed solid solutions of (Mg: M)Hx, (M= Al, Ti, Fe, Ni, Cu, and Nb), and some new phases depend on the nature of metals. For example, Ti and Nb are prominent in reacting with hydrogen to form hydride (like TiHx and NbHx), but Al, Ni, Cu, and Fe do not make their hydrides under standard temperature and pressure. Therefore in this solid solution, the alloy atoms destabilize the Mg-H bond into the Mg lattice. As a result, the Mg-H bond gets weaker resultant the dissociation temperature and re/de-hydrogenation kinetics get improved. The alloy of MgH_2 with metallic additives LaNiFe, PdFe, FeMnTi, and non-metallic additive Si have been synthesized by Reule et al., [30]. They have reported that the alloy with metallic additives shows better desorption kinetics than alloy with non-metallic additives.

In addition to MA, that leads to the formation of fine nanocrystalline particles, which is likely to provide rapid enhanced absorption/desorption kinetics.

4.2. Scaffolding

In a combined, theoretical and experimental study done by Vajeeston et al., [17], they have found that numerous nanoparticles of complex hydride have their particle size nearly order of 2 nm. But it is not very easy to synthesize the nanoparticles of the order ~ 2 nm by using the available experimental tools like milling. So, it is essential to develop a new tool by which these nanoparticles can be synthesized or isolated efficiently.

The scaffold is a technique used for holding and isolating the materials in an elevated work platform and a supporting structure. Nielsen et al., [16] have reported the MgH_2 nanoparticles embedded with nanoporous carbon aerogel scaffold material for hydrogen storage application. Sievert's type instruments were used for the study of hydrogen storage properties. Based on hydrogen de/re-hydrogenation kinetics and thermal desorption study, they have reported that dehydrogenation kinetics depends on the pore size of the scaffold material. The small size pores mediated faster desorption kinetics due to a reduction in the size of scaffold MgH_2. Thus it can be expected that the removal of hydrogen from the center of the scaffold MgH_2 is easier than bulk MgH_2.

4.3. Nanoconfinement

The various ways have been studied to significantly modify the thermodynamic and kinetic properties of the magnesium hydride material. However, along with kinetics and thermodynamics, the reversibility of the material is also an essential property of hydrogen storage materials [31]. In many cases, complex hydride materials have high hydrogen storage capacity but are poor in reversibility under moderate temperature and pressure conditions. Most likely, limited reversibility of the materials is often due to the formation of stable intermediates, by-products, and significant change in its microstructures. This section addresses the nanoconfinement of the MgH_2 for hydrogen storage application.

The nanoconfinement of the MgH_2 in porous materials for improving its hydrogen storage properties (like kinetics, reversibility, and thermodynamics) can be considered [16, 32, 33]. For example, the nanoconfined MgH_2 confined in mesopores have been synthesized using the Mg melt infiltration process followed by a direct hydrogenation process [33] and reported a significant change in kinetics and thermodynamics of the prepared samples. Another study has also done by Utke et al., [34] and reported the $LiBH_4$-MgH_2 nanoconfined in carbon aerogel with enhanced reversible hydrogen storage properties.

4.4. Addition of Additive or Catalyst

For a long time, the research has believed that the dopping or addition of a catalyst(s) is the most general and effective technique for making MgH_2 a viable hydrogen storage material. In the past few years, several catalysts have been investigated by different research groups for catalyzing the MgH_2 [12–14, 35, 36]. Bhatnagar et al., [13] have synthesized the graphene templated Fe_3O_4 catalyst for improving the hydrogen storage properties of MgH_2. They have reported that MgH_2 catalyzed by 5 wt% of graphene templated Fe_3O_4 (MgH_2-Fe_3O_4@GS) shows the lower hydrogen desorption temperature of 262°C is 142°C more inferior than the as-received MgH_2. The Mg-Fe_3O_4@GS absorb 6.20 wt% of hydrogen content within 2.50 minutes, and the formation enthalpy for MgH_2-Fe_3O_4@GS is 60.62 kJ/mol of H_2. Based on kinetics and thermodynamics studies, our previous study [12] revealed that TiH_2@Gr is the best catalyst for catalyzing MgH_2 than TiO_2@Gr and Ti@Gr.

Liang et al., [37] have reported the MgH_2 catalyzed by transition metal (like Ti, V, Mn, Fe, Ni) and found they are the superior catalysts for improving the kinetics of the MgH_2. The MgH_2 catalyzed by 5 wt% of highly crumpled graphene was reported by Liu et al., [38]. They have written that graphene works as a catalyst for enhancing the kinetics of MgH_2 and works as a pulverizer. Our recent study [39] has reported the synthesis of SrF_2@Gr additive, which simultaneously improved the kinetics and thermodynamics of the MgH_2. The MgH_2 catalyzed by SrF_2@Gr has released the 6.20 wt% of hydrogen within 15 minutes at 290°C under 1 atm H_2 pressure with desorption enthalpy 67.60 kJ/mol-H_2.

5. Possible Mechanism for Hydrogen Storage in MgH_2

Different research groups have given various hypotheses on mechanisms to understand the uptake and release of the hydrogen from the magnesium/ magnesium hydride.

Vajeeston et al., [17] have reported a mechanism for dehydrogenation of MgH_2 based on Mulliken population analysis. To understand the mechanism of hydrogen desorption from MgH_2, the bond overlap population (BOP) value ware calculated, providing us information about the character of the bonding interaction between atoms. A high BOP value reports a strong covalent bond, while a low BOP value reports an ionic bond. In the case of MgH_2, the Mg-H bond should be weaker for a fast hydrogen release system. The computed BOP value for the Mg-H bond (bulk phase) is 0.64 (which is smaller than the B-H (borohydride) bond) indicating the ionic interaction. Since ionic bonds are weaker than covalent thus, hydrogen can quickly get released from MgH_2 during dehydrogenation. Bhatnagar et al., [40] have reported the catalyst migration concept using TiH_2 catalyst during hydrogen release and uptake by the MgH_2/Mg. They have said that catalyst (TiH_2) migrated towards the surface of the MgH_2 matrix during rehydrogenation, while during dehydrogenation, it migrated inside the Mg matrix. Our previous study reported forming a non-stoichiometric phase of TiH_2 that is $TiH_{1.924}$, during dehydrogenation of MgH_2, which plays a crucial role in enhancing the de/re-hydrogenation kinetics of MgH_2. The formation of the non-stoichiometric phase is only possible due to the presence of the most oxidizing group (-COOH) attached with the catalyst in the form of graphene. The schematic diagram reported for the mechanism of this study by Verma et al., [12] is given in Figure 2.

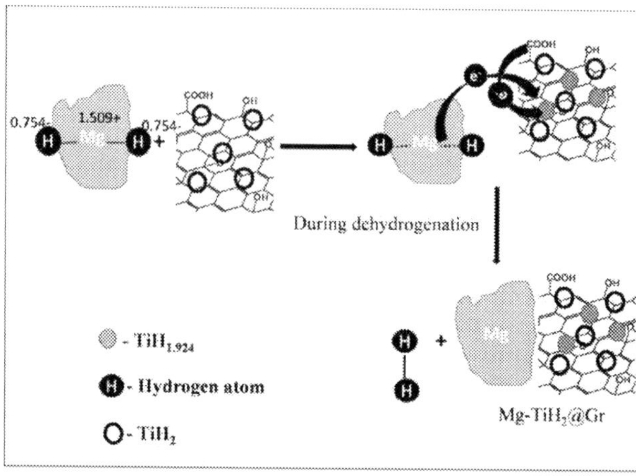

Figure 2. Schematic diagram for the catalytic action of TiH_2@Gr on MgH_2 during dehydrogenation [12].

6. Future Prospects

The area of the research field is very dynamic and broad with anticipation of wide-scale future implementation for developments in particular areas. These specific fields guide the current and future works for solving particular issues. The major problems associated with Mg/MgH_2 material for hydrogen storage applications are high thermodynamics stability, slow kinetics, and poor cyclability. Different research groups have built various approaches, which are discussed in this chapter. Different groups have been reported the MgH_2 catalyzed by graphene and graphene templated metals like Ti, Fe, Cu, etc. The investigations reported by these studies indicate that further research on graphene and possibly other 2D material as a template for the existing catalyst may open new avenues for catalysis in light-weight hydride. There may be an adventure for design, such as a new bi-metallic catalyst, and its templated version on graphene or other 2D materials open new era of research. Although still there is a particular gap between practical applications and theoretical predictions of the nanostructured Mg. The synthesis and modification of nanostructured Mg will be developed rapidly for the attention about the hydrogen storage. Therefore, it is no doubt that Mg-based alloys will play an essential role in utilizing hydrogen energy in the future. The future perspective and research directions of Mg/MgH_2 for hydrogen storage applications are also shown in Figure 3.

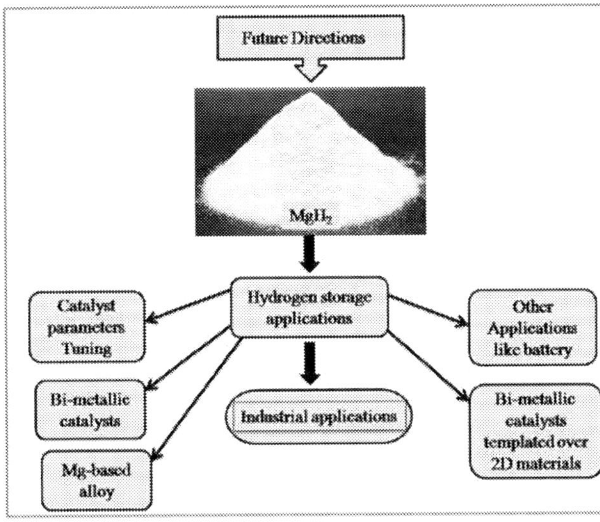

Figure 3. Schematic diagram for future research directions of Mg/MgH_2.

Conclusion

In this chapter, we have summarized the synthesis processes, different properties, and applications of the MgH_2 for hydrogen storage. The low cost, higher availability, higher hydrogen content (7.6 wt% gravimetric and 110 g/L volumetric), cyclic stability, higher reversibility, and lower toxicity of Mg/MgH_2 makes it a front running candidate for hydrogen storage applications. Different advanced synthesis tools can overcome the slow kinetics and high thermodynamic barriers associated with MgH_2. Many research groups have studied 3D Mg/MgH_2 with varying types of catalysts and composite materials, but still, only a few of them were made on other advanced structures such as Mg films, nanowires, and nanoparticles/clusters. Hydrogen production as an energy vector should be considered for clean, green, renewable, and eco-friendly energy systems. The results of different reported studies reveal the nanostructured MgH_2 is the most promising option for the future clean, renewable, alternative, and sustainable energy storage materials. The various nanostructures have exhibited different properties that significantly affect de/re-hydrogenation kinetics and thermodynamics. In addition, the nanoconfined Mg and scaffold nano Mg within porous carbon structures can be considered as key hydrogen storage materials for sustainable and renewable energy applications.

Acknowledgments

Author (S.K.V.) would like to thank the Council of Scientific and Industrial Research (CSIR) New Delhi, India, for providing financial assistance as CSIR-Senior Research Fellowship (Award No. 09/013(0872)/2019-EMR-I).

References

[1] Crivello J-C., Dam B., Denys R. V., Dornheim M., Grant D.M., Huot J., et al., (2016). Review of magnesium hydride-based materials: development and optimisation. *App. Phys. A* 122:97. doi:org/10.1007/s00339-016-9602-0.

[2] Wang H., Wu G., Cao H., Pistidda C., Chaudhary A-L., Garroni S., et al., (2017). Near Ambient Condition Hydrogen Storage in a Synergized Tricomponent Hydride System. *Adv. Energy Mats.* 7:1602456. doi:org/10.1002/aenm.201602456.

[3] Bhatnagar A., Gupta B.K., Tripathi P., Veziroglu A., Hudson M.S.L., Shaz M.A., et al., (2017). Development and Demonstration of Air Stable rGO-EC@AB5 Type Hydrogenated Intermetallic Hybrid for Hydrogen Fuelled Devices. *Adv. Sustainable Systems* 1:1700087. doi:org/10.1002/adsu.201700087.

[4] U.S Deparment of Energy 2017. *DOE Technical Targets for Onboard Hydrogen Storage for Light-Duty Vehicles.* https://energy.gov/eere/fuelcells/doe-technical-targets-onboard-hydrogen-storage-light-duty-vehicles.

[5] Lotoskyy M., Denys R., Yartys V.A., Eriksen J., Goh J., Nyamsi S.N., et al., (2018). An outstanding effect of graphite in nano-MgH2–TiH2 on hydrogen storage performance. *J. Mat. Chem. A* 6:10740–54. doi:org/10.1039/C8TA02969E.

[6] Liu G., Wang Y., Qiu F., Li L., Jiao L., Yuan H. (2012). Synthesis of porous Ni@rGO nanocomposite and its synergetic effect on hydrogen sorption properties of MgH2. *J. Mat. Chem.* 22:22542–9. doi:org/10.1039/C2JM34730J.

[7] Yartys V.A., Lototskyy M. V., Akiba E. Albert R., Antonov V.E., Ares J.R., et al., (2019). Magnesium based materials for hydrogen based energy storage: Past, present and future. *Int. J. Hyd. Energy* 44:7809–59. doi:org/10.1016/j.ijhydene.2018.12.212.

[8] Pandey S., Bhatnagar A., Shahi R., Hudson M.S., Singh M., Srivastava O.N. (2013). Effect of TiO2 Nanoparticles on the Hydrogen Sorption Characteristics of Magnesium Hydride. *J. Nanosci. Nanotech.* 13:5493–9. doi:org/10.1166/jnn.2013.7516.

[9] Wang Y., Li L., An C., Wang Y., Chen C., Jiao L, et al., (2014). Facile synthesis of TiN decorated graphene and its enhanced catalytic effects on dehydrogenation performance of magnesium hydride. *Nanoscale* 6:6684–91. doi:org/10.1039/C4NR00474D.

[10] Lyu J., Lider A., Kudiiarov V. (2019). Using Ball Milling for Modification of the Hydrogenation/Dehydrogenation Process in Magnesium-Based Hydrogen Storage Materials: An Overview. *Metals* 9:768. doi:org/10.3390/met9070768.

[11] Hudson M.S.L., Takahashi K., Ramesh A., Awasthi S., Ghosh A.K., Ravindran P., et al., (2016). Graphene decorated with Fe nanoclusters for improving the hydrogen sorption kinetics of MgH2-Experimental and theoretical evidence. *Cat. Sci. Tech.* 6:261–8. doi:org/10.1039/c5cy01016k.

[12] Verma S.K., Bhatnagar A., Shukla V., Soni P.K., Pandey A.P., Yadav T.P., et al., (2020) Multiple improvements of hydrogen sorption and their mechanism for MgH2 catalyzed through TiH2@Gr. *Int. J. Hyd. Energy* 45:19516–30. doi:org/ 10.1016/ j.ijhydene.2020.05.031.

[13] Bhatnagar A., Pandey S.K., Vishwakarma A.K., Singh S., Shukla V., Soni P.K., et al., (2016). Fe3O4@graphene as a superior catalyst for hydrogen de/absorption from/in MgH2/Mg. *J. Mat. Chem. A* 4:14761–72. doi:org/10.1039/C6TA05998H.

[14] Singh S., Bhatnagar A., Shukla V., Vishwakarma A.K., Soni P.K., Verma S.K., et al., (2020).Ternary transition metal alloy FeCoNi nanoparticles on graphene as new catalyst for hydrogen sorption in MgH2. *Int. J. Hyd. Energy* 45:774–86. doi:org/ 10.1016/J.IJHYDENE.2019.10.204.

[15] Shukla V., Bhatnagar A., Singh S., Soni P.K., Verma S.K., Yadav T.P., et al., (2019). A dual borohydride (Li and Na borohydride) catalyst/additive together with intermetallic FeTi for the optimization of the hydrogen sorption characteristics of $Mg(NH_2)_2$/2LiH. *Dalton Transactions* 48:11391–403. doi:org/10.1039/C9DT0227 0H.

[16] Nielsen T.K., Manickam K., Hirscher M., Besenbacher F., Jensen T.R. (2009). Confinement of MgH2 Nanoclusters within Nanoporous Aerogel Scaffold Materials. *ACS Nano* 3:3521–8. doi:org/10.1021/nn901072w.

[17] Vajeeston P., Sartori S., Ravindran P., Knudsen K.D., Hauback B., Fjellvåg H. (2012). MgH2 in Carbon Scaffolds: A Combined Experimental and Theoretical Investigation. *J. Phys. Chem. C* 116:21139–47. doi:org/10.1021/jp3008199.

[18] Bobet J-L, Chevalier B., Darriet B. (2000). Crystallographic and hydrogen sorption properties of TiMn2 based alloys. *Intermetallics* 8:359–63. doi:org/10.1016/S0966-9795(99)00092-8.

[19] Von Egon Wiberg H.G. and R.B. (1951). Synthesis von magnesium hydrid aus den elementen. *Naturforschung* 6 b:394–5.

[20] Zaluska A., Zaluski L., Ström–Olsen J.O. (1999). Nanocrystalline magnesium for hydrogen storage. *J. Alloys and Comp.* 288:217–25. doi:org/10.1016/S0925-8388 (99)00073-0.

[21] Barbaras G.D., Dillard C., Finholt A.E., Wartik T., Wilzbach K.E., Schlesinger H.I. (1951). The Preparation of the Hydrides of Zinc, Cadmium, Beryllium, Magnesium and Lithium by the Use of Lithium Aluminum Hydride1. *J. of the Ame. Chem. Society* 73:4585–90. doi:org/10.1021/ja01154a025.

[22] Michalczyk M.J. (1992). Synthesis of magnesium hydride by the reaction of phenylsilane and dibutylmagnesium. *Organometallics* 11:2307–9. doi:org/10.1021/ om00042a055.

[23] Bogdanović B., Liao S., Schwickardi M., Sikorsky P., Spliethoff B. (1980). Catalytic Synthesis of Magnesium Hydride under Mild Conditions. *Angewandte Chemie International Edition in English* 19:818–9. doi:org/10.1002/anie.198008 181.

[24] Bhatnagar A., Shaz M.A., Srivastava O.N. (2019). Synthesis of MgH2 using autocatalytic effect of MgH2. *Int. J. Hyd. Energy* 44:6738–47. doi:org/10.1016/j.ijhydene.2019.01.163.

[25] Gygi D., Bloch E.D., Mason J.A., Hudson M.R., Gonzalez M.I., Siegelman R.L., et al., (2016). Hydrogen Storage in the Expanded Pore Metal–Organic Frameworks M2(dobpdc) (M = Mg, Mn, Fe, Co, Ni, Zn). *Chem. of Mat.* 28:1128–38. doi:org/10.1021/acs.chemmater.5b04538.

[26] Wilfried Knott, Klaus-Dieter Klein GK. *United States Patent* 1993. (19) 5,198,207.

[27] Chen Y. (1998). Different oxidation reactions of ilmenite induced by high energy ball milling. *J. Alloys and Comp.* 266:150–4. doi:org/10.1016/S0925-8388(97)00494-5.

[28] Sun D., Enoki H., Gingl F., Akiba E. (1999). New approach for synthesizing Mg-based alloys. *J. Alloys and Comp.* 285:279–83. doi:org/10.1016/S0925-8388(98)01037-8.

[29] Shang C.X., Bououdina M., Song Y., Guo Z.X. (2004). Mechanical alloying and electronic simulations of (MgH2+M) systems (M=Al, Ti, Fe, Ni, Cu and Nb) for hydrogen storage. *Int. J. Hyd. Energy* 29:73–80. doi:org/10.1016/S0360-3199(03)00045-4.

[30] Reule H., Hirscher M., Weißhardt A., Kronmüller H. (2000). Hydrogen desorption properties of mechanically alloyed MgH2 composite materials. *J. Alloys and Comp.* 305:246–52. doi:org/10.1016/S0925-8388(00)00710-6.

[31] Huen P., Paskevicius M., Richter B., Ravnsbæk D.B., Jensen T.R. (2017). Hydrogen Storage Stability of Nanoconfined MgH2 upon Cycling. *Inorganics* 5. doi:org/10.3390/inorganics5030057.

[32] Jia Y., Sun C., Cheng L., Abdul Wahab M., Cui J., Zou J., et al., (2013). Destabilization of Mg–H bonding through nano-interfacial confinement by unsaturated carbon for hydrogen desorption from MgH2. *Phys. Chem. Chem. Phys.* 15:5814–20. doi:org/10.1039/C3CP50515D.

[33] Gosalawit–Utke R., Thiangviriya S., Javadian P., Laipple D., Pistidda C., Bergemann N., et al., (2014). Effective nanoconfinement of 2LiBH4–MgH2 via simply MgH2 premilling for reversible hydrogen storages. *Int. J. Hyd. Energy* 39:15614–26. doi:org/10.1016/j.ijhydene.2014.07.167.

[34] Utke R., Thiangviriya S., Javadian P., Jensen T.R., Milanese C., Klassen T., et al., (2016). 2LiBH4–MgH2 nanoconfined into carbon aerogel scaffold impregnated with ZrCl4 for reversible hydrogen storage. *Mat. Chem. Phys.* 169:136–41. doi:org/10.1016/j.matchemphys.2015.11.040.

[35] Zhou C., Fang Z.Z., Ren C., Li J. (2013). Effect of Ti Intermetallic Catalysts on Hydrogen Storage Properties of Magnesium Hydride. *The J. Phys. Chem. C* 117:12973–80. doi:org/10.1021/jp402770p.

[36] Youn J-S., Phan D-T., Park C-M., Jeon K-J. (2017). Enhancement of hydrogen sorption properties of MgH2 with a MgF2 catalyst. *Int. J. Hyd. Energy* 42:20120–4. doi:org/10.1016/j.ijhydene.2017.06.130.

[37] Liang G., Huot J., Boily S., Van Neste A., Schulz R. (1999). Catalytic effect of transition metals on hydrogen sorption in nanocrystalline ball milled MgH2-Tm (Tm

= Ti, V, Mn, Fe and Ni) systems. *J. Alloys and Comp.*292:247–52. doi:org/10.1016/S0925-8388(99)00442-9.

[38] Liu G., Wang Y., Xu C., Qiu F., An C., Li L., et al., (2013). Excellent catalytic effects of highly crumpled graphene nanosheets on hydrogenation/dehydrogenation of magnesium hydride. *Nanoscale* 5:1074–81. doi:org/10.1039/c2nr33347c.

[39] Shukla V., Bhatnagar A., Verma S.K., Pandey A.P., Vishwakarma A.K., Srivastava P., et al., (2021). Simultaneous improvement of kinetics and thermodynamics based on SrF2 and SrF2@Gr additives on hydrogen sorption in MgH2. *Mat. Adv.* 2:4277–90. doi:org/10.1039/D1MA00012H.

[40] Bhatnagar A., Johnson J.K., Shaz M.A., Srivastava O.N. (2018). TiH2 as a Dynamic Additive for Improving the De/Rehydrogenation Properties of MgH2: A Combined Experimental and Theoretical Mechanistic Investigation. *The J. Phys. Chem. C* 122: 21248–61. doi:org/10.1021/acs.jpcc.8b07640.

Chapter 4

The Magical Green Fuel: Hydrogen

Vivek Shukla[*] and Thakur Prasad Yadav
Hydrogen Energy Center, Department of Physics, Institute of Science, Banaras Hindu University, Varanasi, India

Abstract

The increasing gap between supply and demand of energy and climate change forces us to an abrupt search for sustainable solutions that can also provide long-term perspective. Major issues about which fuel or fuels will emerge and how many alternative fuels would replace gasoline remain unresolved. Renewable energy sources like solar and wind produce intermittent electricity, while large-scale electricity storage and alternative transportation fuel options remain restricted. Hydrogen is the only renewable energy source that may be used for both business and residential purposes. In a comprehensive clean-energy paradigm, hydrogen is the optimal form of energy storage for transit and conversion. This chapter examines hydrogen as an alternative energy source. The goal is to use hydrogen to generate electricity. Despite the apparent benefits of relying on hydrogen to meet our future energy demands, there are several technological, political, and socio-economic hurdles to overcome before hydrogen becomes a viable choice. Hydrogen is more challenging to store, and extracting pure hydrogen, despite its abundance, could be costly. Nevertheless, hydrogen can be utilized to make electric automobiles, boats, electric motors, and various household energy applications.

Keywords: green fuel, hydrogen energy, hydrogen production, hydrogen storage

---

[*] Corresponding Author's Email: vivekshukla8888@gmail.com.

In: Materials Science: Future Aspects
Editors: Kalpana Awasthi, Arti Srivastava and Mridula Tripathi
ISBN: 978-1-68507-843-0

1. Introduction

Energy is inevitable for all living organisms' survival, irrespective of whether it is a single cellular or multicellular organism. Since the industrial revolution, fossil fuels have become the major source of energy for improving the living standard of human life on earth. At present, nearly 80% of world energy demand is satisfied by fossil fuels. However, fossil reserves are limited on the earth's crust, and it is geographically distributed non-uniformly. On the other hand, the more usage of fossil fuels, particularly in the industrial and transportation sectors, adds more greenhouse gas emissions causing adverse effects to the environment. In addition to fossil fuels, there are other energy sources that can be derived from non-renewable sources, such as solar and wind. However, different sectors rely on different forms of energy. For example, the industrial and transportation sector largely depends on fossil fuels such as coal and gasoline, leading to the significant contribution of greenhouse gas emissions. Our energy choices and decisions impact the earth's natural systems that we may not be fully aware of. So we must carefully choose our energy sources [1-2]. It can be anticipated that it is directly related to any country's economy [3]. As we are already witnessing the adverse effect of climate change due to greenhouse gas emissions from fossil fuels, there is an urgent action required to step up from fossil fuels to clean and renewable energy forms [4]. Some of the issues which need to be circumvented are the following.

- Climate change effect (Global warming)
- The gap between demand and supply energy is increasing
- Depletion of fossil fuels
- Economic burden (Large import bills for oils)
- Urban air pollution

Globally, the total primary energy production had increased from 35.54 Quadrillions Btu (Btu is short term of British thermal unit, and 1 Btu = 1.06kJ) in 1950 to 87.60 Quadrillions Btu in 2017. On the other hand, the consumption had far increased from 34.61 Quadrillions Btu in 1950 to 97.59 Quadrillions Btu in 2017. This production and consumption profile estimates that this gap between demand and supply is going to increase further in the forthcoming years. This is confirmed by a report of (Energy Information Administration EIA), Primary Energy Consumption by Source, (http://www.eia.doe.gov/

emeu/aer/txt/ptb0103.html). United States Energy Information Administration (EIA) has predicted that from 2018 to 2050 that world energy consumption will grow by nearly 50% [5-6].

The estimated prediction for global energy consumption from 2010 to 2050 is shown in Figure 1. From this, it clear that daily consumption of energy is growing very fast day by day. Figure 1(b) depicts the year-wise consumption of various natural energy sources. With continuous exhaustion of fossil fuels renewable dependency of energy sources is increasing year by year as fossil fuel, and other natural resources are getting exhausted. The estimated prediction for all-natural resources' exhaustion has been shown in Figure 1(c). Figure 1(e) depicts the global emission of CO_2 from 1990 to 2020. Figure1 (a-e) depicts the ever increasing emission of CO_2.

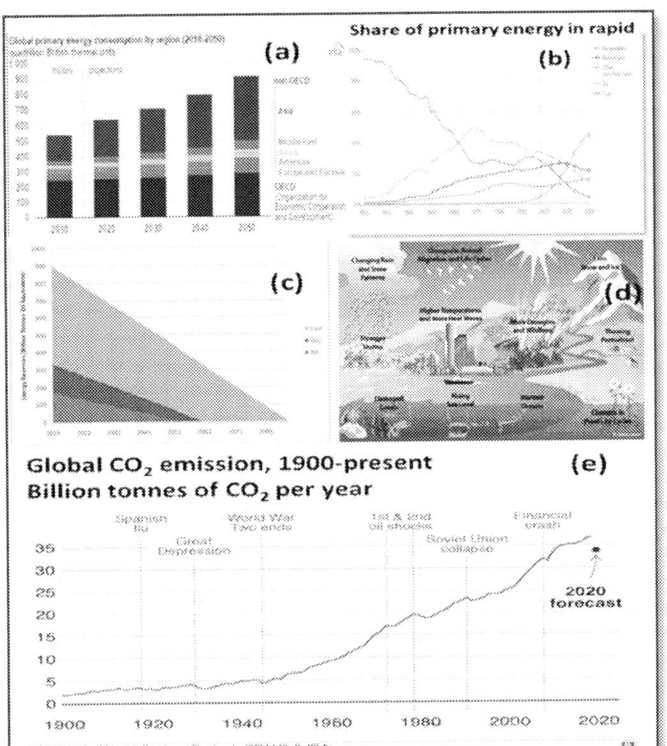

Figure 1. Global energy consumption and CO_2 emission scenario (a) Worldwide primary energy consumption (b) Share of primary energy in rapid (c) Natural resources (Coal, Gas, Oil) limit (d) Effect of global warming (e) CO_2 emission scenario from 1990 to present.

CO_2 emissions from the use of oil, gas, and coal have increased dramatically over the world. While conventional energy sources have transformed the world, the carbon seeping into our atmosphere has driven up global temperatures by just over 1°C since the mid-1850. The major part of the CO_2 that gets emitted by industry, transportation, and cooking is 38% of the total CO_2 emission (Figure 2). Thus, if we continuously increase the emission of CO_2, the earth's surface temperature could rise by 3-4°C by the end of this century, and it will affect our ecosystem. The CO_2 concentration presently in the atmosphere is 413 ppm. It should be cut down by 50%, so that the temperature of the earth does not exceed the 2°C (limit set by Paris convention 2016, UNFCCC (2016)).

Because of the above said scenario, the focus should naturally shift to renewable energies. Decades of research have shown that hydrogen as a fuel is possibly the best option amongst all present sustainable solutions [7-10].

2. Hydrogen Energy

Hydrogen is found in water (H_2O) and is the most abundant element on the planet. Water is composed of two-thirds hydrogen and can be found in conjunction with other elements. It can be used as a fuel to generate electricity once separated. Hydrogen is a powerful energy source that can be used to power ships, vehicles, homes, industries, and rockets. It is completely renewable, can be produced on demand, and emits no toxic gases into the atmosphere. Although hydrogen has approximately three times the energy content of a fossil fuel such as oil, it has a low density and less energy content on a volumetric basis. The problem of hydrogen storage remains unresolved. Hydrides, which are currently being extensively researched, are thought to be an efficient and safe storage mode. However, no hydride with all of the required properties has yet been discovered. Some hydrides, particularly lightweight hydrides, are on the verge of becoming viable.

In fact, hydrogen's ideal means energy is related to clean, completely green, and renewable fuel [11-12]. The element most abundant in the universe is hydrogen. The majority of hydrogen is present in the form of water. The chemical energy content (per mass) of hydrogen is 142 MJ kg^{-1} which is three times higher than that of liquid hydrocarbons (e.g., gasoline it is 47 MJ kg^{-1}) [13-15].

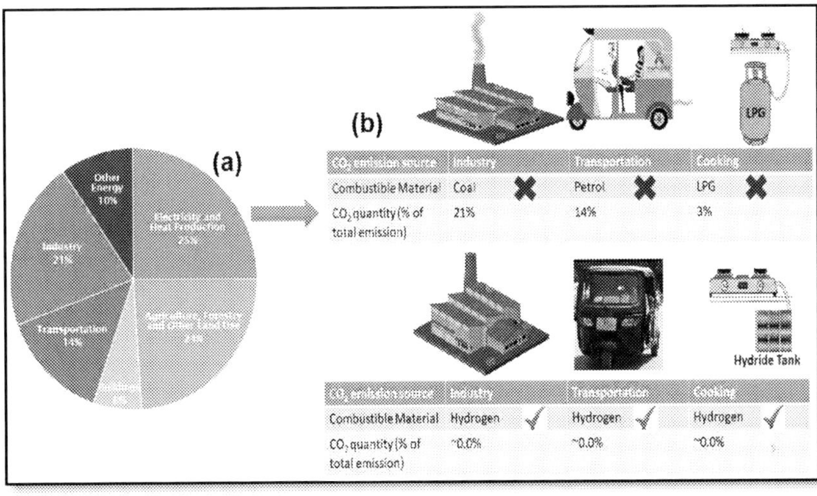

Figure 2. (a) Global CO_2 emission by different sectors (b) Contribution in the emission of CO_2 using natural resources and using hydrogen as fuel in industry, transportation and in cooking.

Hydrogen can be produced from water. It burns back to water through hot combustion in internal combustion (I.C.) engine and cold combustion in a fuel cell [16]. The following equation shows the combustion of hydrogen with oxygen,

$$2H_2\,(g) + O_2\,(g) \rightarrow 2H_2O\,(l) + 572\ kJ \qquad (1)$$

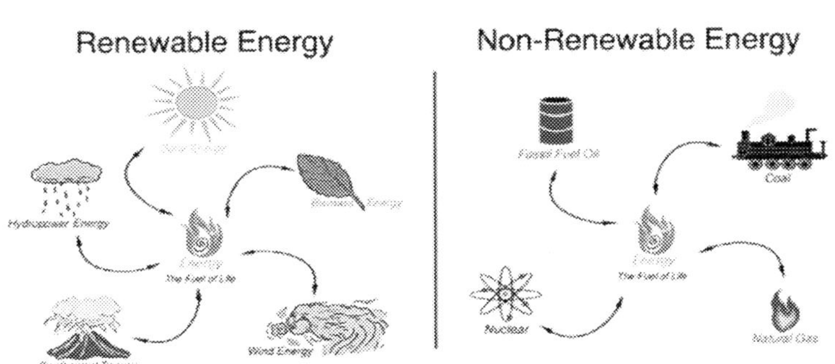

Figure 3. Symbolic representation of renewable as well as non-renewable energy.

The observed enthalpy change for hydrogen combustion is -286 kJmol-1 [17]. On a mass basis, the energy released by hydrogen combustion is greater than that released by any other fuel. Hydrogen's low heating value (LHV) is 2.4, 2.8, and 4 times greater than that of methane, gasoline, and coal, respectively [18]. The cold combustion of hydrogen in fuel cells leads to electrical power generation; however, hot combustion in motor vehicles' IC engines similarly offers motive power [19]. Thus, hydrogen is abundant, indigenous, clean, and climate-friendly fuel. Thus it can solve the quadruple problems of the gap in demand and supply of fossil fuel, climate change, and economic burden [19]. Symbolic representation of renewable as well as non-renewable energy has been given in Figure 3.

2.1. Hydrogen as Energy Carrier

It provides the long-term prospect of abundant clean energy supplies. Hydrogen is regarded as a promising candidate for mobile applications [20-21]. It is clean (pollution-free), renewable, and environmentally friendly, as opposed to petroleum fuels. The vast majority of hydrogen is chemically bound in water as H_2O, and some is bound to liquid or gaseous hydrocarbons. The most environmentally friendly method of producing hydrogen from water is to use sunlight in conjunction with photovoltaic cells to perform water electrolysis. It is the long-term solution to reducing global fossil fuel consumption and combating the effects of climate change [22-23]. Water is emitted during both the cold and hot combustion processes [24-25]. One of the primary benefits of using hydrogen as a fuel is that it produces no CO_2.

In the case of hydrogen, the great attention of this is its pollution-free combustion. The chemical element hydrogen has the symbol H and the atomic number one. With a standard atomic weight of 1.008, hydrogen is the lightest element in the periodic table. Hydrogen is the most abundant chemical element in the universe, accounting for roughly 75% of all baryonic mass. In the form of water, hydrogen is the most abundant element on the planet (H_2O). In the form of molecular hydrogen, it is less than 1% is present in the atmosphere. However, some of the hydrogen is attached to liquid hydrocarbons. Because of its low density, hydrogen was first used to fill balloons and airships. However, it reacts violently with oxygen (to form water), and its future as a filling material for airships ended when the Hindenburg caught fire. The most environmentally friendly method of producing hydrogen from water is to combine photovoltaic cells and water electrolysis with sunlight.

2.2. Hydrogen Cycle

The photovoltaic cell is most commonly used to convert sunlight energy to electricity. Water electrolysis is powered by electricity generated from renewable energy sources. The oxygen is released into the atmosphere during the electrolysis process, and the hydrogen is then stored, transported, and distributed. Finally, hydrogen and oxygen are combusted, releasing energy as heat. After burning, the H_2 releases water and steam into the atmosphere. Henceforth, hydrogen forms a clean and closed cycle.

2.3. Physical and Chemical Properties of Hydrogen

The element hydrogen is the lightest in the periodic table. Hydrogen has three isotopes: Protium, Deuterium, Tritium. The tritium isotope is the most common of the three common isotopes, with only one proton and one electron. The clean combustion of hydrogen has sparked a lot of interest. Hydrogen atoms can easily combine to form H_2 molecules. These molecules are smaller in size than the molecules of most other gases. Hydrogen in its molecular form is colourless, odourless, and tasteless. It is about 14 times lighter than air (density of hydrogen is 0.0000899 g/cm^3 at 1 atm) and diffuses faster (diffusion constant ~ 0.756 cm^2/s at 20°C) than any other gas. The physical properties of hydrogen are given in Table.1.

Because of its high gravimetric energy density of 120 MJ/kg, hydrogen has attracted interest as a potential future energy carrier for mobile and stationary applications (i.e., 2-3 times of natural gas or gasoline); and being a non-polluting fuel (H_2O and energy are the only byproducts) at the point of use [27]. However, on a volume basis, hydrogen has only about a quarter of gasoline's energy density (8 MJ/L for liquid hydrogen versus 32 MJ/L for gasoline. This constitutes a drawback and a challenge for hydrogen to be used efficiently. The following tables give the properties of hydrogen.

The physical and chemical properties of hydrogen together with other liquid hydrocarbon are given below in Table. 2. The great attention of hydrogen is its pollution-free combustion. The combustion of hydrogen is completely clean.

Table 1. Physical properties of hydrogen [28]

	Property	Values and Units
	Molar mass	2.016 kg $kmol^{-1}$
	Particular gas constant	4124 J Kg^{-1} K^{-1}
	Calorific value (gravimetric)	120 MJ kg^{-1} = 33.33 $kWhkg^{-1}$
At triple point	Temperature	-259.35°C (13.80 K)
	Pressure	0.07 bar
	Density gaseous	0.125 kg m^{-3}
	Density liquid	77 kg m^{-3}
	Heat of fusion	58.5 kJ kg^{-1} = 16.25 $kWhkg^{-1}$
At boiling point	Boiling temperature	-252.85°C (20.30 K)
At 1.01325 bar	Heat of vaporization	445.4 kJ kg^{-1}= 123.7 $kWhkg^{-1}$
Liquid phase	Density	70.8 kg m^{-3}
	Calorific value (volumetric)	8.5 MJ dm^{-3} = 2.36 $kWhkg^{-1}$
	Specific heat capacity C_p	9.8 kJ $kg^{-1}K^{-1}$
	Specific heat capacity C_v	5.8 kJ $kg^{-1}K^{-1}$
	Thermal conductivity	0.099 $Wm^{-1}K^{-1}$
	Dynamic viscosity	11.9×10^{-6} N s m^{-2}
	Speed of sound	1089 m s^{-1}
Gaseous phase	Density	1.34 kg m^{-3}
	Calorific value (volumetric)	0.16 MJ dm^{-3} = 0.044 $kWhdm^{-3}$
	Specific heat capacity C_p	12.2 kJ $kg^{-1}K^{-1}$
	Specific heat capacity C_v	6.6 kJ $kg^{-1}K^{-1}$
	Thermal conductivity	0.017 $Wm^{-1}K^{-1}$
	Dynamic viscosity	1.11×10^{-6} N s m^{-2}
	Speed of sound	355 m s^{-1}
At critical point	Temperature	-239.95°C (33.20 K)
	Pressure	13.1 bar
	Density	31.4 kg m^{-3}
At standard conditions (0°C and 1.01325 bar)	Density	0.09 kg m^{-3}
	Calorific value (volumetric)	0.01 MJ dm^{-3} = 2.8 $Whdm^{-3}$
	Specific heat capacity C_p	14.32 kJ $kg^{-1}K^{-1}$
	Specific heat capacity C_v	10.17 kJ $kg^{-1}K^{-1}$
	Thermal conductivity	0.184 $Wm^{-1}K^{-1}$
	Coefficient of diffusion	0.61 cm^2s^{-1}
	Dynamic viscosity	8.91×10^{-6} N s m^{-2}
	Speed of sound	1246 m s^{-1}
Mixtures with air	Lower explosion limit	4 vol% H_2 (λ=10.1)
	Lower detonation limit	18 vol% H_2 (λ=1.9)
	Stoichiometric mixture	29.6 vol% H_2 (λ=1)
	Upper detonation limit	58.9 vol% H_2 (λ=0.29)
	Upper explosion limit	75.6 vol% H_2 (λ=0.13)
	Ignition temperature	585°C (858 K)
	Minimal ignition energy	0.017 mJ
	Maximum Laminar flame speed	~ 3 m s^{-1}
	Adiabatic combustion temperature	~ 2100°C

Table 2. Selected physical properties of hydrogen, methane, propane, and gasoline [29]

Property	Hydrogen	Methane	Propane	Gasoline
Molecular weight (u*)	2.02	16.04	44.06	~107
Density (kg/m^3) at normal conditions	0.084	0.651	1.87	4.4
Buoyancy (density with respect to air)	0.07	0.55	1.52	3.4 to 4.0
Diffusion coefficient (cm^2/s)	0.61	0.16	0.12	0.05
Lean flammability limit in air (% by volume)	4.1	5.3	2.1	1.0
Rich flammability limit in air (% by volume)	75	15	10	7.8
Minimum ignition energy (mJ)	0.02	0.29	0.26	0.24
Minimum self-ignition energy (K)	858	813	760	501 to 744
Lean detonability limit in air (% by volume)	18	6.3	3.1	1.1
Rich detonability limit in air (% by volume)	59	13.5	7.0	3.3
Explosion energy (kg equivalent TNT per m^3 of vapour)	2.02	7.02	20.2	44.2
Molecular weight (u*)	2.02	16.04	44.06	~107
Density (kg/m^3) at normal conditions	0.084	0.651	1.87	4.4
Buoyancy (density with respect to air)	0.07	0.55	1.52	3.4 to 4.0
Diffusion coefficient (cm^2/s)	0.61	0.16	0.12	0.05
Lean flammability limit in air (% by volume)	4.1	5.3	2.1	1.0

*Unified atomic mass unit or atomic mass unit (u) = 1.660538782(83) × 10^{-24}g.

3. Current Hydrogen Energy Scenario

The need for immediate action to address the [19] twin threats of fossil fuel depletion and the onset of extreme climate change is now universally acknowledged. Despite hydrogen's many advantages, a number of issues remain to be addressed before hydrogen can become a feasible option for the future. Significant technological advancements in all aspects will require the transition's implementation of the hydrogen cycle [27, 30]. The three main elements for implementing the hydrogen economy as a viable option in the future are given in the following.

- Hydrogen production
- Storage of hydrogen and its transmission
- Transforming hydrogen into operative power

Hydrogen must be produced since hydrogen is not freely available in nature (the majority is present in the form of water). There are numerous methods for producing hydrogen. However, the majority of hydrogen is currently produced through steam reforming of natural gas or higher hydrocarbons [31]. However, to visualize a viable hydrogen economy, hydrogen needs to be produced based on the electricity from sustainable energy sources or directly from photolysis [32-33]. Since the production of electricity (and thereby H_2) from sustainable sources is intermittent and seldom fits the energy demand. Especially for the transportation sector, the need for an intermediate energy storage medium is evident. It is about 60% of the world's oil consumption. This is equal to 25% of the world's energy consumption [34]. Hydrogen economy can be realized in functional form by storage of hydrogen in metal hydrides and energy conversion in an internal combustion engine or in the fuel cell. The fuel cell provides efficiency twice that of an internal combustion engine [35-36]. Therefore, the fuel cell is preferred due to the Carnot cycle efficiency [37] however, the fuel cell's cost is an issue now a day. The earlier report of US-DOE suggests that the "Hydrogen and Fuel Cells Program Plan" (2010), an improved internal combustion engine (ICE), has also been considered a viable option. It may be mentioned that whereas the internal combustion engine (ICE) is a developed technology, but Fuel Cells are not a pretty industrialized technology yet. The clean and closed hydrogen cycle, describing production, storage, and utilization, is illustrated in Figure 4.

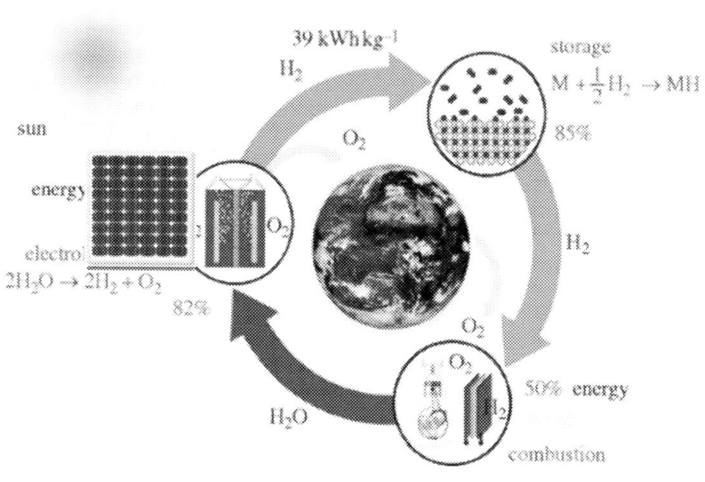

Figure 4. The clean and closed cycle of hydrogen energy [26].

3.1. Hydrogen Production

Hydrogen production systems are a type of industrial method for producing hydrogen gas. Currently, the majority of hydrogen (95 percent) is produced from fossil fuels via steam reforming natural gas, partial oxidation of methane, and coal gasification. These methods embody emission of CO_2. Therefore they cannot form sustainable method of hydrogen production. Other methods of producing hydrogen include biomass gasification and water electrolysis/ photoelectrolysis (Figure 5). Keeping in view the climate change effects, those hydrogen production methods that use renewables as the input energy, e.g., PV-based electrolysis, are the preferred routes of hydrogen production. Because hydrogen is required for many essential chemical processes, hydrogen production is critical in any industrialized society [38]. As of 2019, approximately 70 million tonnes of hydrogen are produced annually worldwide for a variety of uses, including oil refining, the production of ammonia (Haber process) and methanol (carbon monoxide reduction), and as a transportation fuel. In 2017, the hydrogen generation market is expected to be worth US$115.25 billion [35-36].

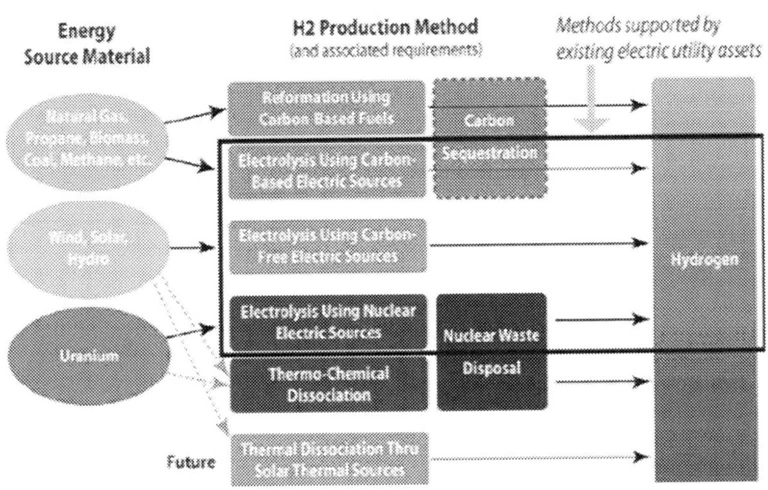

Figure 5. Various method used for Hydrogen production [39].

Hydrogen is an abundant element on earth (15 at.%), and the most widespread element in the universe (93 at.%). The earth's crust's mean hydrogen content is 1.4 gkg^{-1} [39]. Hydrogen is generally found in the form of water or hydrocarbon. Less than 1% of its content as molecular hydrogen is present in the atmosphere.

The source of hydrogen (natural gas or water) and an energy source are needed for producing hydrogen. It may be identical, such as when hydrogen is produced from liquid hydrocarbons. The production of hydrogen by the electrolysis of water has shown in Figure 6. For this, electricity is the energy source, and water is the hydrogen source. Currently, nearly 96% of all hydrogen productions worldwide use fossil fuels (natural gas) (Figure 6). For the production of hydrogen, steam reforming is the most used and cheaper method.

In association with partial oxidation routes, it is likely to use heavier hydrocarbons, such as coal. The need for capital investment is required for the production of hydrogen through electrolysis than steam reforming. One crucial thing about hydrogen produced from electrolysis is that the cost of this method would be down.

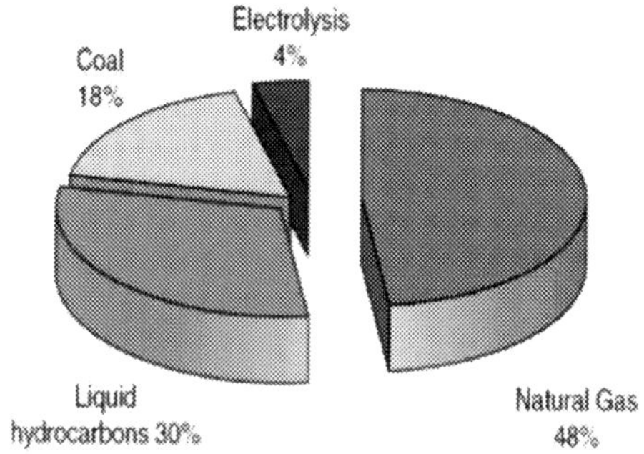

Figure 6. Key energy sources currently used for hydrogen production [40].

3.2. Technical Targets for On-Board Hydrogen Storage Systems

Hydrogen at ambient conditions is a gas (14 times lighter than air). Therefore unlike petroleum it is not self-storable. It has to be stored. For specific purposes e.g., for use of hydrogen in vehicular transport, the storage has to meet certain specifications. The U.S. Department of Energy's technical targets for on-board vehicular hydrogen storage in 2020 and 2025 serve as a useful benchmark for comparing different storage methods [37]. Table. 3 lists several of the current DOE technical targets. Temperature, min/max delivery pressure, re-filling time, cycle life, and fuel purity are all important operational parameters for a storage system's performance.

Table 3. Technical targets for on-board hydrogen storage systems

Storage Parameters	2020	2025	Ultimate
System Gravimetric Capacity (kWh/kg) Usable, specific-energy from H_2 (net useful energy/max system mass)(kg H_2/kg system)	1.5 (0.045)	1.8 (0.055)	2.5 (0.075)
System Volumetric Capacity: Usable energy density from H_2\$/kWh net (\$/kg H_2) \$/gge at pump	 TBD TBD 4	 TBD TBD 4	 TBD TBD 4
Durability/Operability Operating ambient temperature (°C) Min/max delivery temperature(°C) Operational cycle life (1/4 tank to full) (Cycles) Min delivery pressure from storage system; FC=fuel cell, ICE= internal combustion engine [bar (abs)] Max delivery pressure from storage system[bar (abs)] Onboard Efficiency (%) Well" to Power plant Efficiency (%)	 ~40/60(sun) ~40/85 1500 5FC/35ICE 5 90% 60%	 ~40/60 (sun) ~40/85 1500 5FC/35ICE 5 90% 60%	 ~40/60 (sun) ~40/85 1500 5FC/35ICE 5 90% 60%
Charging/discharging Rates System full time (for 5 kg H_2) (min) Minimum full flow rate(kg H_2/min) Start time to full flow (20°C) (s) Start time to full flow (-20°C) (s) Transient response 10%-90% and 90%-0% (s)	 3-5 min 0.02 5 15 0.75	 3-5 min 0.02 5 15 0.75	 3-5 min 0.02 5 15 0.75
Fuel Purity (H_2 from Storage) % H_2	Meet or exceed SAE J2719		
Environmental Health & Safety Permeation & leakage (Scc/h) Toxicity Safety Loss of useable H_2(g/h)kg H_2 stored	Meet or exceed SAE J2719 for system safety		

Storage of hydrogen through physisorption is an attractive way to store the hydrogen. The adsorbed hydrogen does not react chemically during adsorption, so it does not accumulate impurities that can contaminate downstream fuel cells. The physisorption process is also very fast and completely reversible because it does not involve bulk solid diffusion or chemical dissociation, which can satisfy both cycle time and replenishment time.

Table 4. Compares physisorption to the five other basic storage methods [38-41]

Material	Typical framework elements	Surface area (m^2g^{-1})	Porosity (cm^3g^{-1})	Excess hydrogen capacity at 77K and 20 bar	
				(Wt.%)	(KgH_2l^{-1})
Porous carbons	C	3150	1.95	6.9	--
CNTs	C	1160	--	3.8*	--
Zeolites	O, Al, Si	670	--	2.2	0.031
MOFs	H, O, C, TM	2200	0.89	6.1	0.039
PIMs	H, O, C	1050	0.40	2.7**	--
Graphene	C	925	--	1.38*	--

The essential concern with physisorption-primarily based totally garage is that the hydrogen density at ambient situations is simply too low because of the vulnerable binding interplay among the H_2 and the adsorbent surface. Vehicles require a compact, light, safe, and low-priced containment for on-board strength garage. A modern, commercially to be had automobile optimized for mobility with a number four hundred km makes use of approximately 24 kg of petrol in a combustion engine; to cowl the equal distance, an electric powered automobile with a gas mobileular calls for eight kg hydrogen withinside the combustion engine model or four kg hydrogen with inside the electric powered automobile with a gas mobileular. At room temperature and atmospheric pressure, four kg of hydrogen has a quantity of forty five m^3. This corresponds to a balloon with a diameter of five m, that's infrequently a sensible answer for a vehicle [38]. The required technical goals for on-board hydrogen garage structures are proven in Table 3.

3.3. Hydrogen Storage

At 273 K under 1 bar pressure, hydrogen is in the gaseous state with a density of 0.089886 $kgm^{-3,}$ and at 14 K, hydrogen becomes solid. Hydrogen is a liquid between the triple and critical points having a density of 70.8 kgm^{-3} at the temperature of 20 K. Hydrogen gas is described by the Van der Waals (vdW) equation at ambient temperature (298.15 K).

$$P(V) = \frac{nRT}{(V-(nb))} - a\frac{n^2}{V^2} \tag{2}$$

where 'P' denotes the gas pressure, 'V' the volume, 'T' the absolute temperature, 'n' the number of moles, 'R' the universal gas constant, 'a' the repulsion constant or dipole interaction, and 'b' the volume occupied by the hydrogen molecules [44]. The gas's low critical temperature (Tc=33.0 K) is due to the strong repulsive interaction between hydrogen atoms. Between the solids line and the line from the triple point at 21.2 K and the critical point at 32.0 K, there is liquid hydrogen [45].

3.3.1. Storage of Hydrogen in Gaseous State

The most established storage system for hydrogen is physical storage in the form of pressurized hydrogen gas. As previously stated, hydrogen has a very low density of 0.089 kg/m^3, necessitating high pressure or extremely low temperatures for storage [46-47]. Current modern fuel cells require hydrogen to be pressurized to 35-70 MPa. Theoretically, pressurization affects the hydrogen energy content of 11-13% [48]. Due to its extreme lightness, there is a risk of hydrogen leakage from the high-pressure vessel. Steel and aluminum are the traditional materials used in commercial hydrogen storage tanks. When compared to steel or aluminum containers, impact-resistant and strong enough to provide safety in the event of a crash, carbon fiber reinforced plastic containers are an easier solution, but are too expensive and pose different challenges for future cost savings [49-50]. Scientists and engineers are also looking for more cost-effective and practical alternatives. It is assumed that the amount of hydrogen stored is large or the storage time is relatively long. In this case, the compressed hydrogen may be stored in a salt dome or other large underground storage facility in a suitable geological structure. Storing hydrogen in salt caves is a promising prospect for seasonal high-pressure hydrogen storage and can release hydrogen in a reasonable time. Even at high pressure, the salt cave is very resistant to hydrogen and effectively prevents leakage. When solar and wind power are less active on a calm or cloudy day, hydrogen can be extracted from the cave and burned in a combined cycle power plant that produces electricity. For a long time, there was no suitable infrastructure to efficiently burn pure hydrogen, but such infrastructure is now being developed [51]. A recent example is the underground hydrogen storage project in Romania, part of a European assessment project known as the Hy Under project, supported by the Fuel Cells and Hydrogen Joint Venture (FCH JU). It is easier to store enough hydrogen underground for future use, especially in the chemical, transportation and salt industries [52-53]. Despite promising options, the relatively low hydrogen density, extremely high gas pressure, cost, and system safety issues remain major obstacles to this technologically simple and well-proven technology.

3.3.2. Liquid State Storage System

Cryogenic liquids are another way to physically store hydrogen. The higher the density of the liquid, the easier it is to store. At a normal boiling point of 20 K, the density of liquid hydrogen is about 71 g/l. This is about 1.8 times the density of hydrogen at 288 K and pressures up to 70 MPa. Because liquid hydrogen has a low boiling point, the cooling technology requires extremely low temperatures, consuming about 30% of the total energy [54]. As a result, special double-wall vessels with effective thermal insulation systems are required to reduce heat leakage. As a result, more compact and lighter cryogenic pressure vessels offer greater safety benefits than compressed hydrogen vessels. However, the persistent boiloff of hydrogen and the excessive energy required for liquefaction limit the potential use of liquid hydrogen storage systems to applications requiring high energy density as well as uses where hydrogen cost does not matter and consumption occurs in a short period of time, such as air and space and automotive applications [55].

3.3.3. Storage of Hydrogen in the Form of Hydrides

Storage of hydrogen in the structure or on the surface of certain materials can be achieved. However, in the chemical compounds (built-in hydrides) undergo a reaction to release hydrogen. Hydrogen atoms are bonded strongly in the chemical hydrides. However, these are lightweight, having a large hydrogen storage capacity (~3 to 19 wt.%). In the metal and alloys, the hydrogen is chemically absorbed directly into the storage media through absorption. The hydrogen is adsorbed on the surface of storage media for a porous material. Hydrogen is chemically bonded with metal atoms in chemical hydrides, where the dehydrogenation is done through thermal decomposition. Hydrogen storage materials are classified into four general categories based on the phenomenon of hydrogen storage. Metal or Intermetallic hydrides

- Porous materials
- Complex hydrides
- Chemical storage of hydrogen

Through physisorption, hydrogen can be stored on the surface of a porous material as hydrogen molecules (H2) or hydrogen atoms (H). Figure 7(a) depicts the adsorption phenomenon. Hydrogen molecules dissociate into hydrogen atoms during absorption (Figure 7(b) and 7(c)), which are incorporated into the solid lattice framework. This method may allow for the storage of larger amounts of hydrogen in smaller volumes at low pressure and near-room temperature. Finally, as chemical compounds containing hydrogen atoms, hydrogen can be strongly bound within molecular structures (Figure 7(d)).

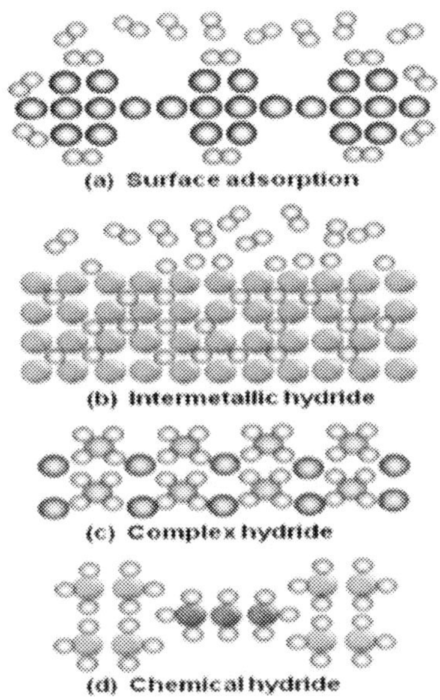

Figure 7. Classification of hydrogen storage materials based on the nature of hydrogen storage [28].

Conclusion

Hydrogen provides a future fuel solution. It enables energy chains that are safe, dependable, and completely sustainable. The great potential of hydrogen is that it can be converted back to water after being used as a fuel in IC engines or fuel cells. In the form of water, clean and green hydrogen is abundant. It has three times the chemical energy content of the other liquid hydrocarbon. Furthermore, the environmental and public health benefits of direct hydrogen fuel are sufficient to justify moving forward based on what we already know about fossil fuels, their consequences, and their limitations.

References

[1] Ley, M. B., Jepsen, L. H., Lee, Y. S., Cho, Y. W., Bellosta von Colbe, J. M., Dornheim, M., Rokni, M., Jensen, J. O., Sloth, M., Filinchuk, Y. (2014), Complex hydrides for hydrogen storage—New perspectives. *Mater. Today* 17: 122–128.

[2] Rude, L. H., Nielsen, T. K., Ravnsbaek, D. B., Bösenberg, U., Ley, M. B., Richter, B., Arnbjerg, L. M., Dornheim, M., Filinchuk, Y., Besenbacher, F. (2011), Tailoring properties of borohydrides for hydrogen storage: A review. *Phys. Status Solidi A* 208: 1754–1773.

[3] Srivastav O. N., Yadav T. P., Shahi R. R., Pandey S. K., Shaz M. A., Bhatnagar A. (2015). Hydrogen energy in India: storage to application, *Proceding of Indian National Science Acadmey* 81: 915-937.

[4] Hudson M. S. L., Dubey P. K., Pukazhselvan D., Pandey S. K., Singh R. K., Raghubanshi H. (2009). Hydrogen energy in changing environmental scenario: Indian context, *Int. J. Hydrogen Energy* 34: 7358–7367.

[5] Bhatnagar A., Pandey S. K., Vishwakarma A. K., Singh S., Shukla V., Soni P. K., Srivastava O. N. (2016). Fe3O4@graphene as a superior catalyst for hydrogen de/absorption from/in MgH2/Mg. *J. Mat. Chem. A* 4: 14761–7.

[6] Paskevicius, M., Jepsen, L. H., Schouwink, P., Cern ˇ ý, R., Ravnsbaek, D. B., Filinchuk, Y., Dornheim, M., Besenbacher, F., Jensen, T. R. (2017). Metal borohydrides and derivatives—Synthesis, structure and properties. *Chem. Soc. Rev.* 46: 1565–1634.

[7] Jepsen, L. H., Ley, M. B., Lee, Y. S., Cho, Y. W., Dornheim, M., Jensen, J. O., Filinchuk, Y., Jørgensen, J. E., Besenbacher, F., Jensen, T. R. (2014). Boron–nitrogen based hydrides and reactive composites for hydrogen storage. *Mater. Today* 17: 129–135.

[8] https://en.wikipedia.org/wiki/Carbon_dioxide_in_Earth%27s_atmosphere.

[9] Bréon F. M., Broquet G., Puygrenier V., Chevallier F., Xueref-Remy I., Ramonet M., Dieudonné E., Lopez M., Schmidt M., Perrussel O., Ciais P. (2015). An attempt at estimating Paris area CO_2 emissions from atmospheric concentration measurements. *Atmospheric Chemistry and Physics*, 15(4): 1707-24.

[10] Taub D. Effects of rising atmospheric concentrations of carbon dioxide on plants. (2010). *Nature Education Knowledge.* 1(8).

[11] Bogdanovic, B., Schwickardi, M. (1997.) Ti-doped alkali metal aluminium hydrides as potential novel reversible hydrogen storage materials. *J. Alloys Compound.* 253–254: 1–9.

[12] Züttel, A., Wenger, P., Rentsch, S., Sudan, P., Mauron, P., Emmenegger, C. (2003). LiBH4 a new hydrogen storage material. *J. Power Sources* 118: 1–7.

[13] Chen, P., Xiong, Z., Luo, J., Lin, J., Tan, K. L. (2002). Interaction of hydrogen with metal nitrides and imides. *Nature* 420: 302–304.

[14] Soulié, J. P., Renaudin, G., Cerný, R., Yvon, K. (2002). Lithium boro-hydride LiBH4: I. Crystal structure. *J. Alloys Compd.* 346: 200–205.

[15] Wang, J., Li, H. W., Chen, P. (2013), Amides and borohydrides for high-capacity solid-state hydrogen storage—Materials design and kinetic improvements. *MRS Bull.* 38: 480–487.

[16] Nakamori, Y., Kitahara, G., Orimo, S. Synthesis and dehydriding studies of Mg–N–H systems. (2004), *J. Power Sources* 138: 309-312.

[17] Smallbone A., Jia B., Atkins P., Roskilly A. P. (2020). The impact of disruptive powertrain technologies on energy consumption and carbon dioxide emissions from heavy-duty vehicles. *Energy Conversion and Management:* 1;6:100030.

[18] Wu Z., Wee V., Ma X., Zhao D. (2021). Adsorbed Natural Gas Storage for Onboard Applications. *Advanced Sustainable Systems.* 5:2000200.

[19] Ouyang L., Chen K., Jiang J., Yang X. S., Zhu M. (2020). Hydrogen storage in light-metal based systems: A review. *J. Alloys and Comp* 829: pp, 154597.

[20] Jena, P. (2011). Materials for hydrogen storage: past, present, and future. *The Journal of Physical Chemistry Letters*, *2*; 3: 206-211.

[21] Rzeszotarska M., Czujko T., Polański M. (2020). Mg_2 (Fe, Cr, Ni) HX complex hydride synthesis from austenitic stainless steel and magnesium hydride. *Int. J. Hydrogen Energy* 31: 19440-54.

[22] Lai Q., Pratthana C., Yang Y., Rawal A., Aguey-Zinsou K. F. (2021). Nanoconfinement of Complex Borohydrides for Hydrogen Storage. *ACS Applied Nano Materials* 4(2):973-8.

[23] Ali, N. A., & Ismail, M. (2021). Modification of NaAlH4 properties using catalysts for solid-state hydrogen storage: A review. *International Journal of Hydrogen Energy*, *46*: 766-782.

[24] Leng, H. Y., Ichikawa, T., Hino, S., Hanada, N., Isobe, S., Fujii, H. (2004), New Metal–N–H System Composed of Mg(NH2)2 and LiH for Hydrogen Storage. *J. Phys. Chem. B* 108: 8763–8765.

[25] Luo, W. (LiNH2–MgH2): A viable hydrogen storage system. (2004). *J. Alloys Compd.* 381: 284–287.

[26] Zuttel A., Borgschulte A., Schlapbach L., *Hydrogen as a Future Energy Carrier.* WILEY-VCH, ISBN: 978-3-527-30817-0, 200.

[27] Xiong, Z., Hu, J.; Wu, G.; Chen, P.; Luo, W.; Gross, K.; Wang, J. (2005). Thermodynamic and kinetic investigations of the hydrogen storage in the Li–Mg–N–H system. *J. Alloys Compd.* 398: 235–239.

[28] Hirscher M., Yartys V. A., Baricco M., von Colbe J. B., Blanchard D., Bowman Jr R. C., Broom D. P., Buckley C. E., Chang F., Chen P., Cho Y. W. (2020). Materials for hydrogen-based energy storage–past, recent progress and future outlook. *J. Alloys Compd.* 827: 153548.

[29] Bennaceur K., Clark, B., Orr Jr F. M., Ramakrishnan T. S., Roulet C., & Stout E. (2005). Hydrogen: a future energy carrier. *Oilfield Review* 17: 30-41. 2005.

[30] Torre, F., Valentoni, A., Milanese, C., Pistidda, C., Marini, A., Dornheim, M., Enzo, S., Mulas, G., Garroni, S. (2015). Kinetic improvement on the CaH2-catalyzed Mg(NH2)2 + 2LiH system. *J. Alloys Compd.* 645: S284–S287.

[31] Shaw, L. L., Ren, R., Markmaitree, T., Osborn, W. (2008). Effects of mechanical activation on dehydrogenation of the lithium amide and lithium hydride system. *J. Alloys Compd.* 448: 263–271.

[32] Liu, Y., Zhong, K., Luo, K., Gao, M., Pan, H., Wang, Q. (2009). Size-Dependent Kinetic Enhancement in Hydrogen Absorption and Desorption of the Li–Mg–N–H System. *J. Am. Chem. Soc.* 131: 1862–1870.

[33] Orimo, S. I., Nakamori, Y., Eliseo, J. R., Züttel, A., & Jensen, C. M. (2007). Complex hydrides for hydrogen storage. *Chemical Reviews*, *107*: 4111-4132.

[34] Züttel, A., Rentsch, S., Fischer, P., Wenger, P. M. C. E. P., Sudan, P. H., Mauron, P., & Emmenegger, C. (2003). Hydrogen storage properties of LiBH4. *Journal of Alloys and Compounds*, *356*:515-520.

[35] Gross A. F., Vajo J. J., Van Atta S. L., Olson G. L. (2008). Enhanced hydrogen storage kinetics of LiBH4 in nanoporous carbon scaffolds. *The Journal of Physical Chemistry C*. 112(14):5651-7.

[36] Jin, S. A., Lee, Y. S., Shim, J. H., & Cho, Y. W. (2008). Reversible hydrogen storage in LiBH4− MH2 (M= Ce, Ca) composites. *The Journal of Physical Chemistry C*, *112*: 9520-9524.

[37] Matsunaga, T., Buchter, F., Miwa, K., Towata, S., Orimo, S., & Züttel, A. (2008). Magnesium borohydride: A new hydrogen storage material. *Renewable Energy*, *33*: 193-196.

[38] Bououdina, M. (2010). Nanomaterials for hydrogen storage: renewable and clean energy. *Int. J. Nanoelectronics and Materials*, *3*:155-167.

[39] Lototskii, M. V., Tarasov B. P. (2007). Hydrogen energetics: Past, present, prospects. *Rus. J of Gen. Chem.* 77: 660-675.

[40] Surer, M. G., Arat H. T. (2018). State of art of hydrogen usage as a fuel on aviation. *European Mechanical Science* 2: 20-30.

[41] Escobar, D. (2007). *Investigation of ZrNi, ZrMn 2 and Zn (BH 4) 2 Metal/Complex Hydrides for Hydrogen Storage*. https://digitalcommons.usf.edu/etd/701/.

[42] PI, J. C. Z., Cui, J., Gao, Y., Kniajansky, S., Lemmon, J., Raber, T., & Soloveichik, G. (2007). *Lightweight Intermetallics for Hydrogen Storage*. https://www.hydrogen.energy.gov/pdfs/review07/st_16_zhao.pdf.

[43] Rönnebro, E., Majzoub, E. H. (2007). Calcium borohydride for hydrogen storage: catalysis and reversibility. *The Journal of Physical Chemistry B.* 111(42):12045-7.

[44] Milanese, C., Garroni, S., Gennari, F., Marini, A., Klassen, T., Dornheim M., Pistidda C. (2018). Solid state hydrogen storage in alanates and alanate-based compounds: A review. *Metals.* 8(8):567.

[45] Fichtner, M., Engel, J., Fuhr, O., Kircher, O., Rubner, O. (2004). Nanocrystalline aluminium hydrides for hydrogen storage. *Materials Science and Engineering: B* 108(1-2):42-7.

[46] Rigas, F., & Sklavounos, S. (2005). Evaluation of hazards associated with hydrogen storage facilities. *Int. J. Hydrogen Energy* 30: 1501-1510.

[47] Abe, J. O., Popoola, A. P. I., Ajenifuja, E., & Popoola, O. M. (2019). Hydrogen energy, economy and storage: review and recommendation. *Int. J. Hydrogen Energy* 44(29): 15072-15086.

[48] Felderhoff, M., Weidenthaler, C., von Helmolt, R., Eberle, U. (2007). Hydrogen storage: the remaining scientific and technological challenges. *Phys. Chem. Chemical Phys.*, 9(21):2643-53.

[49] Bhatnagar, A., Pandey, S. K., Dixit, V., Shukla, V., Shahi, R. R., Shaz, M. A., Srivastava, O. N. (2014). Catalytic effect of carbon nanostructures on the hydrogen storage properties of MgH2–NaAlH4 composite. *Int. J. Hydrogen Energy* 39:14240-6.

[50] Shukla, V., Bhatnagar, A., Pandey, S. K., Shahi, R. R., Yadav, T. P., Shaz, M. A., & Srivastava, O. N. (2015). On the synthesis, characterization and hydrogen storage behavior of ZrFe2 catalyzed Li–Mg–N–H hydrogen storage material. *Int. J of Hydrogen Energy* 40:12294-12302.

[51] Shukla, V., Bhatnagar, A., Singh, S., Soni, P. K., Verma, S. K., Yadav, T. P., Shaz, M. A., Srivastava, O. N., (2019). A dual borohydride (Li and Na borohydride) catalyst/additive together with intermetallic FeTi for the optimization of the hydrogen sorption characteristics of Mg (NH2) 2/2LiH. *Dalton Transactions* 48:1391-403.

[52] Shukla, V., Bhatnagar, A., Soni, P. K., Vishwakarma, A. K., Shaz, M. A., Yadav, T. P., Srivastava, O. N. (2017). Enhanced hydrogen sorption in a Li–Mg–N–H system by the synergistic role of Li 4 (NH 2) 3 BH 4 and ZrFe 2. *Phy. Chem. Chemical Phys.* 19: 9444-56.

[53] Shahi, R. R., Yadav, T. P., Shaz, M. A., Srivastava, O. N. (2010). Studies on dehydrogenation characteristic of Mg (NH2) 2/LiH mixture admixed with vanadium and vanadium based catalysts (V, V2O5 and VCl3). *Int. J. Hydrogen Energy* 35: 238 – 246.

[54] Hino, S., Nakagawa, T., Fujii, H. (2005). Mechanism of hydrogenation reaction in the Li− Mg− N− H system. *The J. Phys. Chem. B.* 109: 10744-8.

[55] Wu, G., Xiong, Z., Liu, T., Liu, Y., Hu, J., Chen, O., Feng, Y., Wee, A. T. S. (2007). Synthesis and Characterisation of a New Ternary Imide – $Li_2Ca(NH)_2$. *Inorganic Chemistry* 46: 517-521.

Chapter 5

High Entropy Materials: An Emerging Material for Battery and Supercapacitive Applications

Amit K. Gupta, Aashish Prakash and Rohit R. Shahi*

Functional Materials Research Laboratory, Department of Physics,
Central University of South Bihar Gaya, India

Abstract

Batteries are an efficient way to store energy. However, these have specific limitations such as low cycle life, low energy density, very high charging time, disruption in power supply, and use of non-eco-friendly electrode materials. Commercially available Li- batteries are relatively expensive, flammable, and have toxic electrode materials. However, aqueous supercapacitors offer higher power density, cyclability, and safety at a lower cost. Thus, the supercapacitor is one of the efficient ways to store the energy and have many triumphs over batteries in terms of cyclic stability (SC > 30,0000 h vs. battery 500 h), power density, fast storage capability (low charge/discharge time; SC: 1-10 sec vs.10-60 min).The main limitations of the supercapacitors are low energy density (10 Wh/kg) and high cost of electrode materials, which prevent them from replacing batteries. According to the literature, the technical challenges are associated with designing and developing new electrode materials that must fulfill high energy density and high power density, low cost with maintaining exceptionally high cycle life. Recently, High Entropy Alloys (HEAs) and High Entropy Oxides are used as emerging electrode materials for battery and supercapacitor. The main advantage of high entropy materials is that we can obtain novel properties by

* Corresponding Author's Email: rohitrshahi@cusb.ac.in, rohitrshahi@gmail.com.

In: Materials Science: Future Aspects
Editors: Kalpana Awasthi, Arti Srivastava and Mridula Tripathi
ISBN: 978-1-68507-843-0

combining different transition elements. These compounds showed interesting, extraordinary high-temperature Li-ion conductivity for solid-state electrodes in TM-HEO. The present chapter summarizes the factors affecting the phases, different synthesis approaches, recently reported results on battery and supercapacitor properties of HEAs and HEOs.

Keywords: high entropy alloys, high entropy oxides, supercapacitor, li-ion battery

1. Introduction

Energy is an essential component of our society. Without this, we can't imagine our life. Batteries are known as one of the efficient ways to store energy. However, batteries have certain limitations, such as low cycle life, low energy density, very high charging time, frequent disruption in power supply, and toxicity to the environment [1]. The supercapacitor is an electrochemical device that fills the gap between batteries and conventional capacitors. Supercapacitors are also known as ultra-capacitor. Nowadays, supercapacitors gained much attention due to their excellent electrochemical performance. Very high specific energy density and specific power with high charging and discharging times of supercapacitors make them unique from other storage devices. Every supercapacitor has collector, polarized electrode, electrolyte, and separator. Table1 summarizes the fundamental difference between capacitor, supercapacitor (SC), and battery [1].

Table 1. Difference between capacitor, supercapacitors and battery [1]

Characteristics	Capacitor	Supercapacitor	Battery
Specific energy ($Wh\ kg^{-1}$)	<0.1	Up to 1091	Up to 1606
Specific power ($W\ kg^{-1}$)	>10000	Up to 196000	<10000
Discharging time	10^{-6}-10^{-3} Sec	Sec to min	0.03 – 3 h
Charging time	10^{-6}-10^{-3} Sec	Sec to min	1-5 h
Columbic efficiency (%)	About 100	Up to 99	70 - 85
Cycle life	Almost infinite	>500,000	About 1000
Charge storage determinant	Electrode Area and dielectric	Microstructure of electrode and electrolyte	Thermo-Dynamic and active mass

2. Classification and Working Principle of Supercapacitor

Based on charge storage mechanism, supercapacitors are classified as a) Electrochemical double-layer capacitors (EDLCs)b) Pseudocapacitor (PCs)

In EDLCs, charge is stored electrostatically (Helmholtz layer), while for Pseudocapacitor (PCs) the charge was stored chemically (Faradic). The specific capacitance of PCs is generally higher than the EDLCs, but its life is much lower than EDLCs [2].We can also combine both the EDLCs and PCs' properties to enhance overall capacitance and life cycles. The combination of EDLCs and PCs is a Hybrid supercapacitor that can store charge both electrostatic and electrochemical. Thus, based on type, we can classify SCs in three different forms: EDLCs, Pseudocapacitor (PCs), and hybrid supercapacitors. Table 2 summarizes the characteristics of different super-capacitors.

The supercapacitor works on the principle of conventional capacitors. As we know, in a conventional capacitor, when a voltage is applied across the terminals, the capacitor gets charge. When we connect it to the load, it will generate energy. From conventional capacitors, we know that capacitance is given by:

$$C=\varepsilon_r\varepsilon_o A/d \quad (1)$$

ε_r : Electrolyte dielectric constant, ε_o : dielectric constant of vacuum, d: the effective thickness of the double layer, A: Electrode surface area.

In supercapacitor, we have two capacitors which combined in series and let the capacitance of two individual capacitors are C_{DL1} and C_{DL2}. Equivalent capacitance can be evaluated by the following equation [4]:

$$C_{eq} = (C_{DL1} \text{ x } C_{DL2})/C_{DL1}+ C_{DL2} \quad (2)$$

Electrical energy stored in the EDLC is evaluated by:

$$E=CV^2_{max}/2 \quad (3)$$

The power density of a supercapacitor is given by:

$$P_{max}= V^2/4 \text{ x } ESR \quad (4)$$

Table 2. Characteristics of different supercapacitors [3]

Parameters	EDLCs	PCs	HSCs
Capacitances	0.1 - 470f	100 - 12000f	300 - 3300f
specific energy (wh/kg)	1.5 - 3.9wh/kg	4 – 9wh/kg	10 – 15 wh/kg
specific power(wh/g)	2 – 10w/g	3 - 10w/g	3 -14 w/g
Temperature range(°c)	40 to+70°c	−20 to +70°c	−20 to +70°c
Recharge cycles(thousand)	100k-1000k	100k - 1000k	20k - 100k

3. Efficiency of Supercapacitors

The efficiency of a supercapacitor depends upon the following factors [5]:

- The surface area of the electrode: Capacitance is directly proportional to the surface area of the electrode. Capacitance is tunned by enhancing the surface area of the electrode.
- The separation between electrodes: The capacitance of SCs is inversely proportional to the separation between electrodes, and by decreasing the separation between the electrodes, the value capacitance can be increased.
- Voltage: The energy stored by the supercapacitor is proportional to the voltagesquare.
- Equivalent Series Resistance (ESR): From equation (4), power is inversely proportional to ESR. Thus by decreasing ESR, we can increase the power.

Along with the factors mentioned above, super capacitance can also depend on the electrode and types of electrolytes. Thus, selecting efficient electrode materials is also one of the critical aspects of an efficient supercapacitor. The literature review confirmed that the significant electrode materials must have the following characteristics [1, 4]:

- High surface area
- Excellent conductivity
- High capacitive retention after 100000 cycles
- High chemical and thermal stability
- Optimized pore size with active surfaces
- Low cost

3.1. Materials Used for Supercapacitors Electrodes

Different types of electrodes have already been studied for high performance of SCs by different research groups. Here we summarized the characteristics of different electrode materials regarding their super capacitive properties.

3.1.1. Carbon-Based Electrode Materials

Carbon-based electrode material got attention due to its low cost, excellent conductivity, high thermal, chemical stability, and the high surface area responsible for its excellent capacitance value [6]. Activated carbons, Carbon nano-tubes, Graphene are carbon-based electrodes are showed very promising results in this area.

Large surface area, good electrical performance, and affordable cost make activated carbon one of the promising materials for SCs application [7]. Ba et al. fabricated a bio-derived cost-effective porous carbon with a surface area of 2000 m^2g^{-1} [8]. Due to its high porosity, they found high capacitance of 340 F/g and 217 F/g at current densities of $0.5Ag^{-1}$ and $20Ag^{-1}$, respectively [8]. The extraordinary properties (lightweight, intrinsic flexibility, high surface area, excellent conductivity, and chemical stability) of carbon-based nano-tubes (CNTs) make them also efficient materials for SCs application [9]. Graphene has a structure of sp^2 bonded atoms in a single honeycombed layer. Graphene was also used for SCs because of its large surface area, high cycle life, chemical and thermal stability, and broad potential window [10-12]. Xiong et al. developed a 3D composite architecture of reduced graphene oxide (rGO)-CNTs grown on carbon fibers and reported capacitance of these electrodes was found to be 203 F/g that was four times greater than pure carbon fiber as electrode [10]. Later Yan et al. developed and demonstrated a high volumetric capacitance of 1040 Fcm^{-3} at a scan rate of $2mVS^{-1}$ and an energy density of 32.6 Wh/L for MXene/rGo electrode [11].

3.1.2. Conducting Polymer as Electrode Material

Conducting Polymer (CP) are organic polymers for which electricity is conducted through a conjugated bond matrix. These materials are unique due to their high energy density than metal oxides and fast reversible faradic redox reactions. Among all CPs, Polyaniline and Polypyrrole are highly used as electrodes for SCs. Polyaniline (PANI) has good stability and high electrical conductivity but low surface area [13, 14]. Luo et al. synthesized a hierarchical graphene@polyaniline nanocomposite which showed a specific capacitance of 488F/g [15]. Miao et al. fabricated hollow PANI nanostructures, which showed capacitance of 601 F/g [16]. Polypyrrole (Ppy) also has high conductivity, high thermal stability, quick charging/discharging, low cost, and high energy density and is used for PCs electrode [17]. The doping can modify the charge storage and specific capacitance and thermal stability of the PPy [18].

3.1.3. Metal Oxides as Electrode Material

Metal oxides, especially the transition metal oxides (TMOs) family, gained vast attention as a Pseudocapacitor electrode because they showed excellent specific energy output compared to EDLCs materials. Nickel oxides (NiO), cobalt oxides, ruthenium oxides, and manganese dioxide are commonly used as electrodes. It has a large specific surface area, and due to its excellent interaction between electrode and electrolyte, its theoretical specific capacitance is very high [19-29]. RuO_2 is one of the promising electrodes for SCs application [20, 22]. It showed capacitive behavior due to its good electrical conductivity and better rate capability, although its preparation was costly [20]. Coaxial RuOx/CNF is a porous macrostructure that displayed a high areal capacitance of 53.76 mF/cm^2 at five mV/s with excellent cycling stability [21]. Fabrications of these materials were very complex and also time-consuming. Hence these materials are not well suited for mass production. Kim et al. proposed a simple electrode preparation method of RuO_2/ACNF [22]. The developed electrode has a high capacitance of 530 F/g at 20mV/s, a large specific surface area (675 m^2/g), and enhanced electrical conductivity (4.5 S/cm) [22]. Manganese oxide MnO_2 has good environmental compatibility, high theoretical specific capacity (1370F/g), extended operating potential, and cost-effectiveness. Hence MnO_2 can be considered as an excellent alternative for RuO_2 as an electrode material. However, it has poor ionic conductivity (10^{-13}s/cm) and electrical conductivity (10^{-5}-10^{-6}S/cm), which limit its commercial applications [23]. Further, Yang et al. fabricated ultra-fine manganese oxide- carbon nanofibers (MnOx-CNF) composite. The reported value of capacitance, cycling durability was found to be 179 F/g and 98% over 5000 cycles [24]. Zhu et al. developed 3D NiO nanowalls and reported excellent areal capacitance of 12.5 mF/cm^2 with 91% retention capability after 3000 cycles [25]. Ren et al. also proposed a hollow tube nanofiber containing NiO with the inclusion of citric acid, which displayed a high specific capacitance of 336 F/g. This hollow structure also provided a shorter electron-diffusion pathway, and due to that, the charge transfer resistance was decreased and found to be 0.23 ohm [26]. Cobalt oxides are the best TMOs for the pseudocapacitor it showed excellent redox reactions, and also it can be synthesized by a simple synthesis route [12]. The charge storage capacity of Co_3O_4 highly depends on its structural morphology and the electronic state of the metal. Iqbal et al. improved the electrical conductivity and electrochemical stability of this TMO Co_3O_4combined with hierarchical porous CNFs to form hybrid SCs and developed CNF–Co hybrid composite [28]. The developed electrode has a high surface area (483 m^2/g), excellent specific capacitance (911F/g) in 1 M H_2SO_4, and increased retention capacity after 1000 cycles [28].

Vanadium oxide (V_2O_5) was also used as promising pseudocapacitor electrodes due to its variable oxidation states, unique layered structure, and broad potential window [29]. Thangappan et al. fabricated GO/V_2O_5 nanofibres. The specific capacitance of the GO/V_2O_5 nanofibres was found to be 453.82 F/g at 10 mV/s in 2 M KOH as the electrolyte, which is higher than those of the individual GO and V_2O_5 components [29].

3.2. Electrolyte

The electrolyte is the combination of salt and solvent, which plays a fundamental role in the formation of EDLC. It provides ionic conductivity and facilitates the charge transfer between both electrodes. In pseudocapacitor, electrolytes are a vital element for storing charges due to redox reactions. The optimized operation of electrolytes determines the performance of the supercapacitor with electrodes. Thus, selecting suitable electrolytes is another crucial aspect of enhancing the efficiency of supercapacitors. For better SCs, electrolytes must have a wide operational voltage window, higher ion conductivity, low ESR, wide range of working temperature chemical and thermal stability, and less toxic and commercial viability. Electrolyte significantly affects the capacitance value, cycle life, energy, and power densities [30].

4. High Entropy Materials

In recent years, it has been found that advanced materials can be developed by thinking out of the box. For instance, Cantor et al., in 2004,developed a multi-component alloy [31]. These alloys in equimolar ratio significantly impact reducing the phase segregation according to the Gibbs phase rule and hence lower the number of phases [31]. For multi-component alloys, configuration entropy of mixing plays a key role, and its value increased as the number of constituent elements increased. Based on this fact, these alloys are termed High Entropy Alloys (HEAs) [31, 32, 33, 34]. It can say that High Entropy Alloys help us to explore the positive effect of constituent elements, which was left unexplored previously. Moreover, changing the number of elements helps control the configurational entropy and tailors the phase and their properties [31]. Nowadays, several researchers have shown keen interest in nonmetallic high entropy materials like nitride and carbides [35-37]. In 2015, Rost et al. reported the first High Entropy Oxide (HEO), which brought a dramatic revolution in high entropy materials [36]. Rost et al. synthesized single-phase oxide (Rocksalt structure) $(Mg_{0.2}Co_{0.2}Ni_{0.2}Zn_{0.2}Cu_{0.2})O$ with 5 different oxides taken in equimolar ratio for entropy-based phase stabilization [36]. Therefore, these materials were termed Entropy Stabilized oxides. Researchers found that Fluorite oxide-like phase was also formed for $(Ce_{0.2}Zr_{0.2}Hf_{0.2}Sn_{0.2}Ti_{0.2})O_2$ HEO, which showed reversible phase transformation at the transition temperature from single-phase (at high temperature) to multiple phases (at low temperature) [38]. HEOs have significant potential to be used in electrochemical energy storage because their configurational entropy helps to stabilize the original crystal structure during the charging/discharging process. The cycle stability of HEO was much higher than available traditional transition-metal oxides. Also, the charge transportation rate for HEO electrodes is much higher than the other reported electrode materials. HEO also has great significance in electrochemical energy storage such as Lithium-Ion batteries (LIB) and supercapacitors due to their better cycling stability than traditional oxide.

4.1. Definition of High Entropy Oxides

High entropy oxide (HEO) is defined as an equiatomic multicomponent oxide containing five or more different cations to formed a single-phase [35-37]. High entropy oxide is a relatively new concept that resembles the idea of High Entropy Alloys. For HEOs, the value of configurational entropy should be more excellent than 1.5 R (R is universal gas constant). The calculation of configuration entropy S_{config} for $A_xB_yO_z$ type oxide can be evaluated by using equation (5) [37]:

$$S_{config} = - R[m\,(\sum_{x=}^{A} m_a * m_a)_{E\text{-sites}} + n\,(\sum_{y=}^{B} n_b * n_b)_{F\text{-sites}} + z\,(\sum_{o=}^{P} z_o * z_o)_{O\text{-sites}}] \quad (5)$$

where A, B, Represents the integer cation present in E, F, O sites and m_a, n_b, z_o is mole fractions of elements in E, F, and O sites. The configurational entropy for O^{2-} the site is generally zero for oxide structure, but in some cases, anion sublattice observed extra configuration entropy because of F^- in Oxyfluorides and oxygen defect [39].

$$S_{config} = - R[m\,(\sum_{x=}^{A} m_a * m_a)]_{cation\text{-}site} \quad (6)$$

The maximum configurational entropy can be achieved from the above when all the constituent elements are in an equimolar ratio like metallic HEAs. The value of S≥1.5R can be obtained only when five or five more cations are taken into account. The equation for the evaluation of configurational entropy is only valid for an ideal solid solution. Murty et al. have classified the material into three parts based on the empirical data collected for the configuration entropy such as ΔS≥1.5 R are termed as "high entropy" materials, 1.5R >S_{config}≥1Ras "medium entropy," S_{config} < 1R as "low entropy" materials [40].

According to Gibb's phase rule

$$\Delta Gmix = \Delta\, Hmix - T\, \Delta Smix \quad (7)$$

From equation (7), we can say that Gibbs free energy of mixing depends on ΔHmix, and ΔSmix at a particular temperature, which has a significant impact on the value of ΔGmix. However, at low temperature phase segregation may occur due to enthalpy [36, 41] Here, high entropy stabilizes the single-phase rather than multiphase, thus enhancing structural stability. The configuration entropy is affected by two factors; a) The S_{config} increased when the constituent elements or cation increased, in particular HEO and b) TheS_{config} is highest for equimolar composition. For entropy stabilized HEOs, the configuration entropy value is reduced when any deviation from equimolar occurs, leading to an increase in transition temperature to a single phase. It is not always necessary that a single phase will produce when the configurational entropy value is greater than 1.5 R, for many cases, the configuration entropy failed to compensate the ΔH mix and the formation of intermetallic phase was reported [40, 42]. For ($Co_{0.2}Cu_{0.2}Mg_{0.2}Ni_{0.2}Zn_{0.2}O$) HEO has a Rock salts structure (R-HEO) and showed reversible phase transformation upon heating and cycled dependency on entropy based phase stabilization [36]. The endothermicity for ($Co_{0.2}Cu_{0.2}Mg_{0.2}Ni_{0.2}Zn_{0.2}O$) R-HEO was related to Zn and Cu cations present in the selected HEO. These cations apply enthalpy penalties for the crystallization of the Rocksalt structure. Thus, single-phase rock salt structure can be formed by removing Zn and Cu cation from the above HEO. However, according to equation (6), removal of any of the cation from the R-HEO leads to multiphase formation due to decrement in configurational entropy value to 1.39R [36]. Along with entropy-based phase stability, other mechanisms are also involved in phase stabilization. For example, in fluorite HEO (F-HEO), the presence of Ce in the composition with Gd, Nd, La, Pr, Sm, Y as a perquisite the necessary condition for the formation of a single phase solid solution compound [43, 44]. It has been reported that multiphase was formed in absence of Ce in any fluorite (F-HEO), irrelevant of the number of elements and synthesis temperature [43, 44]. From this, we can conclude that for F-HEO single phase solid solution compound can only be achieved by the addition of Ce with other elements in the composition. Several HEO like $(Ce_{0.2}La_{0.2}Pr_{0.2}Sm_{0.2}Y_{0.2})O_{2-\gamma}$ and$(Gd_{0.2}La_{0.2}Nd_{0.2}Sm_{0.2}Y_{0.2})FeO_3$ HEOs were not exhibited reversible phase transformation on cyclic heat treatment [45, 46]. In these cases, Hmix value was negative. As a result, the impact of $T\Delta S$ was minimal, and hence ΔG has minimum value [47].

4.2. Classification and Properties of HEOs

Based on the crystal structure, HEO can be classified into 8 groups: rocksalt, fluorite, bixbyite, pyrocholore, magneto plumbite, O3-type layered, perovskite and spinel [37, 48, 49].

Recent studies confirmed that the above crystal structured oxide could be achieved based on cationsionic radius and oxidation states. Moreover, different studies also confirmed that R-HEO and F-HEO showed lower thermal conductivity due to multiple phonon scattering present in these structures [44, 50]. R-HEO also showed the extremely large value of the dielectric susceptibility due to the positive combining effect of the constituent element in R-HEO [37]. HEO also exhibited magnetic properties depending on crystal structure and magnetic ordering of constituent elements such as ferrimagnetic properties in spinel-HEO structure and anti ferrimagnetic behavior in rock salt or perovskite [51]. Figure 1 represents the pictorials representation of structure type, synthesis route, and the properties of HEOs.

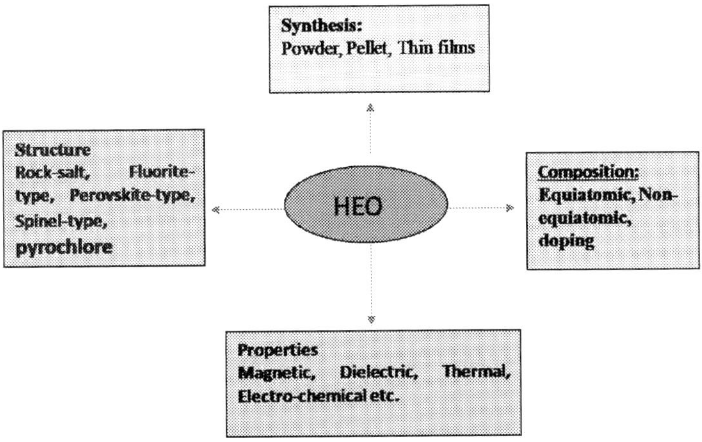

Figure 1. Represents a pictorial representation of HEOs and their synthesis route, structure, and application in a different area.

5. Synthesis Routes for High Entropy Materials

All the existing conventional synthesis routes can also be used to synthesize High Entropy Materials. For the case of $(Ce_{0.2}Zr_{0.2}Hf_{0.2}Sn_{0.2}Ti_{0.2})O_2$ F-HEO was synthesized through the solid-state synthesis route. They found that phase transformation occurred from single-phase (at high temperature) to multiphase (at low temperature), which clearly showed that it was entropy stabilized fluorite oxide [38]. The chart below (Figure 2) represents the various forms of HEOs synthesized by different routes [37].

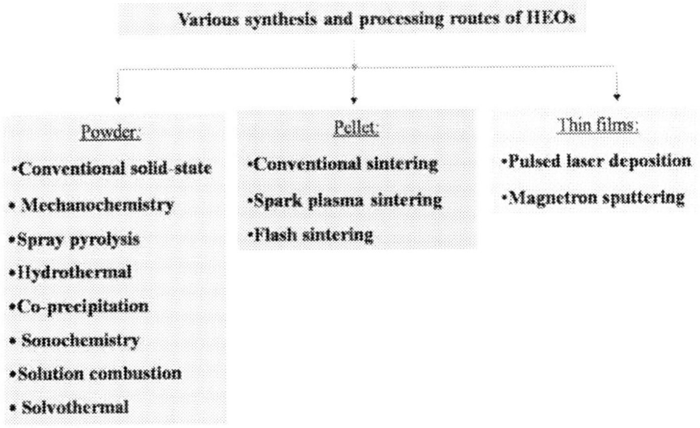

Figure 2. Different synthesis methods of HEOs.

6. Electrochemical Properties of High Entropy Oxide

Some reports are also available for electrochemical properties of Transition Metal based HEO (TM-HEO) [36, 52, 53], Rare Earth-based HEO (RE-HEO) [54], and mixed HEO (TM-RE-HEO) as electrodes for LIB [47]. The reports described that for TM-HEO electrodes, the Li-ion conductivity was found to be very high at room temperature with a high value of dielectric constant [52, 55]. Moreover, it has been found that RE-HEO electrodes facilitated fine-tunable band gaps. The research interest in HEO materials is significantly enhanced in every field due to its easy synthesis and processing technique. Different studies confirmed that HEOs have excellent electrochemical energy storage properties [35]. HEO exhibited different electrochemical properties depending on their structure and the number of elements present in particular HEO. Here we summarized different perspectives for the electrochemical properties of HEOs based on the available reports in the literature.

The room temperature conductivity of Li doped R-HEO was found high as compared to conventional solid-state electrodes due to 6 fold coordination with neighboring ion and the presence of oxygen vacancy in the HEO structure [52, 56]. Moreover, the reversible redox stability of HEOs is also better than conventional electrodes [57]. The conductivity of Li doped in $(Co_{0.2}Cu_{0.2}Mg_{0.2}Ni_{0.2}Zn_{0.2})O$ R-HEO was enhanced from 10^{-8} $Scm^{-1}$to $10^{-3}Scm^{-1}$ with a doping concentration of30% [52]. Also, Na^{+} ion conductivity was found to be 10^{-6} Scm^{-1} for R-HEO [52]. It was reported that ionic conductivity increased when alkali metal elements (Li^{+}, Na^{+}) doped into R-HEO due to oxygen vacancy formation. Thus R-HEO can also be used as an ionic conductor [52], and the R-HEO structure showed stability with different cycles. The structural stability of R-HEO is one of the significant factors for the increased interest of the different researchers to investigate the electrochemical properties of HEO [57]. It has been reported that the decrement of the particle size can further enhance the cyclic stability. It was also reported that after 900 cycles, the specific capacity was found to be greater than 650 $mAhg^{-1}$ and the value of coulombic efficiency was greater than 99.5% at a specific current density of 200 mAg^{-1} micro size particles [57]. However, for R-HEO, particles in the nanometer range have a specific capacity greater than 900 $mAhg^{-1}$ after 300 cycles were recorded [58].

Different studies confirmed that $(Co_{0.2}Cu_{0.2}Mg_{0.2}Ni_{0.2}Zn_{0.2})O$ R-HEO showed better capacity retaining with cycling as compared to R-MEO(-X) (medium entropy oxide) over the first 100 cycles (-X denoted that one cation is removed from the R-HEO compound) [57, 58].Thus as the configurational entropy reduced to 1.39 R from 1.61 R electrochemical performance of R-HEO also reduced. Qiu et al. found that when one of the cation, i.e., Co, was removed from $(Co_{0.2}Cu_{0.2}Mg_{0.2}Ni_{0.2}Zn_{0.2})O$ R-HEO after the first 10 cycles, the cell failed for R-MEO(-Co) clearly represents Co's significance in the R-HEO [56]. Moreover, R-MEO (-Zn) and R-MEO (-Cu) initially showed high specific capacity as compared to R-HEO for the initial 30 - 35 cycles however, after 30 -35 cycles, they observed that specific capacity decreased rapidly. It has been reported that Co and Mg have a tremendous impact on the cyclic stability of $(Co_{0.2}Cu_{0.2}Mg_{0.2}Ni_{0.2}Zn_{0.2})O$ R-HEO [58]. The lithiation potential was also found to be very low when Cu was removed from $(Co_{0.2}Cu_{0.2}Mg_{0.2}Ni_{0.2}Zn_{0.2})O$ R-HEO while during dilithiation two-step oxidation occurred.

Different studies also confirmed that High Entropy Spinel Oxide (HESO) also has excellent electrochemical properties [59]. HESOs are more efficient as electrode materials than HEOs because spinel structure provides 3-dimensionalLi-ion transport than R-HEO and has two Wyckoff sites available due to the coexistence of di and trivalent cations (general formula (AB_2O_4). Also, for HESOs, reversible capacity increased during the lithiation/ dilithiation process due to different valence states. Chen et al. developed an anode of $(Mg_{0.2}Ti_{0.2}Zn_{0.2}Cu_{0.2}Fe_{0.2})_3O_4$ HESO for LIB and found the reversible capacity of 504 $mAhg^{-1}$ after 300cycles at the current density of 100 $mAg^{-1}$and rate capability of 272 $mAhg^{-1}$ at 2000 mAg^{-1} [60]. For $(Co_{0.2}Cr_{0.2}Fe_{0.2}Mn_{0.2}Ni_{0.2})_3O_4$ HESO, it was found that no MgO structure pillar is needed for stability, unlike it was used in rock salt and spinel-type HEO, which ensured good structure and electrode cycling stability [59]. In spinel-type HEO, the multications Cr, Mn, Fe, Co and Ni in non-equimolar were diffused into the two Wyckoff sites of the system, leading to diverse valence states of cations and increasing in oxygen vacancies due to which Li^+ transport phenomenon enhanced [52]. At various specific current rates ranging from 50 to 2000 mAg^{-1}, the HESO electrode exhibited different charging and discharging value. The dilithiation (charging) capacities was found to 1170,1072,979,824, 715,649 and 500 $mAhg^{-1}$at 50,100,200,500,800,1000 and 2000 mAg^{-1} specific current, respectively [61, 62]. Here the electrode delivered a good specific capacity value i.e., 500 $mAhg^{-1}$even at high specific current 2000 mAg^{-1} [61, 62]. From the above discussion, it is clear that HEOs as electrode are more beneficial than conventional oxide electrode materials. For conventional oxides to increase in the specific current rate, the specific capacity decreased rapidly due to low electronic conductivity. However, HESO electrode exhibited higher charging and discharging capacity for 200 cycles. Its value was found to be 1235mAhg^{-1} at specific current 500 mAg^{-1} and has capacity retention of 90% with a coulomb efficiency ~ 99%with 0.05% slowly disappeared capacity per cycle [59]. These unique behaviors of HESO confirmed why HEOs are better than traditional type oxide electrodes. For $(FeCoCrMnZn)_3O_4$ HESOs, metals are dispersed throughout the lattice, which enhances the transportation rate. It was observed that its gravimetric capacitance was 340.3 F/g at a current density of 1A/g in 1M of KOH electrolyte [63]. It was found that specific capacitance at current densities 1A/g, 2A/g,5A/g was 340.3 F/g, 287.9 F/g, 301.5F/g, and at 10 A/g it was 281.8 F/g along with 82.8% of capacitance retention at a high current density of 10A/g [63]. The cyclic performance at a current density of 5A/g indicates 69% of capacitance retention even after 1000 cycles for $(FeCoCrMnZn)_3O_4$ HESOs [63].

For $(Co_{0.2}Cu_{0.2}Mg_{0.2}Ni_{0.2}Zn_{0.2})O$ electrodes using 63 wt% of the TM-HEO as active material was used in secondary Li-based cells. It was found that the specific capacity after hundred cycles was 770 mAhg^{-1} and it decreased when the specific current was increased from 0.05 Ag^{-1} to 3 Ag^{-1} [64, 65]. For atypical conversion type material, it has been reported that the particular capacity value increased with the cycle number because activation occurs in electrodes with large particle sizes. It has been found that although with micrometer-sized particles during the conversion reaction, the cell exhibited good cycling stability at high capacity value, without changing the other components of the cell i.e., electrode composition, binder, and electrolyte. The specific capacity tends to increase during the first 75 cycles. Initially, the cell was discharged with a capacity of 980 mAhg^{-1}, and at the third cycle, the cell stabilized with a capacity value ~600 mAhg^{-1}, the capacity value even increased after 70 cycles. The fluctuation in the value of capacity was found in between 75 to 150 cycles after the 150th cycle the capacity value was stabilized due to structural transformation [64, 66, 67].

Summary and Conclusion

This chapter provides detailed information on the electrochemical properties of HEOs. The content of this chapter can be summarized as follows;

- In HEO, thermodynamic parameters role on phase formation and structure are quite significant, unlike in conventional oxide. In HEO, the synergetic (positive combining effect of constituent element) effect helps us achieve higher performance value in electrochemical, dielectric, thermal, catalytic, and other properties. However, in conventional oxide, such an effect is not pronounced.

- HEO shows better cycle stability as compared to conventional conversion type oxide. In traditional type oxide anode, after a few cycles of lithiation and delithiation the crystalline phase is lost and almost changed to amorphous.
- In traditional type oxide anode, the specific capacity value declined rapidly as the specific current increased. This happens due to low electronic conductivity and restriction in the intercalation process. The electrochemical behavior of HEO is far better than conventional oxide.
- High Entropy Material can be easily synthesized using conventional synthesis routes, which are low-cost, non-toxic materials.
- Among all HEO, HESO electrodes exhibit highest charging and discharging capacity can be found with high coulombic efficiency due to the existence of two Wyckoff sites availability which allows the coexistence of divalent and trivalent cations with general formula (AB_2O_4).
- The effect of particle size on the electrochemical performance that the cyclic stability can be further enhanced withthe decrement of particle size.
- In F-HEO, the presence of Ce in the composition as a prerequisite is the necessary condition for the formation of a single-phase solid solution compound.
- Removal of any element from the composition the cell failed completely, and the specific capacity declined rapidly for R-MEO (-X). Thus, we can tune the electrochemical performance of HEOs by selecting proper transition elements.

Acknowledgement

The authors would like to acknowledge the research funding received from the Science and Engineering Research Board (SERB), Government of India.

References

[1] Raza, W., Ali, F., Raza, N., Luo, Y., Kim, Yi. Li, Yang, J., Kumar, S., Mehmood, A. and Kwon, E.E. (2018).Recent advancements in supercapacitor technology. *Nano Energy* 52: 441-473. https:/doi.org/10.1016/j.nanoen.2018.08.013.

[2] Liu, J., Wang, J., Xu, C., Jiang, H., Li, C., Zhang, L., Lin, J. and Shen, Z.X. (2018). Advanced Energy Storage Devices: Basic Principles, Analytical Methods, and Rational Materials Design. *Adv Sci.* 5: 1700322-1700340. DOI:10.1002/advs.2017 00322.

[3] Conway, B.E. (1999). *Electrochemical Supercapacitors, scientific fundamentals and technological application*, Kluwer Academic/Plenum publisher New Work.

[4] Saikia, B.K., Benoy, S.M., Bora, M., Tamuly, J., Pandey, M. and Bhattacharya, D. (2020). A brief review on supercapacitor energy storage devices and utilization of natural carbon resources as their electrode materials. *Fuel* 282: 118796 - 118812.https://doi.org/10.1016/j.fuel.2020.118796.

[5] Show, Y. (2012). Research article on electric double-layer capacitor fabricated with addition of carbon nano-tube to polarizable electrode. *J. Nanomater 2012:* 1-8. https://doi.org/10.1155/2012/929343.

[6] Cakici, M., Kakarla, R.R. and Alonso-Marroquin, F. (2017). Advanced electro-chemical energy storage supercapacitors based on the flexible carbon fiber fabric-coated with uniform coral-like MnO 2 structured electrodes. *Chem. Eng. J.* 309: 151-158. http://dx.doi.org/10.1016/j.cej.2016.10.012.

[7] Qu, D. andShi, H. (1998). Studies of activated carbons used in double-layer capacitors. *J. Power Sources* 74: 99-107. DOI:10.1016/S0378-7753(98)00038-X.

[8] Ba, H., Wang, W., Pronkin, S., Romero, T., Baaziz, W., Nguyen-Dinh, L., Chu,W.,Ersen, O. and Pham-Huu, C. (2018).Biosourced foam like activated carbon materials as high-performance supercapacitors. *Adv. Sustain. Syst. 2: 1700123-1700134.*DOI: 10.1002/adsu.201700123.

[9] Issi, J.-P., Langer, L., Heremans, J. andOlk, C. (1995). Electronic properties of carbon nano-tubes: experimental results. *Carbon* 33: 941-948. DOI:10.1016/0008-6223(95)00023-7.

[10] Xiong, C., Li, T., Zhao, T., Dang, A., Li, H., Ji, X., Jin,W., Jiao,S.,Shang, Y. and Zhang,Y. (2017).Reduced graphene oxidecarbon nano-tube grown on carbon fiber as binder-free electrode for flexible high performance fiber supercapacitors. *Compos. Part B: Eng.* 116: 7-15. https://doi.org/10.1016/j.compositesb.2017.02. 028.

[11] Wang, F., Wang, C., Zhao, Y., Liu, Z., Chang, Z., Fu, L., et al. (2016). A quasi solid state Li ion capacitor based on porous TiO_2 hollow microspheres wrapped with graphene nanosheets. *Small* 12: 6207–6213. https://doi.org/10.1002/smll.2016023 31.

[12] Abdah, M.A.A.M., Azam, N.H.N., Kulandaivalu, S. and Sulaiman, Y.(2020). Review of the use of transition-metal-oxide and conducting polymer-based fibers for high-performance supercapacitors. *Mat. and Design* 186: 108199-108218. DOI: 10.1016/j.matdes.2019.108199.

[13] Cao, X., Zeng, H.-Y., Xu, S., Yuan, J., Han, J. and Xiao, G.-F. (2019). Facile fabrication of the Polyaniline/layered double hydroxide nanosheet composite for supercapacitors. *Appl. Clay Sci.* 168: 175-183. https://doi.org/10.1016/j.clay.2018.11.011.

[14] Li, X.,Yang, L., Lei, Y., Gu, L. and Xiao, D. (2014). Microwave-assisted chemical-vapor-induced In situ polymerization of polyaniline nanofibers on graphite electrode for high-Performance supercapacitor. *ACS Appl. Mater. Interfaces* 6: 19978–19989. https://doi.org/10.1021/am505533c.

[15] Jose, R.and Misnon, II. (2017). Synthesis and electrochemical evaluation of the PANI/δ-MnO2 electrode for high performing asymmetric supercapacitors. *New J. Chem.* 41: 6574–6584 https://doi.org/10.1039/C7NJ00679A.

[16] Miao, Y.-E., Fan, W., Chen, D. andLiu, T. (2013). High-performance supercapacitors based on Hollow polyaniline nanofibers by electrospinning. *ACS Appl. Mater. Interfaces* 5: 4423–4428. DOI: 10.1021/am4008352.

[17] Fan,L.-Z. andMaier, J. (2006). High-performance polypyrrole electrode materials for redox Supercapacitors. *Electrochem. Commun.* 8: 937–940. https://doi.org/10.1016/j.elecom.2006.03.035.

[18] Asen, P. andShahrokhian, S. (2017).A high performance supercapacitor based on graphene/Polypyrrole/Cu2O–Cu (OH) 2 ternary nanocomposite coated on nickel foam. *J. Phys. Chem. C* 121: 6508–6519. https://doi.org/10.1021/acs.jpcc.7b00534.

[19] Zheng, J.H., Zhang, R.M., Yu, P.F. and Wang, X.G. (2019). Binary transition metal oxides (BTMO) (Co-Zn, Co-Cu) synthesis and high supercapacitor performance. *J. Alloys Comp.* 772: 359-365.*DOI:*10.1016/j.jallcom.2018.09.067.

[20] Tang, C., Tang, Z. and Gong, H. (2012). Hierarchically porous Ni-co oxide for high reversibility asymmetric full-cell supercapacitors. *J. Electrochem. Soc.* 159: A651–A656. https://doi.org/10.1149/2.074205jes.

[21] Jhao, J.-J., Lin, C.-H., Yeh, T.-K., Wu, H.-C., Tsai, M.-C. and Hsieh, C.-K. (2017). The coaxial nanostructure of ruthenium oxide thin films coated onto the vertically grown graphiticanofibers for electrochemical supercapacitor. *Surf. Coat. Technol.* 320: 263–269. DOI:10.1016/j.surfcoat.2017.01.006

[22] Kim,C., Joo, Y.H., Kim, J.H., Lee, W.J. and Yang, K.S. (2006).Performances of electrochemical hybrid supercapacitor of RuO2/activated carbon nanofibres from electrospinning. *WSEAS Transactions on Systems* 5 (8): 1971–1975.

[23] Qu, J., Shi, L., He, C., Gao, F., Li, B., Zhou, Q., Hu, H., Shao, G., Wang, X. and Qiu, J. (2014).Highly efficient synthesis of graphene/MnO2 hybrids and their application for ultrafast oxidative decomposition of methylene blue. *Carbon* 66: 485–492. Doi:10.1016/j.carbon.2013.09.025.

[24] Yang, Y., Lee, S.,Brown, D.E.,Zhao,H., Li, X., Jiang, D., Hao, S., Zhao, Y., Cong, D., Zhang, X. andRen, Y. (2016).Fabrication of ultrafine manganese oxide-decorated carbon nanofibres for high-performance electrochemical capacitors. *Electrochim. Acta* 211:524-532. https://doi.org/10.1016/j.electacta.2016.06.012.

[25] Zhu, G., Chen, J., Zhang, Z., Kang, Q., Feng, X., Li, Y., Huang, Z., Wang, L. andMa, Y. (2016). NiO nanowall-assisted growth of thick carbon nanofibre layers on metal wires for fibre supercapacitors. *Chem. Commun.* 52: 2721-2724. DOI:10.1039/C5CC10113A.

[26] Ren, B., Fan, M., Liu, Q., Wang, J., Song, D. and Bai, X. (2013). Hollow NiO nanofibres modified by citric acid and the performances as supercapacitor electrode. *Electrochim. Acta* 92: 197–204 DOI:10.1016/j.electacta.2013.01.009.

[27] Zang, L., Zhu, J. and Xia, Y. (2014). Facile synthesis of porous NiO nanofibers for highperformance supercapacitors. *J. Mater. Eng. Perform.*23: 679-683. DOI: 10.1007/s11665-013-0797-3.

[28] Iqbal, N., Wang, X., Ge, J., Yu, J., Kim, H.-Y., Al-Deyab, S.S.,El-Newehy, M. and Ding, B. (2016). Cobalt oxide nanoparticles embedded in flexible carbon nanofibres: attractive material for supercapacitor electrodes and CO_2 adsorption. *RSC Adv.* 6: 52171-52179.https://doi.org/10.1039/C6RA06077C.

[29] Thangappan, R., Kalaiselvam, S., Elayaperumal, A. and Jayavel, R. (2014). Synthesis of graphene oxide/vanadium pentoxide composite nanofibres by electrospinning for supercapacitor applications, *Solid State Ionics* 268: 321-325.DOI:10.1016/j.ssi.2014.10.025.

[30] Simon, P. and Gogotsi, Y. (2008). Materials for electrochemical capacitors, *Nature materials* 7: 845-854. DOI: 10.1038/nmat2297.

[31] Cantor, B. (2021). Multi-component high-entropy Cantor alloys. *Prog Mater Sci* 120: 100754-1000789. https://doi.org/10.1016/j.pmatsci.2020.100754.

[32] Tsai, MH. and Yeh, JW.(2014). High-entropy alloys: A critical review. *Mater. Res. Lett.* 2: 107–123. http://dx.doi.org/10.1080/21663831.2014.912690.

[33] Yadav, T.P., Mukhopadhyay, S., Mishra, S.S., Mukhopadhyay, N.K. and Srivastava, O.N. (2018).Synthesis of a single phase of high-entropy Laves intermetallics in the Ti–Zr–V–Cr–Ni equiatomic alloy. *Philosophical Magazine Letters* 97: 494-503.https://doi.org/10.1080/09500839.2017.1418539.

[34] Mishra, S.S., Mukhopadhyay, S., Yadav, T.P., Mukhopadhyay, N.K. and Srivastava, O.N. (2019).Synthesis and characterization of hexanary Ti–Zr–V– Cr–Ni–Fe high-entropy Laves phase. *Journal of Materials Research* 34: 807-818.https://doi.org/10.1557/jmr.2018.502.

[35] Sarkar, A., Wang, Q., Schiele, A., Chellali, M.R., Bhattacharya, S.S., Wang, D., Brezesinski, T., Hahn, H., Velasco, L. and Breitung, B. (2019). High-Entropy Oxides: Fundamental Aspects and Electrochemical Properties. *Adv Mater.* 31: 1806236-1806244.https://doi.org/10.1002/adma.201806236.

[36] Rost, C.M., Sachet, E., Borman, T., Moballegh, A.,. Dickey, E.C., Hou, D., Jones, J.L., Curtarolo, S. and Mar, J.-P. (2015). Entropy stabilized oxides. *Nat. Commun.* 6: 8485-8492. DOI: 10.1038/ncomms9485.

[37] Sarkar, A., Breitung, B. andHahn, H. (2020). High entropy oxides: The role of entropy, enthalpy and synergy. *Scr Mater.* 187: 43–48. DOI: 10.1016/j.scriptamat. 2020.05.019.

[38] Chen, K., Pei, X., Tang, L., Cheng, H., Li, Z., Li, C., Zhang, X. and An, L. (2018). A five-component entropy-stabilized fluorite oxide. *J Eur Ceram Soc.*38: 4161–4164. DOI:10.1016/j.jeurceramsoc.2018.04.063.

[39] Wang, Q., Sarkar, A., Wang, D., Velasco, L., Azmi, R., Bhattacharya, S.S., Bergfeldt, T., Düvel, A., Heitjans, P., Brezesinski, T., Hahn, H. and Breitung, B. (2019).Multi-anionic and -cationic compounds: New high entropy materials for

advanced Li-ion batteries. *Energy Environ. Sci.* 12: 2433–2442. DOI: 10.1039/C9EE00368A.

[40] Murty, B.S. et al. (2019). *High-Entropy Alloys,* Elsevier, Amsterdam.

[41] Biesuz, M., Spiridigliozzi, L., Agli, G.D., Bortolotti, M. and Sglavo, V.M. (2018). Synthesis and sintering of (Mg, Co, Ni, Cu, Zn)O entropy-stabilized oxides obtained by wet chemical methods. *J. Mater Sci.* 53: 8074–8085. https://doi.org/10.1007/s10853-018-2168-9.

[42] Miracle, D.B. and Senkov, O.N.(2017). A critical review of high entropy alloys and related concepts. *Acta Mater.*122: 448-511. https://doi.org/10.1016/j.actamat.2016.08.081.

[43] Chellali, M.R., Sarkar, A., Nandam, S.H., Bhattacharya, S.S., Breitung, B., Hahn, H.and Velasco, L. (2019). On the homogeneity of high entropy oxides: An investigation at the atomic scale. *Scr. Mater.* 166: 58–63. https://doi.org/10.1016/j.scriptamat.2019.02.039.

[44] Gild, J., Samiee, M., Braun, J.L., Harrington, T., Vega, H., Hopkins, P.E., Vecchio, K. and Luo, J. (2018). "High-entropy fluorite oxide. *J. Eur. Ceram. Soc.* 38: 3578–3584. https://doi.org/10.1016/j.jeurceramsoc.2018.04.010.

[45] Jiang, S., Hu, T., Gild, J., Zhou, N., Nie, J., Qin, M., Harrington, T., Vecchio, K. and Luo, J. (2018). A new class of high-entropy perovskite oxide. *Scr. Mater.*142: 116-120. https://doi.org/10.1016/j.scriptamat.2017.08.040.

[46] Djenadic, R., Sarkar, A., Clemens, O., Loho, C., Botros, M., Chakravadhanula, V.S.K., Kübel, C., Bhattacharya, S.S., Gandhi, A.S. and Hahn, H. (2017). Multi-component equiatomic rare earth oxides *Mater. Res. Lett.*, 5: 102-109. https://doi.org/10.1080/21663831.2016.1220433.

[47] Sarkar, A., Djenadic, R., Wang, D., Hein, C., Kautenburger, R., Clemens, O. and Hahn, H. (2018). Rare earth and transition metal based entropy stabilised perovskite type oxides. *J. Eur. Ceram. Soc.* 38: 2318-2327. https://doi.org/10.1016/j.jeurceramsoc.2017.12.058.

[48] Musicó, B., Wright, Q., Ward, T.Z., Grutter, A., Arenholz, E., Gilbert, D., Mandrus, D. and Keppens, V. (2019). Tunable magnetic ordering through cation selection in entropic spinel oxides. *Phys. Rev. Mater.3*: 104416-104425. DOI:https://doi.org/10.1103/PhysRevMaterials.3.104416.

[49] Abrowa, J.D., Stygar, M., Mikuła,A., Knapik, A., Mroczka, K., Tejchman, W., Danielewski, M. and Martin, M. (2018). Synthesis and microstructure of the (Co,Cr,Fe,Mn,Ni) 3O4 high entropy oxide characterized by spinel structure. *Mater. Lett.* 216: 32–36. https://doi.org/10.1016/j.matlet.2017.12.148.

[50] Braun, J.L., Rost, C.M., Lim, M., Giri,A., Olson, D.H., Kotsonis, G.N., Stan, G., Brenner, D.W., Maria, J.-P. and Hopkins, P.E. (2018). Charge-induced disorder controls the thermal conductivity of entropy-stabilized oxides. *Adv. Mater.* 30: 1805004-1805011. DOI: 10.1002/adma.201805004.

[51] Mao, A., Xie, H.X., Xiang, H.Z., Zhang, Z.G., Zhang, H. and Ran, S. (2020). A novel six-component spinel-structure high-entropy oxide with ferrimagnetic property. *J. Magn. Magn. Mater.* 503: 166594-166598. https://doi.org/10.1016/j.jmmm.2020.166594.

[52] Bérardan, D., Franger, S., Meena, A.K. and Dragoe, N.(2016). Room temperature lithium superionic conductivity in high entropy oxides. *J. Mater. Chem. A* 4: 9536-9541, https://doi.org/10.1039/C6TA03249D.

[53] Sarkar, A., Djenadic, R., Usharani, N.J., Sanghvi, K.P., Chakravadhanula, V.S., Gandhi, A.S., Hahn, H. and Bhattacharya, SS. (2017).Nanocrystalline multi-component entropy stabilised transition metal oxides. *J. Eur. Ceram. Soc.* 37: 747–754, DOI:10.1016/j.jeurceramsoc.2016.09.018.

[54] Sarkar, A., Loho, C., Velasco, L., Thomas, T., Bhattacharya, S.S., Hahn, H., and Djenadic, R.(2017). Multi-component equiatomic rare earth oxides with a narrow band gap and associated praseodymium multivalency. *Dalton Trans.* 46: 12167–12176, https://doi.org/10.1039/C7DT02077E.

[55] Bérardan, D., Franger, S., Dragoe, D., Meena, A.K. and Dragoe, N. (2016). Colossal dielectric constant in high entropy oxides. *Phys. Status Solidi Res. Lett.* 10: 328–333. https://doi.org/10.1002/pssr.201600043.

[56] Osenciat, N., Bérardan, D., Dragoe, D., Léridon, B., Holé, S., Meena, A.K., Franger,S. and Dragoe, N. (2019). Charge compensation mechanism s in Li - substituted high -entropy oxides and influence on Li superionic conductivity. *J. Am. Ceram. Soc.102: 6156-6162*. https://doi.org/10.1111/jace.16511.

[57] Sarkar, A., Velasco, L., Wang, D., Wang, Q., Talasila, G., de Biasi, L., Kübel, C., Brezesinski, T., Bhattacharya, S.S., Hahn, H. and Breitung, B.(2018).High entropy oxides reversible energy storage. *Nat. Commun.* 9: 3400-3408. Doi: 10.1038/s41467-018-05774-5.

[58] Qiu, N., Chen, H., Yang, Z., Sun, S., Wang, Y. and Cui, Y.(2019).A high entropy oxide ($Mg_{0.2}Co_{0.2}Ni_{0.2}Cu_{0.2}Zn_{0.2}O$) with superior lithium storage perform. *J. Alloys Comp.*777: 767-774. https://doi.org/10.1016/j.jallcom.2018.11.049.

[59] Nguyen, T.X., Patra, J., Chang, J.K. and Ting, J.M. (2020). High entropy spinel oxide nanoparticles for superior lithiation-delithiation performance. *J Mater Chem A*. 8: 18963–18973. https://doi.org/10.1039/D0TA04844E.

[60] Chen, H., Qiu, N., Wu, B., Yang, Z., Sun, S. and Wang Y. (2020). A new spinel high-entropy oxide (Mg0.2Ti0.2Zn0.2Cu0.2Fe0.2)3O4 with fast reaction kinetics and excellent stability as an anode material for lithium-ion batteries. *RSC Adv*. 10: 9736–9744 https://doi.org/10.1039/D0RA00255K.

[61] Lu, Y., Yu, L. and Lou, X.W.D. (2018). Nanostructured conversion-type anode materials for advanced Lithium-ion batteries. *Chem.* 4: 972–996. DOI 10.1016/j.chempr.2018.01.003.

[62] Cabana, J., Monconduit, L., Larcher, D. and Palacin, M.R. (2010). Beyond Intercalation-Based Li-Ion Batteries: The State of the Art and Challenges of Electrode Materials Reacting Through Conversion Reactions. *Adv. Mater.* 22: E170–E192. https://doi.org/10.1002/adma.201000717.

[63] Liang, B., Ai, Y., Wang, Y., Liu, C. and Ouyang, S. (2020). Spinel-Type (FeCoCrMnZn)3O4 High-Entropy Oxide: Facile Preparation and Supercapacitor. Performance13:5798-5807. DOI:10.3390/ma13245798.

[64] Poizot, P., Laruelle, S., Grugeon, S., Dupont, L. and Tarascon, J.M. (2000).Nano-sized transition-metal oxides as negative-electrode materials for lithium-ion batteries. *Nature* 407: 496–499 DOI:10.1038/35035045.

[65] Ding, C., et al. (2016). A bubble-template approach for assembling Ni–Co oxide hollow microspheres with enhanced electrochemical performance as an anode forlithium ion batteries. *Phys. Chem. Chem. Phys.* 18: 25879–25886. https://doi.org/10.1039/C6CP04097G.

[66] Liu, H., Wang, G., Liu, J., Qiao, S. and Ahn, H. (2011). Highly ordered mesoporous NiO anode material for lithium ion batteries with an excellent electrochemical performance. *J. Mater. Chem.* 21: 3046-3052. https://doi.org/10.1039/C0JM03132A.

[67] Goriparti, S. et al. (2014). Review on recent progress of nanostructured anode materials for Li-ion batteries. *J. Power Sources* 257: 421–443 https://doi.org/10.1016/j.jpowsour.2013.11.103.

Chapter 6

A New Prospective of High Entropy Ceramics: Properties and Remarkable Applications

Subhash, Pinki and Ashu Chaudhary*

Department of Chemistry, Kurukshetra University, Kurukshetra, Haryana, India

Abstract

High-entropy ceramics (HECs) have immediately picked up consideration since 2015. Disordered multicomponent frameworks, possessing the generally unknown focuses of stage graphs, were proposed in 2004 as creative materials with promising applications. The thought was to amplify the configurational entropy to settle the equimolar mixture and accomplish more vigorous frameworks, known as high-entropy materials. Until this point, about all work has zeroed in on five-segment, equimolar arrangements. This viewpoint article quickly surveys various groups of HECs and chooses properties. High entropy ceramics are novel materials with no under four different cations or anions. The advancement of high entropy ceramics production follows the 'configurational entropy balanced out single stage' idea, which was the first exhibited for high entropy metal combinations. The upsides of high entropy ceramics production are their compositional and structural decent variety, and huge numbers of them have a band hole, which makes them expected useful materials for a wide scope of utilizations. Initial research-intensive principally on metal combinations and nitride films. entropy stabilization was shown in a blend of oxides. Other high-entropy scattered ceramics quickly followed, animating the expansion of more segments to acquire materials communicating a mix of properties, regularly profoundly improved. The frameworks were before long

* Corresponding Author's Email: ashuchaudhary21@gmail.com.

In: Materials Science: Future Aspects
Editors: Kalpana Awasthi, Arti Srivastava and Mridula Tripathi
ISBN: 978-1-68507-843-0

demonstrated to be helpful in wide-running innovations, including thermal obstruction coatings, thermo-electrics, catalysis, batteries, and wear-resistant and corrosion-resistant coatings. Here we talk about the present status of the cluttered ceramics field by analyzing the applications and the high-entropy highlights fulfilling them, covering both hypothetical expectations and exploratory outcomes. The impact of entropy is unavoidable and can never again be disregarded. In the space of ceramics, it prompts new materials that, both as bulk and thin films, will play significant parts in innovation in the decades to come.

Keywords: HECs, lithium-ion batteries, catalysis, thermochemical water splitting, optical lenses, thin-film materials

1. Introduction

The expression "Ceramics" originates from the Greek word Keramikos, which signifies "Burnt stuff" or drinking vessel, demonstrating that attractive properties of these materials are regularly accomplished through a high-temperature heat treatment measure called Firing, however, was later applied by the Greeks to every single terminated item. Ceramic materials are inorganic, non-metallic materials and things produced using them. They can be Crystalline or somewhat translucent. They are shaped by the activity of warmth and resulting cooling. Most ceramics are compounds among metallic and non-metallic components for which the interatomic bonds are either absolutely ionic or prevalently ionic however having the same covalent character. Clay was perhaps the most punctual material used to create ceramics however various pottery materials are presently utilized in homegrown, modern, and building items. A wide-running group of materials whose fixings are the clay, sand, and felspar [1].

High entropy ceramics (HECs), here and there additionally alluded to as high entropy mixes, are single-stage ceramics production with no under four types of cations or anions. The idea of high entropy ceramics production is acquired from the field of high entropy alloys (HEAs). The two groups of materials are on a very basic level different, and henceforth their applications are likewise different [2, 3]. From the early instruments of crude people to the modern upheaval, advancement of railways and space investigation, progress has depended on the presentation of accessible materials. Diverting properties through blend alloying is a well-established strategy to improve materials.

Table 1. High-entropy ceramics for bulk materials applications

Year	Material	Description	Ref.
Bulk structural materials			
2015	(MgCoNiCuZn)O	fcc lattice, entropy-stabilized, antiferromagnetic TN = 106-140 K	39
2016	Refractory borides	Hexagonal lattice, H = 17–27 GPa, fracture toughness = 3.64–4.47 MPa $m^{1/2}$	49
2017	(Ce{RE})O_2	Three to six cations: fluorite (CaF2-type) structure; seven cations: cubic structure (space group Ia3 #206); Egap = 1.95–2.14 eV direct and 1.42–1.64 eV indirect	50
2018	Perovskite oxides	Two cation sublattices; ({Sr,Ba})({M})O_3: cubic structure (space group Pm3m #221); ({RE})({TM})O_3: orthorhombic structure (space group Pbnm #62), complex antiferromagnetic-dominated magnetic state	51
2018	$(CoCrFeMnNi)_3O_4$	Spinel structure retains ferrimagnetic behavior of other MFe2O4 spinels	52
2018	(HfNbTaTiZr)C	fcc lattice, H = 15–40.6 GPa, E = 443–552 GPa, κ = 5.42–6.45 W m^{-1} K^{-1}, oxidation resistant to 800°C (versus 200°C for binary carbides)	27
2018	(MoNbTaVW)C	fcc lattice, highest EFA despite two non-fcc precursors, H = 27 GPa, E = 533 GPa	53
2018	(HfTaTiWZr)C	fcc lattice, H = 33 GPa, E = 473 GPa	53
2018	(HfNbTaZr)C	fcc lattice, H = 36.1 GPa, E = 598 GPa	48
2019	$(GdTbDyHoEr)_2O_3$	Bixbyite structure, non-reversible transformation to single-phase, paramagnet	54
2019	$(CoCrFeMnZn)_3O_4$ and $(CoCrFeNiZn)_3O_4$	Spinel structure, ferromagnets	55
2019	(NbTaZr)C	fcc lattice, flexural strength = 366–496 MPa, toughness = 2.9 MPa $m^{1/2}$, E = 563 GPa	56
2019	(MoNbTaTiVW)C(N)	fcc lattice, H = 28 GPa, E = 554.9 GPa, fracture toughness = 8.4 MPa $m^{1/2}$	57
2019	(HfMoTaTi)(BC)	Three phases (two fcc, one hexagonal), H = 27.4 GPa, E = 505.8 GPa	58
2019	(HfMoTaTi)(BC)–SiC	Composite (hcp high-entropy phase and SiC compound), H = 35.4 GPa, E = 472.4 GPa	59
Energy-storage materials			
2016	(MgCoNiCuZn)O and derivatives	fcc lattice, anode material, specific capacity (I = 0.1 A g^{-1}, 300 cycles) = 920 mAh g^{-1}, approximately full capacity retention after 300 cycles; with Lix: colossal εr ≈ 2 × 105, superionic Li conductivity >10^{-3} S cm^{-1}; with LixFx: cathode material, working potential (versus Li^+/Li) = 3.4 V, specific capacity (I = 20 mA g^{-1}, 300 cycles) ≈ 120 mAh g^{-1}, 90% coulombic efficiencies after 300 cycles	[31]
2018	(CrMoNbVZr)N	fcc lattice, C = 78 F g^{-1} at 100 mV s^{-1} scan rate	[48]

Table 1. (Continued)

Year	Material	Description	Ref.
Thermochemical water-splitting materials.			
2018	Pt/Ru-loaded (MgCoNiCuZn)O	Mixed-phase (rock-salt and spinel), H_2 yields (1300°C) = 10.1 ± 0.5 ml-H_2 g^{-1}, H2 yields (1100°C) = 1.4 ± 0.5 ml-H_2 g^{-1}	[44]
Materials for thermal and environmental protection			
2018	(MgCoNiCuZn)O	fcc lattice, E = 108–152 GPa, bulk modulus = 187.7 ± 3.5 GPa, bending strength = 323 ± 19 MPa, κ = 2.95 ± 0.25 W m^{-1} K^{-1}, volumetric heat capacity = 3.01 ± 0.49 MJ m^{-3} K^{-1}, thermally stable ~450°C and mechanically stable to ~50 GPa	[46]
2018	(HfZrCe)({M}) $O_2^{-\delta}$	Fluorite (CaF_2-type) structure, H = 12.3–13.6 GPa, κ = 1.1–1.81 W m^{-1} K^{-1}	[78, 79]
2019	$(YbYLuScGd)_2Si_2O_7$	Monoclinic structure, stable up to 1300°C, α = 3.7–5.7 × 10^{-6} K^{-1}, coated on Cf/SiC enduring a 50% H2O–50% O2 environment at 1250°C for 300 h: residual flexural strength = 352 ± 26 MPa, strength retention = 86%	[28]
2019	$(YHoErYb)_2SiO_5$	Monoclinic structure, E (300–1600°C) = 150–165 GPa, κ(300–1300°C) ≈ 2 W m^{-1} K^{-1}, α(650–1450°C) = 5.5–6.5 × 10^{-6} K^{-1}	[29]
2019	$(TiZrHf)P_2O_7$	Metal pyrophosphate structure, κ(300 K) = 0.78 W m^{-1} K^{-1}, higher thermal stability than single-component phases (>1550°C)	[31]
2018	Pt/Ru-loaded (MgCoNiCuZn)O	CO oxidation: completely converted at 155°C, no catalytic activity reduction after >40 h at 135°C; CO_2 hydrogenation at 500°C: CO yields and CO_2 conversions >45%, CO selectivities >95%	[39]
2018	(AgBiGe)Se	fcc lattice, κ = 0.43 W m^{-1} K^{-1}, power factor (773 K) = 3.8 μW cm^{-1} K^{-2}, ZT (677 K) = 0.45	[43]
2018	$(Cu_5GeMgSnZn)\ S_9$	Tetragonal lattice, κ = 1.0-1.9 W m^{-1} K^{-1}, power factor (773 K) = 8 μW cm^{-1} K–2, ZT (723 K) = 0.58	[42]

Where: Cf, carbon fiber; CO, carbon monoxide; CO_2, carbon dioxide; EFA, entropy-forming ability; fcc, face-centered cubic; hcp, hexagonal close-packed; M, metal; RE, rare-earth metal; TM, transition metal; ZT, thermoelectric figure of merit. Electronic properties: C, specific capacitance; Egap, bandgap; I, specific current; εr, dielectric constant. Magnetic properties: TN, Neel temperature. Mechanical properties: E, elastic modulus; H, hardness. Thermal properties: α, coefficient of thermal expansion; κ, thermal conductivity.

Instinct directs that a mixture should yield a normal of its segments' properties however, at times upgraded or in any case new highlights can emerge. In profoundly disordered, multicomponent frameworks [4-7] high entropy produces alluring characteristics, including an inclination for single-phase solid solutions with straightforward crystal structures, drowsy energy, lattice contortions and an assortment of properties outstripping those of the component materials [8]. Despite continuous discussion concerning the significance of high-entropy impacts in metal alloys [8, 9], they give off an impression of being significant in high-entropy ceramics, which stay single phase under extraordinary temperatures [10, 11], pressures and chemical environments [12, 13] exhibiting security and strength across applications (Table 1). Because of the escalated research on HEAs, HECs have now become another concentration in the quest for novel materials. Like HEAs, HECs comprise multi-part components in a solitary stage, for which their huge configurational entropy adds to their formation [14]. Rather than metallic HEAs, HECs are commonly semiconductors or insulators with a band hole, which makes them conceivably helpful practical materials. Transition metal nitrides and carbides are ordinarily considered for high-temperature application, attributable to their high melting point and magnificent chemical steadiness [15, 16]. Specifically, ZrC, ZrN, TiC and TiN with the rock salt structures are highlighted with high Vickers hardness, high Young's modulus and high thermal conductivity, which make them the promising possibility for atomic application [17]. In addition to the fact that this presented the main affirmed instance of an entropy-driven change to a homogeneous, single-phase framework however it additionally presented another group of materials: entropy-balanced out oxides. Since their revelation in 2015, high-entropy earthenware production, which contain at least four metal components in required atomic proportions or if nothing else with each metal component being somewhere in the range of 5 and 35 at.% in a solitary phase structure, have stimulated concentrated exploration interests lately for their immense composition space, extraordinary microstructure, customizable properties, and different potential applications [18]. To date, broad endeavors have been committed to investigating an assortment of high-entropy ceramics, including oxides [19], carbides [20], what's more, diborides [21], with wide-going applications in the auxiliary and useful fields, which have noteworthy properties prompting applications in thermal and ecological security, thermoelectricity, water parting, catalysis and energy storage (Figure 1).

The universe of high-entropy materials is adaptable: by including species, sudden properties rise, which would then be able to be tweaked by fitting concentrations. The tremendous measure of opportunities for finding specially appointed disordered materials can be tended to with proficient high-throughput approaches [22] and quickened man-made consciousness methods [23]. Albeit scattered systems have been known for quite a long time, an unambiguous showing of entropy stabilization grabbed the eye of the network in 2015 [24].

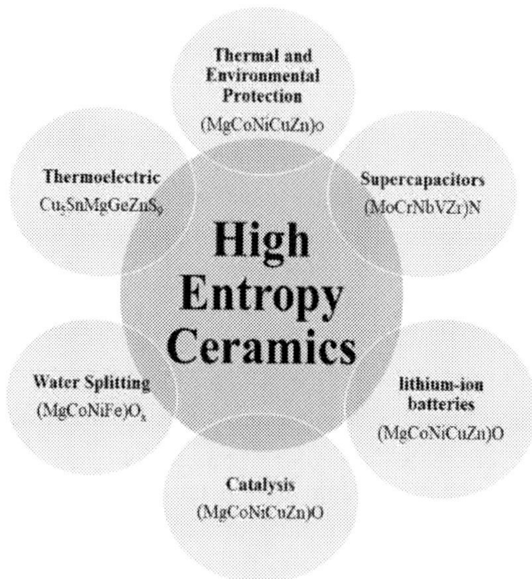

Figure 1. Applications of high entropy ceramics.

This chapter deals with the idea of entropy stabilization are summed up first, trailed by a review of theoretical reads for synthesizability and properties expectation. At that point, a few material sciences (oxides, borides, carbides, nitrides, and silicides) are presented along with the applications previously investigated (batteries, supercapacitors, thermoelectric, catalysis, thermo-chemical water parting, and thermal and environmental protection). disordered thin films are then talked about for wear-resistant, corrosion-resistant and biocompatible coatings, microelectronic dispersion barriers, electronic ceramics, spintronic layers and thermal insulators. We finish up with open doors for high-entropy ceramics production as far as applications, sciences and strategies.

2. Properties and Applications of High Entropy Ceramics

The advantages of high entropy ceramics are their compositional and structural diversity, and many of them have a bandgap, which makes them potential functional materials for a wide range of previously investigated applications like in batteries, supercapacitors, thermoelectric, catalysis, thermochemical water parting, and thermal and environmental protection.

2.1. Lithium-Ion Batteries

Lithium-ion batteries are significant for a wide assortment of uses going from consumer electronics to automotive drive to stationary load-leveling for intermittent power generation, for example, from wind or sun-oriented vitality. Battery execution relies basically upon the materials utilized, so the advancement of new materials is significant for propelling battery innovation [25, 26]. One materials challenge is the advancement of electrode materials with expanded energy density, quicker discharge kinetics and better dependability. Another materials challenge is the advancement of more secure and more reliable electrolytes to supplant the right now utilized natural carbonate liquid solutions. Specifically, the properties, particularly conductivity, of solid electrolytes for lithium-ion batteries are looked into.

The utilization of a solid electrolyte kills the requirement for control of the liquid electrolyte, which streamlines the cell configuration, just as improves safety and strength. The high flexible moduli of ceramics make them more appropriate for rigid battery plans as in, for instance, thin film-based gadgets. Ceramics are more appropriate for high temperatures or other aggressive situations. One of the most significant properties of electrolyte materials (Figure2) is ionic conductivity.

The superionic conductivity saw in (MgCoNiCuZn)O started ensuing examinations of this material for use in Li-ion batteries [27, 29]. As an anode (negative terminal) material, (MgCoNiCuZn)O conveys high Li storage limit maintenance and great cycling steadiness that improves with molecule size reduction. For reference, transition metal binary oxides offer restricted limit maintenance and efficiency, and four-cation variations which require post-annealing treatment to be synthesized experience the ill effects of strength and execution deficiencies. The cycling stability is likely an aftereffect of the conservation of the rock salt structure all through the whole redox measure, in which a portion of the cations, (for example, Co and Cu) take an interest in the conversion reactions while others, (for example, Mg) keep the structure intact and stifle dynamic nanograin aggregation. The capacity to present imperfections during conversion reactions without inciting stage partition is ascribed to entropy stabilization, which is a thermodynamic inclination for the homogeneous five-cation framework. As a functional application, full cells with a $LiNi_{1/3}Co_{1/3}Mn_{1/3}O_2$ cathode (positive terminal) were assembled [30]. Pocket cells were additionally constructed and utilized to control light-radiating diodes. Through the consolidation of both lithium and fluorine, (MgCoNiCuZn)O can likewise fill in as a lithium source, that is, as cathode dynamic material [31]. Oxyfluorides are significant cathode materials for their capacity to smother oxygen misfortune during cycling, with fluorine offering opposition against hydrogen fluoride etching. X-ray diffraction validates the arrangement of a rock salt stage with lithium and fluorine atoms haphazardly disseminated on the cation and anion sublattices, individually. A reversible change from a multiphase to a solitary stage state shows entropy stabilization.

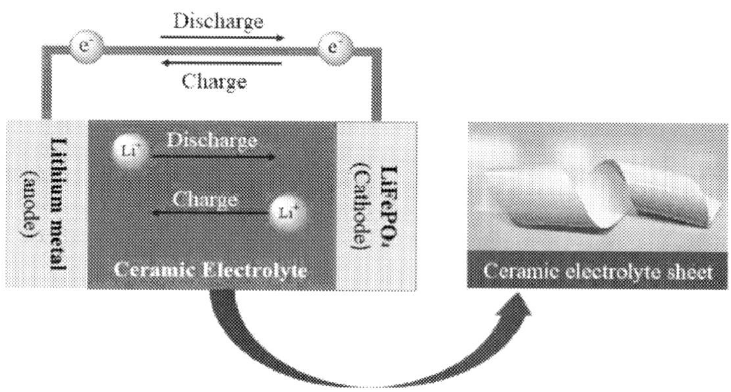

Figure 2. Ceramics are more appropriately used as electrolyte materials in lithium-ion batteries.

(MgCoNiCuZn)O has additionally been researched as a polysulfide anchor for the sulfur cathode in lithium-sulfur batteries, alleviating the shuttle impact brought about by the disintegration of lithium polysulfides in the electrolyte and dissemination to and fro between the electrodes [28]. The communication among (MgCoNiCuZn)O and lithium polysulfides makes Li–O and S–Ni bonds adding to the immobilization of lithium poly-sulfides in the sulfur anode. Coin cells utilizing (MgCoNiCuZn)O as a polysulfide anchor show serious reversible limit, extraordinary cycling stability and low-capacity decay.

2.2. Catalysis

The solid-solution metal oxide (MgCoNiCuZn)O is the primary known high-entropy (HE) metal oxide combined, framing a perfect example of the rising high-entropy oxide materials, which is gotten from high-temperature synthesis techniques (>900°C). Which demonstrated great stability at high temperature and delivered a high catalytic action in the hydrogenation of atmospheric CO_2

to CO. The last work exhibited the extraordinary bit of leeway of utilizing HE materials to disperse catalysis focuses.

For catalysis, high-entropy metal combinations have been read for some time and have been appeared to have helpful characteristics, including corrosion resistance, which limits conventional transition metal compounds as electrocatalysts in acidic or alkaline situations [32]; resistance to poisoning [33]; an enormous number of special restricting destinations giving an almost constant dispersion of adsorption energies [34]; synergetic and unexpected activity improvements; and expanded miscibility of components permitting the streamlining of restricting quality for higher activity [35-37]. In reality, high-entropy alloys have been demonstrated to be compelling in a few reactions, for example, oxidation (methanol, ammonia, carbon monoxide), advancement (oxygen, hydrogen), disintegration (ammonia), reduction (oxygen) and degradation (azo dye) [38]. The combination of bond chemistries in high-entropy ceramics can prompt promising roads for catalysis.

Valuable metal-stacked (MgCoNiCuZn)O was examined for CO oxidation and CO_2 hydrogenation [39]. The disarranged oxide was found to advance profoundly scattered Pt/Ru up to 5 wt.%, upgrading action, and offer protection from valuable metal sintering during high-heat treatments (up to 900°C for CO oxidation and 700°C for CO_2 hydrogenation) and at reaction temperatures (CO was changed over at 155°C Over the Pt-stacked catalyst, which held great reusability). At a reaction temperature of 500°C, CO yields and CO_2 changes of over 45% were accounted for 5 wt.% Pt/Ru-stacked catalysts. CO selectivities for Pt/Ru-stacked CO_2 hydrogenation catalysts were over 95% [39]. mesoporous (MgCoNiCuFe)Ox–Al_2O_3 was examined for CO oxidation 130. Complete CO transformation happened at 260°C. The ceramic showed immaterial degradation after 48 h 130, and better sulfur resistance than CuO–Al_2O_3. The system containing nitrogen, carbon, oxygen and boron were recognized as promising possibilities for oxygen reduction [34, 40]. Regardless of the great activity of such systems, performance enhancements are expected to rival platinum and state-of-the-art alloys. High-entropy ceramics production may introduce an ideal convergence of critical properties, particularly in non-trivial geometries as well as if their pore structure is tuned [41] to improve their surface region. Given their promising properties, we anticipate plenty of studies on catalysis dependent on high-entropy ceramics in the coming years.

2.3. Thermoelectric

High-entropy ceramics with compositional intricacy can be planned as new thermoelectric materials. Their decreased lattice thermal conductivity makes high-entropy semiconductors promising materials for thermoelectric applications. The presentation of Cu–S-based high-entropy materials were researched by orchestrating $Cu_5SnMgGeZnS_9$ and $Cu_3SnMgInZnS_7$ utilizing ball-milling and SPS [42]. A thermoelectric figure of merit (ZT) of 0.58 at 723 K was estimated for the synthesis $Cu_5Sn_{1.2}MgGeZnS_9$, like the qualities for requested ternary and quaternary precious diamond-like sulfides. Little improvement was gotten by presenting disorder, probably because of the effectively low thermal conductivity of Cu–S materials.

An all the more encouraging methodology may be the presentation of disorder into materials that have the capability of getting great thermoelectric if their thermal conductivity can be brought down. The high-entropy selenide (AgBiGe)Se was incorporated by a melting reaction of GeSe and $AgBiSe_2$ in a fixed vacuum tube [43], shaping a homogeneous rock-salt structure for half $AgBiSe_2$. Its capacity factor was 3.8 $\mu W\ cm^{-1}\ K^{-2}$ at 677 K, and its thermal conductivity was 0.43 $W\ m^{-1}\ K^{-1}$ at 300 K. ZT arrived at a limit of 0.45 at 677 K.

2.4. Thermochemical Water Splitting

Almost certainly, poly-cation oxides PCOs with complex cation structures will offer new open doors for both key examinations of redox thermochemistry just as adaptable H_2 creation utilizing foundation viable chemical systems. (MgCoNiCuZn)O enlivened the improvement of the $(MgCoNiFe)O_x$ ($x \approx 1.2$) polycation oxide, which uses a mixed solid-phase state (rock-salt and spinel) to perform two-step thermochemical water splitting at decreased temperatures pertinent for present-day enormous scope chemical infrastructure (≤1100°C) [44]. $(MgCoNiFe)O_x$ beats state-of-the-art materials in a few limits: H_2 yields of 10.1 ± 0.5 ml-H_2 g^{-1} at 1300°C and 1.4 ± 0.5 ml-H_2 g^{-1} at 1100°C, though ceria and spinel ferrites require >1300°C to yield considerable H_2; better protection from H_2 oxidation invert reaction during water splitting than Mn-based perovskites; and practically no presentation degradation more than 10 cycles for both temperature conditions, likely a consequence of high mixing entropy forestalling phase separation into binary oxides. The reactivity of CeO_2-based ceramics in the O_2-delivering step could be upgraded by doping the pottery with cations with a higher valence and a smaller powerful ionic range [45]. CeO_2-based ceramics for hydrogen creation employing two-step water-splitting cycles (Figure 3).

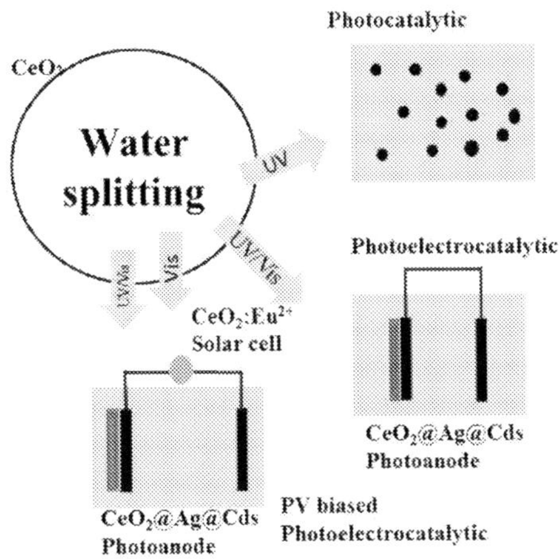

Figure 3. CeO_2-based ceramics for hydrogen creation utilizing two-step water-splitting cycles.

2.5. Thermal and Environmental Protection

The (MgCoNiCuZn)O shows amorphous-like thermal conductivity low qualities expanding with temperature without compromising its mechanical stiffness, a typical compromise in non-metals, in which phonons are the predominant heat transporters [46]. Rayleigh scattering (instead of momentum-destroying Umklapp measures) was discovered to be the prevailing phonon dispersing mechanism directing thermal conductivity in high-entropy ceramics. The thermal conductivity can be additionally decreased with the option of a 6th cation (Sc, Sb, Sn, Cr, or Ge), which is conceivable in thin films developed by pulsed laser deposition [47]. The subsequent drop in thermal conductivity by a factor of two is to a great extent autonomous of mass, yet determined rather by nearby ionic charge disorder, which strains the oxygen sublattice and prompts huge varieties in the interatomic force constants. High-entropy oxides display proportions of versatile modulus to thermal conductivity among the most elevated details at room temperature, surpassing those of noticeable thermal barrier coatings, for example, zirconate materials (Table 1).

2.6. Supercapacitors

Crystalline high-entropy ceramics production (CHC), another class of crystalline solid arrangements that contain at least five natural species, has been pulled into noteworthy consideration because of their exceptional physical properties and potential applications. The high-entropy nitride (MoCrNbVZr)N [48] has been explored for use in supercapacitor applications. (MoCrNbVZr)N was incorporated utilizing a mechanochemical delicate urea approach and had explicit capacitances running from 230 to 54 F g^{-1} for filter rates somewhere in the range of 10 and 200 mV s^{-1}.

3. High-Entropy Ceramics for Thin-Film Materials Applications

High-entropy nitride, carbide, boride, and oxide thin films have been stored on Si, steel, Ti_6Al_4V compound, quartz glass, WC–Co established carbide, MgO and $SrTiO_3$ substrates utilizing faltering of natural or high-entropy amalgam focuses in an Ar + N_2, Ar + CH_4 or Ar + O_2 environment, just as utilizing a cathodic arc and pulsed laser deposition. Pulsed laser deposition extends the dissolvability of cations by contributing a 'viable temperature' through episode molecule motor energy, empowering the consolidation of Ca, Sc, Cr, Ge, Sn, and Sb [60]. High-entropy ceramic films have been researched for likely use as wear-resistant, corrosion-resistant, and/or oxidation-resistant coatings, dispersion obstructions for microelectronics, electronic pottery, biocompatible coatings, antiferromagnetic layers for spintronics, and thermal protectors (Table 2).

The primary nitride films depended on high-entropy combinations shaping on body-centered cubic or face-centered cubic (fcc) cross-sections: the films would in general be amorphous or multiphase, not framing the single-phase glasslike solid solutions that describe high-entropy alloy s [61]. This was credited to the presence of metals that don't shape nitrides, for example, Cu, prompting the arrangement of isolated metallic stages. Utilizing great nitride formers, for example, Al, Cr, Ta, Ti, and Zr, prompts the development of single-phase fcc solid-solution films [62]. Hexagonal AlN, specifically, breaks up effectively into the fcc grid, with (AlTi)N solid-solutions having an fcc cross-section up to~60% AlN, and Al upgrading the oxidation obstruction of coatings. Like nitrides, high-entropy oxides were initially investigated as thin films of high-entropy composites (metals) kept at high oxygen pressures, frequently yielding high-symmetry structures, for example, hexagonal close-packed, spinel and fcc [63].

3.1. Diffusion Barriers for Microelectronic Applications

The progressing scaling down of microelectronic hardware down to the nanoscale requires new materials to forestall interdiffusion at the interface among Cu and Si components and the arrangement of Cu_3Si, which increments electrical obstruction, decreases proficiency and meddles with the usefulness of the gadget. Truly, such diffusion barriers have incorporated the SiO_2 native oxide alongside nitrides, for example, Si_3N_4 and TiN. High-entropy nitrides are especially encouraging for such applications, because the lattice distortion and expanded packing density coming about because of the diverse basic radii bring about drowsy dispersion even through layers only a couple of nanometers thick (Table 2).

The effectiveness of expanding the natural segments to improve the dispersion hindrance was shown by an examination of nitrides with the number of metals going from one to six [67]. The disappointment temperature expanded monotonically with the number of components from 550°C for TiN to 900°C for (AlCrRuTaTiZr)N.

3.2. Wear-Resistant, Corrosion-Resistant and Oxidation-Resistant Coatings

High-entropy carbides, nitrides, carbo-nitrides and oxides show upgraded hardness due to solid-solution reinforcing, just as great oxidation and corrosion resistance because of their slow dissemination, which limits the entrance of oxygen and different species (Table 2). They are important for applications as defensive coatings impervious to wear, oxidation and additional corrosion for machine parts, for example, cutting instruments and drill bits, working at high temperatures in destructive situations. Properties including the hardness, versatile modulus, wear rate, coefficient of contact, resistivity, mass addition during strengthening in air, and corrosion currents and possibilities in destructive arrangements were estimated for a scope of carbide, nitride and oxide coatings (Table 2). Although nitrides are regularly harder, the comparing carbides can display a lower wear rate because of their lower coefficients of grating, which can be clarified by the development of a free carbon layer on a superficial level going about as a lubricant [68]. Al_2(CoCrCuFeNi)O and (AlCrTaTiZr)O films accomplish hardness esteems among the most noteworthy showed by available oxide coatings; twofold oxide films seldom exceed 20 GPa [69]. Hardness by and large increments with faltering substrate inclination and nitrogen content up to immersion: super-saturated films can create voids that diminish hardness. Higher aluminum and oxygen concentrations [70] and tempering additionally upgrade hardness, which can be ascribed to the fuse of strong oxygen–metal bonding, end of film cracks and voids, and development of nanocrystalline phases.

Table 2. High-entropy ceramics for thin-film materials applications

Year	Material	Description	Refs.
Diffusion barriers for microelectronics			
2008	(AlMoNbSiTaTiVZr)N	Amorphous, barrier intact to 850°C, failure at 900°C	[75]
2009	(AlCrTaTiZr)N	fcc nanocrystallites in amorphous matrix, barrier intact to 900°C	[77]
2012	(AlCrRuTaTiZr)N	Amorphous (4 nm film thickness) or fcc (102–103 nm film thickness), barrier intact to 800°C, failure at 900°C	[76]
2012	(CrHfTiVZr)N	Amorphous (10 nm film thickness) or fcc (102–103 nm film thickness), barrier intact to 800°C, failure at 900°C	[77]
Wear-resistant and corrosion/oxidation-resistant coatings			
2007	Al2(CoCrCuFeNi)O	Spinel structure, H = 22.6 ± 1.6 GPa, Egap = 1.56–1.68 eV	[81]
2010	(AlCrTaTiZr)O	Crystallizes at 900°C (mixed phase), H ≈ 20 GPa, E ≈ 230 GPa, ρ = 1012 μΩ-cm	[80]
2011	(AlCrMoTaTiZr)N	fcc lattice, H = 25–40.2 GPa, E = 370–420 GPa, μ = 0.74–0.80, wear rate = 2.8–2.9 × 10^{-6} mm^3 N^{-1} m^{-1}	[86]
2011	(AlCrTaTiZr)CN	fcc lattice, H = 20–35 GPa, elastic modulus = 242–280 GPa	[87]
2012	(CrTaTiVZr)N	fcc lattice, H = 11.3–36.4 GPa, E = 200.3–273.8, ρ = 131– 446 μΩ-cm Amorphous with fcc nanocrystallites, H = 16–17 GPa, E = 220–230 GPa, corrosion current = 3.1–19.1 μA/cm2	[88]
2012	(AlCrSiTiZr)N	Amorphous with fcc nanocrystallites, H = 16–17 GPa, E = 220–230 GPa, corrosion current = 3.1–19.1 μA/cm^2	[78]
2012	(HfNbTiVZr)N	fcc lattice, H = 36–70 GPa, E = 281–384 GPa, μ = 1.19, wear rate = 0.39 × 10^{-6} mm^3 N^{-1} m^{-1}	[79]
2012	(AlBCrSiTi)N	fcc crystallization at 800°C, H = 23 GPa, E = 256.6 GPa	[80]

Year	Material	Description	Refs.
2012	(HfNbTaTiZr)C	fcc lattice, H = 27.5 GPa, μ = 0.15, wear rate = 0.8 × 10^{-6} mm^3 N^{-1} m^{-1}	[68]
2012	(HfNbTaTiZr)N	fcc lattice, H = 32.9 GPa, μ = 0.96, wear rate = 2.9 × 10^{-6} mm^3 N^{-1} m^{-1}	[68]
2013	(CuSiTiYZr)C	Amorphous; H = 20.7–29.5 GPa, μ = 0.15, wear rate = 1.89 × 10^{-6} mm^3 N^{-1} m^{-1}, corrosion current = 0.535–0.841 μA/cm^2	[81]
2013	(AlCrNbSiTi)N	fcc lattice, H = 16–36.7 GPa, E = 300–450 GPa, good oxidation resistance at 900°C	[40]
2013	(AlCrMoTaTi)N	fcc lattice, H = 20.6–30.6 GPa, E = 260–290 GPa, ρ = 536–8212 μΩ-cm, oxidation temperature = 1073 K, increases to >1173 K with Si addition	[82]
2017	(HfNbTaTiVZr)C	fcc lattice, H = 43–48 GPa, E = 337 GPa	[83]
2018	(CrNbTaTiW)C	fcc lattice (bct for Ta and W rich), H = 11.7 GPa (Nb rich)–35.5 GPa (Ta and W rich), E = 367 GPa (~equimolar)–568 GPa (Ta and W rich)	[84]
2018	(CrNbSiTiZr)C	fcc lattice, H = 22.3–32.8 GPa, E = 191–358 GPa, μ = 0.07–0.4, wear rate = 0.2–3.3 × 10^{-6}mm3 N^{-1} m^{-1}	[85]
2018	$(HfTaTiVZr)B_2$	H = 47.2 GPa, E = 540.1 GPa	[86]
Antiferromagnetic layers for spintronics			
2017	(MgCoxNiCuZn)O	fcc lattice, 10-fold exchange-bias enhancement over permalloy/CoO heterostructures	[71]
Electronic ceramics			
2011	(Ti/Al)FeCoNi)O	fcc lattice, ρ = 19 μΩ-cm; with Ti: ρ = 28–35 μΩ-cm; with Al: ρ = 18 μΩ-cm	[63]
Thermal insulators			
2018	(MgCoNiCuZn)O	fcc lattice, E = 151.0–236.7 GPa, κ = 1.41–1.68 W m^{-1} K^{-1}, volumetric heat capacity = 3.29–3.96 MJ m^{-3} K^{-1}	[46]
2018	$Ba(ZrSnTiHfNb)O_3$	Perovskite structure, κ = 0.54–0.58 W m^{-1} K^{-1}	[72]
Biocompatible coatings			
2012	(HfNbTaTiZr)C	fcc lattice, H = 22.4–32.1 GPa, μ = 0.12–0.32, wear rate = 0.2–0.9 × 10–6 mm^3 N^{-1} m^{-1}	[74]
2012	(HfNbTaTiZr)N	fcc lattice, H = 30.9 GPa, μ = 0.17, wear rate = 0.29 × 10–6 mm^3 N^{-1} m^{-1}	[73]
2016	(NbSiTaTiZr)C	fcc lattice, corrosion current = 0.025 μA/cm^2	[87]

Where: bct, body-centred tetragonal; fcc, face-centred cubic. Electronic properties: Egap, bandgap; ρ, electrical resistivity. Mechanical properties: E, elastic modulus; H, hardness; μ, coefficient of friction. Thermal properties: κ, thermal conductivity.

3.3. Antiferromagnetic Layers for Spintronics

Entropy-stabilized materials are balanced out by the configurational entropy of the constituents, as opposed to the enthalpy of formation of the compound. A special benefit to entropy-stabilized materials is the expanded solvency of components, which opens a wide compositional space, with ensuing local synthetic and basic disorder coming about because of various nuclear sizes and favored coordination of the constituents.

The anti-ferromagnetic conduct of (MgCoNiCuZn)O was first explored in a heterostructure with ferromagnet permalloy [71]. Exchange coupling at the ferromagnet/antiferromagnet interface can be advanced by adjusting the chemical and the level of magnetic particles (it increments with Co concentrations). By tuning the Co concentration, the exchange inclination can be improved to arrive at a worth a significant degree higher than that of permalloy/CoO heterostructures.

3.4. Electronic Ceramics

Oxidation can have the opposite effect on resistivity relying upon the alloy system. The resistivity of (AlCrTaTiZr)O films increment with the oxygen concentration (up to 1012 $\mu\Omega$ -cm) [69], while (TiFeCoNi) O films display their least resistivities (35 ± 3 $\mu\Omega$ -cm) after oxidation at high temperatures, which eliminates little voids at amorphous grain boundaries [63]. Indeed, the resistivity of (TiFeCoNi)O films keeps on diminishing with annealing temperatures dipping under that of bulk TiFeCoNi even after the resistivity in TiFeCoNi films begins to increment as the film respond with the SiO_2/Si substrate forming high-resistivity silicide's. The resistivity of (TiFeCoNi)O films is equivalent to that of single-crystal RuO_2 and lower than that of tin-doped indium oxide thin films (150 $\mu\Omega$ -cm). Although this is as yet higher than the resistivity of numerous unadulterated metals, it is low contrasted and that of ceramic semiconductors at room temperature [63].

3.5. Biocompatible Coatings

Because of their protection from wear and corrosion, high-entropy carbides and nitrides are promising materials for coatings for biomedical inserts, on the off chance that they can be exhibited to be biocompatible and non-poisonous. Carbide and nitride coatings demonstrated preferred corrosion resistance over the bare substrate in recreated body fluid. Biocompatibility was tried utilizing osteoblast [73] or osteosarcoma cells: no cytotoxic reactions were watched, with most cells still alive after 72 h. Si-containing coatings were found to improve cell connection and practicality, with (HfNbSiTiZr)C being the most biocompatible material tested [74]. A potential clarification for the improved cell connection comes from the expanded negative surface charge of Si-containing coatings: positively-charged proteins in the arrangement cling to the negatively-charged covering and afterward pull in negatively-charged cells, mooring them to the surface.

Conclusion

Disorder prompts unexpected properties. High-entropy ceramics show surprising properties symmetrical to those of high-entropy metal combinations, for example, high hardness and melting temperatures, the steadiness of structure in outrageous conditions, amorphous like thermal conductivity, wear and corrosion opposition, and affirmed entropy stabilization. Contrasted with metals, in which electrons are delocalized, ceramics can have articulated impacts beginning from disorder-induced charge vacillations. Potential applications are bountiful, among them, energy storage, water splitting, catalysis, thermoelectricity, electronic device materials, spintronics, thermal and ecological protection, wear resistance and biocompatible coatings. The augmentation to corresponding systems, in which configurational disorder additionally shows up in the anion sublattice, further builds entropy and gives a bigger compositional space to find new materials with streamlined properties. The outside of disorder materials has scarcely been damaged: the immense universe of structures and compositions requires successful joint trial, hypothetical and computational endeavors, all the while tending to the synthesizability, stability and applicability of these materials. Entropy is the platform for significant future materials-space investigations.

Acknowledgments

The authors Mr. Subhash and Pinki wish to express gratitude to the University Grants Commission (UGC) (Ref. No.- 92(CSIR-UGC NET DEC. 2018) and Council of Scientific and Industrial Research (CSIR) (Ref. No.- 16/06/2019(i) EU-V.), New Delhi, India New Delhi for financial assistance in the form of JRF.

References

[1] Ashby, M. F. and D. R. H. Jones, (2005). Engineering Materials 1, *An Introduction to Their Properties and Applications*, 3rd edition, Butterworth-Heinemann, Woburn, UK.

[2] Yeh, J. W., Chen, S. K., Lin, S. J., Gan, J. Y., Chin, T. S., Shun, T. T., Tsau, C. H. and Chang, S. Y. (2004). *Adv. Eng. Mater.* 6: 299–303.

[3] Cantor, B., Chang, I. T. H., Knight, P. and Vincent, A. J. B. (2004). *Mater. Sci. Eng. A*, 375–377: 213–218.

[4] Li, F, Zhou, L, Liu, JX, et al. High-entropy pyrochlores with low thermal conductivity for thermal barrier coating materials. *J Adv Ceram* 2019, 8: 576–582.

[5] Rui, Feng, Liaw, Peter K., Gao, Michael C. and Widom, Michael (2017). First-principles prediction of high-entropy-alloy stability, *Computational Materials,* 3, 1-7.

[6] Widom, M. (2018). Modeling the structure and thermodynamics of high-entropy alloys. *J. Mater. Res.* 33: 2881–2898.

[7] George, E. P., Raabe, D. and Ritchie, R. O. (2019). High-entropy alloys. *Nat. Rev. Mater.* 4:515–534.

[8] Yeh, J. W. (2006). Recent progress in high-entropy alloys. *Ann. Chim. Sci. Mat.* 31: 633–648.

[9] Miracle, D. B. (2017). High-entropy alloys: a current evaluation of founding ideas and core effects and exploring nonlinear alloys. *JOM* 69: 2130–2136.

[10] Grzesik, Z. Smoła, G., Stygar, M., Dąbrowa, J., Zajusz, M., Mroczka, K. and Danielewski M. (2019). Defect structure and transport properties in (CoCuMgNiZn)O high entropy oxide. *J. Eur. Ceram. Soc.* 39: 4292–4298.

[11] Dong, Y. ren, K., Yonghong, l., Liu, J and wang, Y. (2019). High-entropy environmental barrier coating for the ceramic matrix composites. *J. Eur. Ceram. Soc.* 39: 2574–2579.

[12] Chen, J., Liu, W., Liu, J., Zhang, X., Yuan, M., Zhao, Y., Yan, J., Zhang, H. and Yin. Z. (2019). Stability and compressibility of cation-doped high-entropy oxide MgCoNiCuZnO5. *J. Phys. Chem. C.* 123: 17735-17744.

[13] Sarkar, A. Velasco, L., Wang, D., Wang, Q., Talasila, G., Biasi, L. d., Kübel, C., Brezesinski, T., Bhattacharya, S. S., Hahn, H. and Breitung, B. (2018). High entropy oxides for reversible energy storage. *Nat. Commun.* 9: 3400.

[14] Zheng, Y. Yi, Y., Fan, M., Liu, H., Li, X., Zhang, R. and Li, M. (2019). A high-entropy metal oxide as chemical anchor of polysulfide for lithium-sulfur batteries. *Energy Storage Mater.* 23: 678-683.

[15] Berardan, D., Franger, S., Meena, A. K., Dragoe, N. Room. (2016). Temperature Lithium Superionic Conductivity in High Entropy Oxides. *J. Mater. Chem. A.*, 4, 9536–9541.

[16] Harrison, R. W. and Lee, W. E. (2016). Processing and properties of ZrC, ZrN and ZrCN ceramics *Adv. Appl. Ceram.* 115: 294–307.

[17] Chen, K., Kamran, S. (2013). *Model Numer. Simul. Mater. Sci.* 3: 7–11.

[18] Bao, W., Robertson, S., Liu, J. X., Zhang, G. J., Xu, F. and Wu, H. (2018). Structural integrity and characteristics at lattice and nanometer levels of ZrN polycrystalline irradiated by 4 MeV Au ions. *J. Eur. Ceram. Soc.* 38: 4373-4383.

[19] Liu, D., Liu, H., Ning, S., Ye, B., Chu, Y. (2019). Synthesis of high-purity high-entropy metal diboride powders by boro/carbothermal reduction. *J Am Ceram Soc.* 102: 7071-7076.

[20] Hong, W., Chen, F., Shen, Q., Han, Y., Fahrenholtz W. G. and Zhang, L. (2019). Microstructural evolution and mechanical properties of (MgCoNiCuZn)O high-entropy ceramics. *J Am Ceram Soc.* 102: 2228-2237.

[21] Wei, X., Liu, J., Li, F., Qin Y., Liang, Y., Zhang, G. (2019). High entropy car-bide ceramics from different starting materials. *J Eur Ceram Soc.* 39: 2989-2994.

[22] Liu, D., Wen, T., Ye, B., Chu, Y. (2019). Synthesis of superfine high-entropy metal diboride powders. *Scripta Mater.* 167: 110-114.

[23] Curtarolo, S. Hart, G. L. W., Nardelli, M. B. Mingo, N. Sanvito, S. and Ohad. (2013). The high-throughput highway to computational materials design. *Nat. Mater.* 12: 191-201.

[24] Schmidt, J., Marques, M. R. G., Botti, S. and Marques, M. A. L. (2019). Recent advances and applications of machine learning in solid-state materials science. *npj Comput. Mater.* 5, 83.

[25] Rost, C. M. Sachet, E., Borman, T., Moballegh, A., Dickey, E. C., Hou, D., Jones, J. L., Curtarolo, S. and Maria, J. P. (2015). Entropy-stabilized oxides. *Nat. Commun.* 6: 8485.

[26] Scrosati, B. and Garche, J. (2010). Lithium batteries: Status, prospects and future. *J. Power Sources.* 195: 2419-2430.

[27] Goodenough, J. B. and Kim, Y. (2010). Challenges for Rechargeable Li Batteries. *Chem. Mater.* 22: 587-603.

[28] Wang, Q., Sarkar, A., Li, Z., Lu, Y., Velasco, L., Bhattacharya, S. S., Brezesinski, T., Hahn, H., Breitung, B. (2019). High Entropy Oxides as Anode Material for Li-Ion Battery Applications: A Practical Approach. *Electrochem. Commun.* 100, 121–125.

[29] Zheng, Y. Yib, Y., Fana, M., Liuc, H., Lic, X., Zhanga, R., Lib, M. and Qiaoa, Z. (2019). A high-entropy metal oxide as chemical anchor of polysulfide for lithium-sulfur batteries. *Energy Storage Mater.* 23: 678-683.

[30] Qiu, N. Chenab, H., Yanga, Z., Suna, S., Wanga, Y. and Cui, Y. (2019). A high entropy oxide ($Mg_{0.2}Co_{0.2}Ni_{0.2}Cu_{0.2}Zn_{0.2}O$) with superior lithium storage performance. *J. Alloys Compd.* 777: 767-774.

[31] Wang, Q. Sarkara, A., Lia, Z., Lu, Y., Velascoa, L., Bhattacharya, S. S., Brezesinskia, T., Hahnac, H. and Breitunga, B. (2019). High entropy oxides as anode material for Li-ion battery applications: a practical approach. *Electrochem. Commun.* 100: 121-125.

[32] Wang, Q. Sarkar, A. Wang, D., Velasco, L., Azmi, R. Bhattacharya, S. S., Bergfeldt, T. and Düvel, A. (2019). Multi-anionic and -cationic compounds: new high entropy materials for advanced Li-ion batteries. *Energy Environ. Sci.* 12: 2433-2442.

[33] Zhang, G. Mingb, K., Kanga, J., Huanga, Q., Zhanga, Z. and Zhengc, X. (2018). High entropy alloy is a highly active and stable electrocatalyst for hydrogen evolution reactions. *Electrochim. Acta* 279: 19–23.

[34] Wang, A. L., Wan, H. C., Xu, H., Tong, Y. X. and Li, G. R. (2014). Quinary PdNiCoCuFe alloy nanotube arrays as efficient electrocatalysts for methanol oxidation. *Electrochim. Acta* 127: 448-453.

[35] Löffler, T. Meyer, H., Savan, A., Wilde, P., Garzón, A. and Yen, M. (2018). Discovery of a multinary noble metal-free oxygen reduction catalyst. *Adv. Energy Mater.* 8: 180-269.

[36] Batchelor, T. A. A. Pedersen, J. K., Winther, S. H., Castelli, I. E., Jacobsen, K. W. and Rossmeisl, J. (2019). High-entropy alloys as a discovery platform for electrocatalysis. *Joule 3*, 834–845.

[37] Cui, X., Zhang, B., Zeng, C. and Guo, S. (2018). Electrocatalytic activity of high-entropy alloys toward oxygen evolution reaction. *MRS Commun.* 8: 1230-1235.

[38] Mao, A., Xiang, H.-Z., Zhang, Z.-G., Kuramoto, K., Yu, H., Ran, S. (2019). Solution Combustion Synthesis and Magnetic Property of Rock-Salt High-Entropy Oxide Nanocrystalline Powder. *J. Magn. Magn. Mater.* 2019, 484, 245–252.

[39] Li, Z., Pradeep, K. G., Deng, Y., Raabe, D., Tasan, C. C. (2016). Metastable high-entropy dual-phase alloys overcome the strength-ductility trade-off. *Nature* 2016, 534, 227–230.

[40] Chen, H. Lin, W., Zhang, Z., Jie, K., Mullins, D. R., Sang, X., Yang, S., Jafta, C. J., Bridges, C. A., Hu, X., Unocic, R. R., Fu, J., Zhang, P. and Dai, S. (2019). Mechanochemical synthesis of high entropy oxide materials under ambient conditions: dispersion of catalysts via entropy maximization. *ACS Mater. Lett.* 1: 83-88.

[41] Gong, K., Du, F., Xia, Z., Durstock, M. and Dai, L. (2009). Nitrogen-doped carbon nanotube arrays with high electrocatalytic activity for oxygen reduction. *Science* 323: 760-764.

[42] Lefèvre, M., Proietti, E., Jaouen, F. and Dodelet, J. P. (2009). Iron-based catalysts with improved oxygen reduction activity in polymer electrolyte fuel cells. *Science* 324: 71-74.

[43] Zhang, R.-Z., Gucci, F., Zhu, H., Chen, K. and Reece, M. J. (2018). Data-driven design of eco-friendly thermoelectric high-entropy sulfides. *Inorg. Chem.* 57: 13027-13033.

[44] Roychowdhury, S., Ghosh, T., Arora, R., Waghmare, U. V. and Biswas, K. (2018). Stabilizing n-type cubic GeSe by entropy-driven alloying of $AgBiSe_2$: ultralow thermal conductivity and promising thermoelectric performance. *Angew. Chem. Int. Ed.* 57: 15167-15171.

[45] Zhai, S. Rojas, J., Ahlborg, N., Lim, K., Toney, M. F., Jin, H., Chueh, W. C., and Majumdar. A. (2018). The use of poly-cation oxides to lower the temperature of two-step thermochemical water splitting. *Energy Environ. Sci.* 11: 2172-2178.

[46] Menga, P. Q., Lee, C., Ishihara, T., Kaneko, H. and Tamaura, Y. (2011). Reactivity of CeO_2-based ceramics for solar hydrogen production via a two-step water-splitting cycle with concentrated solar energy. *International Journal of Hydrogen Energy.* 36: 13435-13441.

[47] Senkov, O. N., Scott, J. M., Senkova, S. V., Miracle, D. B., Woodward, C. F. (2011). Microstructure and Room Temperature Properties of a High-Entropy TaNbHfZrTi Alloy. *J. Alloys Compd.* 509, 6043–6048.

[48] Okejiri, F., Zhang, Z., Liu, J., Liu, M., Yang, S., Dai, S. Room-Temperature Synthesis of High-Entropy Perovskite Oxide Nanoparticle Catalysts through Ultrasonication-Based Method. *ChemSusChem*. 2020, 13, 111–115.

[49] Feuerbacher, M., Heidelmann, M., Thomas, C. (2015). Hexagonal High-Entropy Alloys. *Mater. Res. Lett*. 3, 1–6.

[50] Gludovatz, B. Hohenwarter, A., Catoor, D., Chang, E. H., George, E. P., Ritchie, R. O. (2014). A FractureResistant High-Entropy Alloy for Cryogenic Applications. *Science* 345, 1153–1158.

[51] Djenadic, R. Sarkar, A., Clemens, O., Loho, C., Botros, M., Chakravadhanula, V. S. K. and Kübel, C. (2017). Multicomponent equiatomic rare earth oxides. *Mater. Res. Lett.* 5, 102–109.

[52] Jiang, S., Hu, T., Gild, J., Zhou, N., Nie, J., Qin, M. and Harrington, T. (2018). A new class of high-entropy perovskite oxides. *Scr. Mater.* 142: 116–120.

[53] Dąbrowa, J., Stygar, M., Mikuła, A., Knapik, A., Mroczka, K., Tejchman, W., Danielewski, M. and Martin, M. (2018). Synthesis and microstructure of the $(CoCrFeMnNi)_3O_4$ high entropy oxide characterized by spinel structure. *Mater. Lett.* 216: 32-36.

[54] Jiang, S., Hu, T., Gild, J., Zhou, N., Nie, J., Qin, M., Harrington, T., Vecchio, K., Luo, J. (2018). A New Class of High-Entropy Perovskite Oxides. *Scr. Mater*. 142, 116–120.

[55] Tseng, K. P., Yang, Q., McCormack, S. J. and Kriven, W. M. (2019). High-entropy, phase-constrained, lanthanide sesquioxide. *J. Amer. Ceram. Soc.* 103: 569-576.

[56] Zhao, Z. Chen, H. Xiang, H. Dai, F. Z. Wang, X. Xu, W. Sun, K. Peng, Z. Zhou, Y. (2020). High Entropy Defective Fluorite Structured Rare-Earth Niobates and Tantalates for Thermal Barrier Applications. *J. Adv. Ceram.* 9, 303–311.

[57] Demirskyi, D., Borodianska, H., Suzuki, T. S., Sakka, Y., Yoshimi, K. and Vasylkiv, O. (2019). High-temperature flexural strength performance of ternary high-entropy carbide consolidated via spark plasma sintering of TaC, ZrC and NbC. *Scr. Mater.* 164: 12-16.

[58] Liu, Y., Jia, D., Zhou, Y., Zhou, Y., Zhao, J., Nian, H., Liu, B. (2020). A Non Equimolar Multicomponent Perovskite Ceramic with Low Thermal Conductivity. *J. Eur. Ceram. Soc.* 40, 6272–6277.

[59] Teng, Z., Zhu, L., Tan, Y., Zeng, S., Xia, Y., Wang, Y., Zhang, H. (2020). Synthesis and Structures of High-Entropy Pyrochlore Oxides. *J. Eur. Ceram. Soc.* 40, 1639–1643.

[60] Zhang, H., Hedman, D., Feng, P., Han, G. and Akhtar, F. A. (2019). high-entropy $B_4(HfMo_2TaTi)$ C and SiC ceramic composite. *Dalton Trans*. 48: 5161-5167.

[61] Liu, D., Wen, T., Ye, B., Chu, Y. (2019). Synthesis of Superfine High-Entropy Metal Diboride Powders. *Scr. Mater.* 167, 110–114.

[62] Chen, T. K., Shun, T. T., Yeh, J. W. and Wong, M. S. (2004). Nanostructured nitride films of multi-element high-entropy alloys by reactive DC sputtering. *Surf. Coat. Technol.* 188–189: 193-200.

[63] Lai, C. H., Lin, S. J., Yeh, J. W. and Chang, S. Y. (2006). Preparation and characterization of AlCrTaTiZr multi-element nitride coatings. *Surf. Coat. Technol.* 201: 3275-3280.

[64] Yang, Y. C., Tsau, C. H. and Yeh, J. W. (2011). TiFeCoNi oxide thin film a new composition with extremely low electrical resistivity at room temperature. *Scr. Mater.* 64: 173-176.

[65] Anand G, Wynn AP, Handley CM, (2018). Phase stability and distortion in high-entropy oxides. *Acta Mater* 146: 119-125.

[66] Chang, S. Y., Li, C. E., Chiang, S. C. and Huang, Y. C. (2012). 4-nm thick multilayer structure of multi-component (AlCrRuTaTiZr)Nx as a robust diffusion barrier for Cu interconnects. *J. Alloys Compd.* 515: 4-7.

[67] Chang, S. Y., Chen, M. K. and Chen, D. S. (2009). Multiprincipal- element AlCrTaTiZr–nitride nanocomposite film of extremely high thermal stability as diffusion barrier for Cu metallization. *J. Electrochem. Soc.* 156: G37-G42.

[68] Chang, S. Y., Huang, Y. C., Li, C., Hsu, H. F., Yeh, J. and Lin, S. J. (2013). Improved diffusion-resistant ability of multicomponent nitrides: from unitary TiN to senary high-entropy (TiTaCrZrAlRu)N. *JOM* 65: 1790-1796.

[69] Braic, V., Vladescu, A., Balaceanu, M., Luculescu, C. R. and Braic, M. (2012). Nanostructured multi-element (TiZrNbHfTa) N and (TiZrNbHfTa)C hard coatings. *Surf. Coat. Technol.* 211: 117-121.

[70] Lin, M. I., Tsai, M. H., Shen, W. J. and Yeh, J. W. (2010). Evolution of structure and properties of multi-component (AlCrTaTiZr)Ox films. *Thin Solid Films* 518: 2732-2737.

[71] Chen, T. K. and Wong, M. S. (2007). Structure and properties of reactively-sputtered AlxCoCrCuFeNi oxide films. *Thin Solid. Films* 516: 141-146.

[72] Ding Q., Zhang Y., Chen X., et al. (2019). Tuning element distribution, structure, and properties by composition in high-entropy alloys. *Nature* 574: 223–227.

[73] Yan X. L., Constantin L., Lu Y. F., et al. $(Hf_{0.2}Zr_{0.2}Ta_{0.2}Nb_{0.2}Ti_{0.2})C$ high-entropy ceramics with low thermal conductivity. *J Am Ceram Soc* 2018, 101: 4486–4491.

[74] Braic, V. Balaceanu, M., Braic, M., Vladescu, A., Panserib, S. and Russo, A. (2012). Characterization of multi-principal element (TiZrNbHfTa)N and (TiZrNbHfTa)C coatings for biomedical applications. *J. Mech. Behav. Biomed. Mater.* 10: 197-205.

[75] Sarkar, A., Djenadic, R., Wang, D. (2018). Rare earth and transition metal-based entropy stabilized perovskite-type oxides. *J Eur Ceram Soc.* 38: 2318–2327.

[76] Cheng, K. H., Lai, C. H., Lin, S. J. and Yeh, J. W. (2011). Structural and mechanical properties of multi-element $(AlCrMoTaTiZr)N_x$ coatings by reactive magnetron sputtering. *Thin Solid. Films* 519: 3185-3190.

[77] Chang, S. Y., Lin, S. Y. and Huang, Y. C. (2011). Microstructures and mechanical properties of multi-component $(AlCrTaTiZr)N_xC_y$ nanocomposite coatings. *Thin Solid. Films* 519: 4865-4869.

[78] Lee, C. T. Cho, W. H., Shiao, M. H. and C. N. Hsiao, (2012). Effects of DC bias on the microstructure, residual stress and hardness properties of TiVCrZrTaN films by reactive RF magnetron sputtering. *Proc. Eng.* 36, 316–321.

[79] Hsueh, H. T., Shen, W. J., Tsai, M. H. and Yeh, J. W. (2012). Effect of nitrogen content and substrate bias on mechanical and corrosion properties of high-entropy films (AlCrSiTiZr)100−xNx. *Surf. Coat. Technol.* 206: 4106-4112.

[80] Sobol, O. V., Andreev, A. A., Gorban, V. F., Krapivka, N. A. and Stolbovoi, V. A. (2012). Reproducibility of the single-phase structural state of the multielement high-entropy Ti–V–Zr–Nb–Hf system and related super hard nitrides formed by the vacuum-arc method. *Tech. Phys. Lett.* 38: 616-619.

[81] Tsai, C. W., Lai, S. W., Cheng, K. H. and Tsai, M. H. (2012). Strong amorphization of high-entropy AlBCrSiTi nitride film. *Thin Solid. Films* 520: 2613-2618.

[82] Braic, M., Balaceanu, M., Vladescu, A., Zoita, C. N. and Braic, V. (2013). Deposition and characterization of multi-principal-element (CuSiTiYZr)C coatings. *Appl. Surf. Sci.* 284: 671-678.

[83] Tsai, D. C. Chang, Z. C., Kuoa, B., Shiaoc, M. H. and Changa, S. Y. (2013). Structural morphology and characterization of (AlCrMoTaTi)N coating deposited via magnetron sputtering. *Appl. Surf. Sci.* 282: 789-797.

[84] Gorban, V. F., Andreyev, A. A., Kartmazov, G. N., Chikryzhov, A. M., Karpets, M. V., Dolomanov, A. V., Ostroverkh. A. A. and Kantsyr, E. V. (2017). Production and mechanical properties of high-entropic carbide based on the TiZrHfVNbTa multicomponent alloy. *J. Superhard Mater.* 39: 166-171.

[85] Malinovskis, P., Fritze, S., Riekehr, L., Fieandt, L., Cedervall, J., Rehnlund, D. and Nyholm, L. (2018). Synthesis and characterization of multicomponent (CrNbTaTiW)C films for increased hardness and corrosion resistance. *Mater. Des.* 149: 51-62.

[86] Jhong, Y. S., Huang, C. W. and Lin, S. J. (2018). Effects of CH4 flow ratio on the structure and properties of reactively sputtered (CrNbSiTiZr)Cx coatings. *Mater. Chem. Phys.* 210: 348-352.

[87] Mayrhofer, P. H., Kirnbauer, A., Ertelthaler, Ph. and Koller, C. M. (2018). High-entropy ceramic thin films; a case study on transition metal diborides. *Scr. Mater.* 149: 93-97.

[88] Dinu, M., Pana, I., Braic, V., Miculescu, F., Balaceanu, M., Vladescu, A. and Braic, M. (2016). In vitro corrosion resistance of Si containing multi-principal element carbide coatings. *Mater. Corros.* 67: 908-914.

Chapter 7

History and Developments of Heusler Alloys

S. S. Mishra[1,2,*] and T. P. Yadav[2]

[1]Department of Materials Science and Engineering, Indian Institute of Technology Kanpur, Kanpur, India
[2]Hydrogen Energy Centre, Department of Physics, Institute of Science, Banaras Hindu University, Varanasi, India

Abstract

Heusler alloys have an ordered structure while retaining the majority of metal properties. Because of their unique phenomena and potential applications, Heusler alloys are exciting research areas for theoretical and experimental researchers. Half-metallicity, in which the majority spin band exhibits typical metallic nature, and the minority spin band reveals insulating or semiconducting behavior, is one of the three basic unusual properties of Heusler alloy. The inverse magnetocaloric effect, which occurs when a material is exposed to a varying magnetic field, exhibits an opposite pattern of temperature alteration. The magnetic shape memory effect, the material undergoes structural transformations as the magnetic field changes. Along with these three unusual properties, this class of compounds gives the fundamental aspects for magnetism in complex systems. The historical research and development of Heusler alloys have been summarized in this book chapter.

Keywords: Heusler alloy, $L2_1$ structure, half metallicity, magnetic ordering, spin polarization

[*] Corresponding Author's Email: smishra747@gmail.com.

In: Materials Science: Future Aspects
Editors: Kalpana Awasthi, Arti Srivastava and Mridula Tripathi
ISBN: 978-1-68507-843-0

1. Introduction

More than 100 years before, the Heusler alloys were discovered by Friedrich Heusler [1]. He found that when metalloids (Al, In, Sn, Sb, or Bi) are added to the Cu-Mn alloy, it becomes ferromagnetic, despite the fact that the alloy constituents are not ferromagnetic [1, 2]. After this discovery, the basic knowledge of the composition and crystal structure remained unknown for a long time. Potter measured X-rays on Cu-Mn-Al alloy in 1929 and demonstrated that the alloy constituents are well ordered and arranged on an FCC superlattice [3]. Using anomalous scattering and X-ray, the detailed investigation on the Cu-Mn-Al system has been performed by Bradley and Rodgers and found a correlation between the chemical ordering and composition [4]. Many studies were made after the complete knowledge of the crystal structure. The unit cell of Heusler alloys is formed by the fusion of two ordered binary (XY and XZ) B2 compounds with CsCl-prototype. For example, CuMn and CuAl give raise Cu_2MnAl. It has also been observed that if one sublattice remains unoccupied, it gives $C1_b$ crystal structure and is called semi or half Heusler alloy, whereas the $L2_1$ structure is named full-Heusler alloys. Comprehensive experimental investigations indicate that in stoichiometric composition, the Heusler compounds are ferromagnetically ordered. Stoichiometry, crystal structure, and heat treatment have all been identified to be effective features in evaluating magnetic response.

Full Heusler alloys (X_2YZ) and half Heusler alloys (XYZ) crystallize in $L2_1$ and $C1_b$ structures, respectively, at the stoichiometric configuration (Figure 1). The transition metal group comprises the X and Y constituents, while the III-V group comprises the Z constituent.

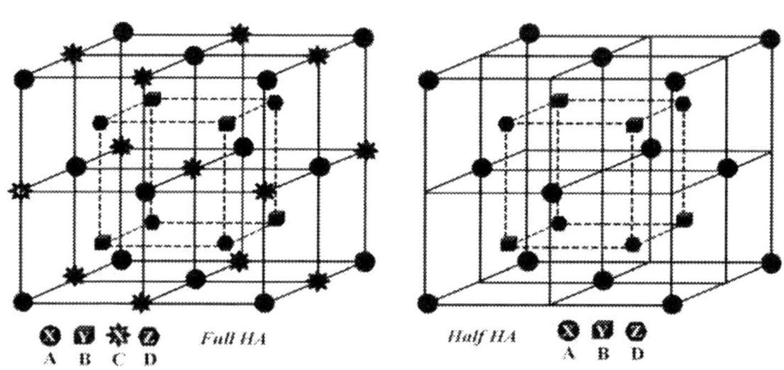

Figure 1. The full- and half-Heusler alloys adapted $L2_1$ and $C1_b$ structures. The lattice is made up of four interconnected FCC lattices. One of the four sub-lattices in the half-Heusler alloys is vacant [5].

Table 1. Stoichiometry, type of magnetism, and structure of Heusler compounds [13].

Y	X	Z	Magnetic order	Crystal structure
V	Mn	Al, Ga	FM*	$L2_1$
	Fe	Al, Ga	FM	$L2_1$
	Fe	Si	PM	$L2_1$
	Co	Al, Ga, Sn	FM	$L2_1$
Cr	Co	Al, Ga	FM	$L2_1$
	Fe	Al, Ga	FM	$L2_1$
Fe	Fe	Al, Si	FM	DO_3
	Co	Al, Si, Ga	FM	$L2_1$
Co	Fe	Ga	FM	$L2_1$
Mn	Cu	Al, In, Sn	FM	$L2_1$
	Cu	Sb	AFM	$C1_b$
	Ni	Al	AFM	B2
	Ni	Sb	FM	$C1_b$
	Ni	Al, Ga, In, Sn, Sb	FM	$L2_1$
	Co	Al, Si, Ga, Ge, Sn	FM	$L2_1$
	Co	Sb	FM*	$C1_b$
	Fe	Al, Si	FM	$L2_1$
	Pd	Al	AFM	B2
	Pd	In	AFM	$L2_1$-B2
	Pd	Ge, Sn, Sb	FM	$L2_1$
	Pd	Sb	FM	$C1_b$
	Pd	Te	AFM	$C1_b$
	Rh	Al, Ga, In	FM	B2
	Rh	Ge, Sn, Pb	FM	$L2_1$
	Rh	Sb	FM	$C1_b$
	Ru	Ga	FM	$C1_b$
	Au	Zn, Cu	AFM	B2
	Au	Al, Ga, In	AFM	$L2_1$
	Au	Sb	FM	$C1_b$
	Pt	Al, Ga	AFM	$L2_1$
	Pt	Ga	FM	$C1_b$
	Ir		AFM	$L2_1$
	Ir		AFM	$C1_b$
Ni	Fe	Al, Ga	PM	$L2_1$

FM* - Ferrimagnetic

Table 1 summarizes the elements associated with the X, Y, and Z. Mn element finds the entry in most of the Heusler alloys [6], whereas the compound with Mn at X position is rare. For instance, Mn_2VGa [7] and Mn_2VAl [8] are the only Heusler compounds that have been experimentally investigated so for. At the stoichiometric configuration, the chaos can take the form of fractional atom exchange in distinct sub-lattices. The $L2_1$-B2 type disorder arises when Y and Z atoms partially populated the lattice sites of one another. Allowing the interchange of half of the Y and Z atoms in their positions B2-type structure can be found. The $L2_1$/B2 phase ratio is affected by heat treatment. An antiferromagnetic ordering is more energetically advantageous because of the relatively small lattice spacing in the B2-type configuration. It is exciting in these compounds that the same family of alloys exhibits very diverse magnetic phenomena [9-11]. The magnetic shape memory effect, half-metallic behaviour, and an inverse magnetocaloric effect of the Heusler alloys have sparked a great deal of experimental and theoretical interest. Groot proposed half-metallicity in the NiMnSb half-Heusler alloy, which gives it the potential to become a material of intense research [12].

In addition to strong spin polarization, half-metallic alloys should have a crystal structure compatible with commercially available zinc blend semiconductors. Furthermore, because of the high Curie temperatures of this class of alloys, they can be used in room-temperature device applications. Thus, the Heusler alloys are also a promising system in this perspective. Combining the structural and magnetic features, Heusler alloys shows many other interesting properties.

2. Classification of Heusler Alloys

Depending on the lattice sites occupied by different constituents (or elements), Heusler alloys are categorized into four groups as follows:

2.1. Full Heusler Alloy

The full-Heusler (X_2YZ) alloy unit cell is made up of the eight tightly packed body-centered cubic (BCC) lattices depicted in Figure 1. The exterior sublattice, which is made up of eight cubic lattices, is populated by X atoms, whereas the inner cubic sublattice, which is made up of body-centered sites from each BCC lattice, is inhabited by Y and Z atoms on a regular basis. So, each Y or *Z* atom was surrounded by eight X atoms occupying the octahedral position, whereas each X was surrounded by four Y and *Z* atoms. As a result, the crystal's geometry is simplified to tetrahedral because the X atoms populate two separate sub-lattices, and the surroundings of the first sub-lattice are just like the surroundings of the second sublattice with the rotated by 90^0, and these sub-lattices are chemically equivalent. Kandpal et al. reported that the proportion of valance electrons strongly influences the site preference of the transition metal elements in Heusler compounds [14]. Full-Heusler alloys have a Cu_2MnAl-type crystal structure (space group 225, Fm-3m) in which two X atoms ate settled at A and C sites, while Y and Z atoms, respectively, populate B and D sites. There exist two kinds of disordering on the Full-Heusler alloys. The structure reduces to the B2 type when the B and D sub-lattices are randomly captured by Y and Z atoms. Moreover, when the disordering between the A and B, D sub-lattices takes place, the structure is transformed to the A2 type. Usually, such structural ordering and disordering can be distinguished by x-ray diffraction (XRD) analysis. The odd superlattice diffraction lines (e.g., (111) and (333) etc.) were eliminated in the first type of disordering (which gives the B2 phase). In the second type of disordering, even superlattice diffraction line (e.g., (200)) vanish. Whereas the primordial diffraction lines ($(h + k+l)/2 = 2n$, for example, (220)) are unaffected by structural ordering. To explain the magnetic properties, the interaction of X atoms is important as X atoms occupy the second neighbor site in the $L2_1$ structure to explain the magnetic properties. Galanakis et al. reported that orbital magnetism plays a very negligible role in these compounds when discussing their magnetic properties [15]. Kübler et al. have investigated the mechanism, sustaining the ferro- or the antiferromagnetism in Full-Heusler alloys [16]. The first time, Ishida et al. have investigated the presence of half-metallicity in these alloys using *ab-initio* electronic structure calculations, and they discovered the same property in Fe_2MnZ (Z = Si, Ge) compounds [17].

2.2. Half Heusler Alloy

Half-Heusler or Semi-Heusler alloys consist of two different TMs, X and Y, and Z is a *sp* valence element and crystallized in $C1_b$ structure. This crystal structure consists of one vacant and three filled interpenetrating FCC sub-lattices. From the body diagonal (rocksalt structure), a one-fourth shift of the third FCC sublattice was found in the unit cell [18]. Rather than 2, the 8 valence electrons are allocated among 3 atoms. The octahedral vacancies are occupied by the third atom in the Zinc-blende lattice. This fact spontaneously results in the evolution of a rocksalt-like sublattice [19]. Full and Half- Heusler alloys are structurally comparable as both structures are comprised of four interpenetrating FCC sub-lattices, but in the previous case, alternative sub-lattices are occupied by X atoms. It took almost 50 years for scientists to realize that it is possible that out of the four sub-lattices, one sublattice may remain vacant/unoccupied by the atoms (generally by X) to yield XYZ alloys that crystallize into Cl_b-type structure [1]. In 1952 Castelliz et al. synthesized an alloy whose crystal structure was very identical to that of the Heusler alloys with composition CuMnSb [20]. Forster et al. (1968) prepared and studied the CuMnSb half- Heusler using X-ray and neutron diffraction techniques and reported that the alloy has an ordered MgAgAs-type structure [21]. A pioneering discovery was made by Groot et al. (1983), which demonstrated the half-metallic ferromagnetic (HMF) behavior in NiMnSb half-Heusler alloy [12]. Such HMFs is considered to be lie between metal and semiconductor. The fast developments in the field of magneto-electronics increase the importance of these compounds because the effectiveness of any spin-dependent device scales with the spin polarization at Fermi level (E_F).

2.3. Inverse Heusler Alloy

Along with the conventional full-Heusler compounds, the compounds also exist termed inverse full-Heusler alloys. These compounds have similar chemical formula (X_2YZ) as that of full-Heusler alloys. However, in the preceding case, the number of valence electrons of the Y atom is more than that of the X atom. As a result, the inverse Heusler alloys emerged in Hg_2TiCu-type structure and so-called *XA* or *Xα* structure, having a sequence of constituents X-X-Y-Z, obtained by replacing X atom with Y atom, Figure 1 [22]. In literature, many inverse Heusler alloys have been investigated using *ab initio* calculations [23-25]. In all studies, it has been observed that the$L2_1$ structure was less energetically compatible than the XA structure in the case of standard full-Heusler compounds. The experiments performed on the Mn_2CoGa and Mn_2CoSn films confirm these results [26-28]. However, in the case of Mn_2NiSb alloy, the preparation method influences the ordering of atoms at different lattice positions [29]. Large Curie temperature (can be above 1000K) and coherent growth on semiconductors make the Inverse Heusler alloys interesting for applications [30]. Skaftouros et al. have performed an *ab-initio* investigation on the inverse full-Heusler alloys X_2YZ where (X = Sc, Ti, V, Cr or Mn, Z = Al, Si or As) and the Y expends from Ti to Zn [31]. Out of these, many alloys exhibit half-metallic magnetic behavior. In all scenarios, the occurrence of half-metallicity is correlated with the Slater-Pauling behavior of the entire spin-magnetic moment.

2.4. Quaternary Heusler Alloy

In addition to the full-Heusler and inverse full-Heusler alloys, there is another full-Heuslers community of the LiMgPdSn-type crystal structure known as Quaternary Heusler alloys [32]. The stoichiometry of this new Heusler compound is XX'YZ (1:1:1:1), where X, X', and Y are d-block elements, and Z is the p-block element. Quaternary Heuslers crystallizes into Y-type (LiMgPdSn-prototype) structure with space group F-43m. The valance of X is greater than both X' and Y, and the valance of X' is greater than Y. The order of the atoms is X-Y-X'-Z along the cubic FCC diagonal that is preferably the most stable structure [28, 33]. In the crystal structure, the X and X' atoms occupy the (000) (A) and (0.5 0.5 0.5) (C) position, respectively, the Y element occupy the (0.25 0.25 0.25) (B) sites, III-IV group element Z lies at the (0.75 0.75 0.75) (D) sites. This occupation rule's applicability to quaternary Heusler alloys has been demonstrated, and the formation of a highly ordered structure has been demonstrated [33, 34]. Some new half-metallic LiMgPdSn-type compounds have been investigated [32, 35, 36]. In 2013 Özdogan et al. had investigated the magnetic and electronic characteristics of the sixty LiMgPdSn-prototype quaternary Heusler alloys. They demonstrated that most of the alloys are half-metallic with few exceptions, following the same Slater-Pauling rule applicable for full-Heusler alloys [37].

3. Properties of Heusler Alloys

The extraordinary tunability of the Heusler compounds makes them favorable to design almost any functionality. For instance, the high spin-polarization in spin-resolved photoemission and tunnel junction devices has been revealed by Co_2-based Heusler compounds. Other Mn-rich Heusler alloys have stimulated the curiosity of researchers working on the spin-Hall effect, spin-transfer torque, and RE metal-free hard magnets. Several Mn-rich Heusler alloys crystallize into inverse structures (XA), distinguished by the antiparallel coupling of magnetic moments of Mn atoms. The absence of inversion symmetry and ferromagnetic ordering introduces many new characteristics that are not present in the centrosymmetric ($L2_1$) ferromagnetic Heusler structures, e.g., topological Hall Effect, non-collinear magnetism and Skyrmions. Here we are discussing, in short, some interesting properties of Heusler alloys.

3.1. Thermoelectric Properties

Theoretical band structure calculations reveal that Heusler alloys are half-metallic ferromagnets. Therefore, the electric current is mainly due to the up-spin electrons. Thermoelectric materials can directly convert thermal energy into electrical energy and hence efficient material for the development of alternative power engineering devices. On the examination of $Fe_{2-x}V_{1+x}Al$ ($0 \leq x \leq 0.11$) alloys, it has been noticed that the concentration of Fe and V remarkably affects the thermoelectric property in the low-temperature range 10- 300K [38]. The low-temperature electrical conductivity and the Seebeck coefficient enhanced sharply on replacing Fe by V, whereas the conduction type changed from p to n [38]. It has also been observed that the Seebeck coefficient was found to be sensitive to Al fraction in $(Fe_{2/3}V_{1/3})_{100-y}Al_y$ ($23.8 \leq y \leq 25.8$) alloys [39, 40]. Nishino et al. (2006) investigated the influence of Si and Ge substitutions on the Al site in the $Fe2VAl_{1-x}Si_x$ ($0 \leq x \leq 0.50$) and $Fe2VAl_{1-x}Ge_x$ ($0 \leq x \leq 0.20$) alloys [41], respectively, and observed a significant increase in the Seebeck coefficient and electrical conductivity. Another investigation on $(Fe_{1-x}Co_x)_2TiAl$ and $(Fe_{2-x}Co_x)(V_{1-y}Ti_y)Al$ alloys demonstrated that Co is a viable replacing constituent in the Fe_2VAl system [42, 43]. The *ab-initio* method was used to estimate the band structure of these Heusler compounds, and it is found that the value of the Seebeck coefficient obtained was 300 μV/K at 300K at a charge-carrier concentration from 1020 to 1021 cm^3 [44].

3.2. Magnetic Properties

The presence of ferromagnetism in the first Heusler alloys (Cu_2MnAl, Cu_2MnSn and CuMnSb) attracts attention though the constituents are non-magnetic. In these alloys, the Mn atom plays a significant role. Generally, the magnetic Heusler alloys carry the Mn or some rare earth (RE) elements (MnNiSb, GdNiS, PtMnBi or YbPtBi). The rare-earth atoms have RE^{3+} valance electrons, *i.e.*, they donate three electrons. Mn shows the same behavior as the rare earth element; for example, in the NiMnSb, Mn has d^4 configuration, causing a very strong magnetic moment of about 4 μB, while the contribution of Ni is zero [45]. Therefore, localized type magnetism is emerged by Mn and/or RE element at the Y position in the lattice in the Heusler structure. Independently Groot et al. (1983) and Kübler et al. (2006) reported a unique band structure in NiMnSb and Co_2MnZ Heusler alloys [12,

16, 46]. Kübler was the first who recognized the delocalization of the local magnetic moment of cobalt at X sites and localized at Y sites [16]. The Co_2TiSn alloy exhibits a Curie temperature of 350K, reveals the ferromagnetic feature, having magnetic moment 2 μB/formula unit [14]. Co_2(Cr, Fe)Al was discovered to have a high magnetoresistance at ambient temperature because of high spin-polarization [47, 48]. In the Heusler alloy Co_2MnSi, with an increase in the proportion of manganese, the highest tunnel magnetoresistance of (>2000%) was observed [49]. In the same alloy (Co_2MnSi), a high spin-polarization was found directly by spin-polarized photoemission [50]. The off-stoichiometric Heusler alloys were also found to be revealing unique magnetic response [51, 52].

3.3. Mechanical Properties

In recent years, significant progress in developing materials with efficient thermoelectric power production or refrigerator capability is noticed. These properties are closely related to the chemical stability, material degradation temperatures, and mechanical robustness of the material but also to resist cracking or failure from vibrations. For instance, to develop a thermoelectric module, the elastic stiffness which governs its mechanical behavior becomes a crucial parameter concerning its design. It is beneficial if a thermoelectric couple's legs (n-type and p-type) have a comparable stiffness value. The fracture toughness, wear resistance of brittle materials, and machinability is closely related to material hardness [53, 54]. Furthermore, the hardness and strength of any material are inextricably linked. Hardness and stiffness as a function of elemental concentration were investigated for a variety of Heusler compounds [55, 56]. Kawaharada et al. discovered that by replacing Sn with Sb in $NiZrSn_{1-x}Sb_x$ (x = 0.01-0.28) half-Heusler alloys, the elastic modulus decreases from 111 GPa to 56.9 GPa [55]. Rogl et al. found that in $Ti_{1-x}Zr_xNiSn$ and $Ti_{1-x}Hf_xNiSn$ half-Heusler alloys, the hardness increases with any extra amount of Ni but decreases with increasing Ti/Hf content [57]. In the low-pressure region (<12 GPa), the spin polarization reduces by 97.28%, and the total magnetic moment decreases slowly for NiMnSb Heusler alloys. In high pressure (>12 GPa), a dramatic depletion of the spin polarization is noticed while the decrease of the magnetization is more visible [58]. In the $Zr_{0.5}Hf_{05}Co_{1-x}Ir_xSb_{0.99}Sn_{0.01}$ half-Heusler compound, incorporation of iridium had no discernible impact on the stiffness or hardness of the material [59].

4. Applications of Heusler Alloys

Many Heusler compounds, such as Co_2NbSn, Ni_2MnGa, undergo the structural transformation from cubic austenitic to martensitic phase at low temperature and, if magnetic in this case, then shows the magnetic shape memory effect (MSM). As a result, they can be used as actuators and sensors in a variety of fields. Furthermore, because of their diverse magnetic behavior and complex crystal structure, Heusler alloys provide critical characteristics for magnetism in complex systems in addition to these three exciting and unique properties.

Heusler alloys possess several extraordinary properties along with easy tunability that makes them favorable to design almost any functionality. This alloy family has a lot of applications. For example, in developing the thermoelectric devices, spintronic devices, giant magnetoresistive devices, solar cells, topological insulators, noble catalyst and magnetic tunnel junctions etc. A. Fert and P. Grünberg investigated the giant magnetoresistive effect, which is a well-known phenomenon in the field of spintronics and served as a great incentive for many subsequent thorough investigations due to its historically significant applications [60, 61]. The half-metallic Heusler alloys such as Co_2MnSi is emerging compound for significantly improved magnetoresistive properties due to its excellent resistivity and enhanced spin-polarization [62]. This increase in the magneto-resistance ratio at Co_2MnSi/Ag interface was described by the following large interface spin-asymmetry [63]. Due to the spin-transfer torque feature, the half-metallic Heusler alloys reveal several magnetization dynamics in a nanometer scale and become a potential candidate for several applications like nanoscale radio frequency oscillator [64-66]. The $Co_2Cr_{0.6}Fe_{0.4}Al$ Heusler compound has shown the >30% enhancement in the tunnel magnetoresistance at ambient temperature [47]. The $Co_2MnSi/AlO/Co_2MnSi$ sample deposited on MgO(100) substrate demonstrates a massive tunnel magnetoresistance (>550%) at 2K [67]. There are many benefits, such as enhanced data processing speed, lower power consumption, non-volatility, and increased integration densities on incorporating the spin degree of freedom into standard electrical appliances [68]. Because of its interesting physical properties, the HMFs, like NiMnSb half-Heusler alloys have got attention of the scientific world for their technological applications, especially in the area of spintronics, *i.e.*, spin valves, tunnel junctions etc. [69, 70].

Conclusion

In this chapter, we have summarized the convenience of the Heusler alloys for future application. The Heusler alloys with stoichiometry X_2YZ, where X and Y stand for d-block elements and Z for p-block element, exhibit a wide range of exotic electronic, magnetic, and transport properties.

Acknowledgment

This book chapter is dedicated to the beloved and distinguished researcher Prof. O. N. Srivastava, who departed prematurely from this world on 24th April 2021 due to COVID-19.

References

[1] FR, H. J. G. Verh. 1903. *Deutsch. Phys.*5:219.

[2] Heusler, F.S., W. Haupt, E. 190. Über die synthese ferromagnetischer Manganlegierungen [On the synthesis of ferromagnetic manganese alloys]. *Verh Deut Phys Ges*. 19.

[3] Potter, H. H. (1929). *Proceedings of the Physical Society*.;41.

[4] Bradley, A. J., Rodgers, J. (1934). The crystal structure of the Heusler alloys. *Proceedings of the royal society of london Series A, Containing Papers of a Mathematical Physical Character*. 144:340-59.

[5] Mishra, S. S., Yadava, T. P., Mukhopadhyay, S., Yadav, R. M., Subrahmanyam, V. S., Mukhopadhyay, N. K., et al. (2015). Rapidly Quenched Ni45Fe5Mn40Sn10 Heusler Alloys. *Mat. Res. Exp.*18:101-5.

[6] Mishra, S. S., Yadav, T. P., Yadav, R. M., Puthirath, A. B., Deng, L., Adnani, M. et al. (2021). Effect of Ga substitution on structural and magnetic properties of Fe50Mn25Al25-xGax Heusler alloys. *J. Alloys and Comp.* 854:156756.

[7] Ramesh Kumar, K., Arout Chelvane J., Markandeyulu G., Malik, S. K., Harish Kumar, N. (2010). Effect of Co substitution on the magnetic and transport properties of the half-metallic ferrimagnet Mn2VGa. *Solid State Comm.* 150:70-3.

[8] Itoh, H., Nakamichi, T., Yamaguchi, Y., Kazama, N., 1983. Neutron diffraction study of Heusler type alloy Mn0.47V0.28Al0.25. *Trans. of the Japan institute of metals*. 24:265-71.

[9] Pierre, J., Skolozdra, R. V., Tobola, J., Kaprzyk, S., Hordequin, C., Kouacou, M. A. et al. (1997). Properties on request in semi-Heusler phases. *J. Alloys and Comp.* 262-263:101-7.

[10] Toboła, J., Kaprzyk, S., Pecheur, P., (2003). Theoretical search for magnetic half-Heusler semiconductors. *Physica status solidi*. 236:531-5.

[11] Gilleßen M., Dronskowski, R. (2010). A combinatorial study of inverse Heusler alloys by first-principles computational methods. *Journal of computational chemistry.* 31:612-9.

[12] de Groot, R. A., Mueller F. M., Engen P., Gv. Buschow, K. H. J., (1983). New Class of Materials: Half-Metallic Ferromagnets. *Phys. Rev. Lett.*. 50:2024-7.

[13] Kurt, H., Coey, J., *Heusler Alloys: Properties, Growth, Applications.* 2016; Springer Series in Materials Science 222.

[14] Kandpal, H. C., Fecher, G. H., Felser, C., (2007).Calculated electronic and magnetic properties of the half-metallic, transition metal based Heusler compounds. *J. of Phy. D: Appl. Phys.* 40:1507-23.

[15] Galanakis, I., (2005).Orbital magnetism in the half-metallic Heusler alloys. *Phys. Rev. B.* 71:012413.

[16] Kübler, J., William, A. R., Sommers, C. B., (1983). Formation and coupling of magnetic moments in Heusler alloys. *Phy.l Rev. B.* 28:1745-55.

[17] Fujii, S., Ishida, S., Asano, S., (1995). A Half-Metallic Band Structure and Fe2MnZ (Z = Al, Si, P). *J. of the Phys. Society of Japan.* 64:185-91.

[18] Tritt, T. M., Black, M., Broido, D., Chen, G., Cronin, S., Venkatasubramanian, R. et al. Recent Trends in Thermoelectric Materials Research III. Semicon-ductors and Semimetals. *Recent Trends in Thermoelectric Materials Research III.* 2001.

[19] Casper, F., Graf, T., Chadov, S., Balke, B., Felser, C., (2012). Half-Heusler compounds: novel materials for energy and spintronic applications. *Semicond. Sci. and Tech.* 27:063001.

[20] Castelliz, L., (1952). Studies on the heusler alloys—III. The antiferro-magnetic phase in the Cu-Mn-Sb system. *Mh Chem.*;82:1314.

[21] Forster, R. H., Johnston, G. B., Wheeler, D. A., (1968). Studies on the heusler alloys—III. The antiferro-magnetic phase in the Cu-Mn-Sb system. *J. Phys. Chem. Solids*. 29:855-61.

[22] Özdog˜an, K., Galanakis, I., (2009). First-principles electronic and magnetic properties of the half-metallic antiferromagnet Cr2MnSb. *J. Magn. Magnetic Mat.* 321:L34-L6.

[23] Kervan, N., Kervan, S., (2012). A first-principle study of half-metallic ferrimagnetism in the Ti2CoGa Heusler compound. *J. Magn. Mag. Mat.*324:645-8.

[24] Kervan, N., Kervan, S., (2012). A first-principle study of half-metallic ferrimagnetism in the Ti_2CoGa Heusler compound. *J. Mag. Mag. Mat.* 324:645-8.

[25] Bayar, E., Kervan, N., Kervan, S., (2011). Half-metallic ferrimagnetism in the Ti2CoAl Heusler compound. *J. Magnet. Magnetic Mat.* 323:2945-8.

[26] Meinert, M., Schmalhorst, J-M., Reiss, G., (2011). Exchange interactions and Curie temperatures of Mn_2CoZ compounds. *J. Phys.: Condensed Matter*. 23:116005.

[27] Meinert, M., Schmalhorst, J-M., Klewe, C., Reiss, G., Arenholz, E., Böhnert, T., Nilson T., (2011). Itinerant and localized magnetic moments in ferrimagnetic Mn_2CoGa thin films probed by x-ray magnetic linear dichroism: Experiment and ab initio theory. *Phy. Rev. B.* 84:132405(1-4).

[28] Alijani, V., Winterlik, J., Fecher, G. H., Felser, C., (2011). Tuning the magnetism of the Heusler alloys $Mn_{3-x}Co_xGa$ from soft and half-metallic to hard-magnetic for spin-transfer torque applications. *Appl. Phys. Lett.* 99:222510(1-3).

[29] Luo, H. Zhu, W. Ma, L. Liu, G. Li, Y. Zhu, X. et al. (2009). Effect of rapid solidification on the site preference of Heusler alloy Mn_2NiSb. *J. Phys. D: Applied Physics*. 42:095001(1-6).

[30] Galanakis, I. Şaşıoğlu, E. (2011). Stability of ferromagnetism against doping in half-metallic alloys. *J. Appl. Phys.* 109:113912(1-5).

[31] Skaftouros, S. Özdoğan, K. Şaşıoğlu, E. Galanakis, I. (2013). Generalized Slater-Pauling rule for the inverse Heusler compounds. *Phys. Rev. B*. 87:024420(1-7).

[32] Dai, X. Liu, G. Fecher, G.H. Felser, C. Li, Y. Liu, H. (2009). New quarternary half metallic material CoFeMnSi. *J. Appl. Phys.* 105:07E901(1-3).

[33] Alijani, V., Ouardi, S. Fecher, G.H. Winterlik, J. Naghavi, S.S. Kozina, X. et al. 2011. Electronic, structural, and magnetic properties of the half-metallic ferromagnetic quaternary Heusler compounds CoFeMnZ (Z = Al, Ga, Si, Ge). *Phys. Rev. B*. 84:224416(1-10).

[34] Gao, G.Y. Hu, L. Yao, K.L. Luo, B. Liu, N. (2013). Large half-metallic gaps in the quaternary Heusler alloys CoFeCrZ (Z = Al, Si, Ga, Ge): A first-principles study. *J. Alloys and Comp.* 551:539-43.

[35] Gökoğlu, G. (2012). Ab initio electronic structure of NiCoCrGa half-metallic quaternary Heusler compound. *Solid State Sci.* 14:1273-6.

[36] Izadi, S. Nourbakhsh, Z. 2011. Structural, Electronic and Magnetic Properties of Nanolayer and Bulk of MnCo2Si and MnFeCoSi Compounds. *J. Superconductivity and Novel Magnetism*. 24:825-31.

[37] Özdoğan, K. Şaşıoğlu, E. Galanakis, I. (2013). Slater-Pauling behavior in LiMgPdSn-type multifunctional quaternary Heusler materials: Half-metallicity, spin-gapless and magnetic semiconductors. *J. Appl. Phys.* 113:193903(1-5).

[38] Lue, C.S. Kuo, Y.K. (2002). Thermoelectric properties of the semimetallic Heusler compounds $Fe_{2-x}V_{1+x}M$ (M = Al, Ga). *Phys. Rev. B*. 66:085121(1-5).

[39] Soda, K. Murayama, H. Shimba, K. Yagi, S. Yuhara, J. Takeuchi, T. Mizutani U. Sumi H. Kato M. Kato H. Nishinu Y. Sekiymaya A. Suga S. Matsushita T. Saitoh Y. (2005). High-resolution soft x-ray photoelectron study of density of states and thermoelectric properties of the Heusler-type alloys $(Fe_{2/3}V_{1/3})_{100-x}Al_y$. *Phys. Rev. B.*;71:245112(1-6).

[40] Lue, C.S. Chen, C.F. Lin, J.Y. Yu, Y.T. Kuo, Y.K. (2007). Thermoelectric properties of quaternary Heusler alloys Fe2VAl1-xSix. *Phys. Rev. B*. 75:064204(1-6).

[41] Nishino, Y. Deguchi, S. Mizutani, U. (2006). Thermal and transport properties of the Heusler-type $Fe_2VAl_{1-x}Ge_x$ ($0 \leq x \leq 0.2$) alloys: Effect of doping on lattice thermal conductivity, electrical resistivity, and Seebeck coefficient. *Phys. Rev. B*. 74:115115(1-6).

[42] Ishida, S. Fujii, S. Kashiwagi S, Asano S. (1995). Search for Half-Metallic Compounds in Co2MnZ (Z = IIIb, IVb, Vb Element). *J. Phys. Society of Japan*. 64:2152-7.

[43] Nakayama, H. Ide N. Nishino, Y. (2008). Thermoelectric properties of p-type Heusler compounds $(Fe_2{-}xCox)(V1{-}yTiy)Al$. *Mat. Trans.* 49:1858-62.

[44] Yabuuchi, S. Okamoto, M. Nishide, A. Kurosaki, Y. Hayakawa, J. (2013). Large Seebeck Coefficients of Fe_2TiSn and Fe_2TiSi: First-Principles Study. *Appl. Phys. Express*. 6:025504(1-3).

[45] Graf, T. Felser, C. Parkin, S.S.P. (2011). Simple rules for the understanding of Heusler compounds. *Progress in Solid State Chemistry*. 39:1-50.

[46] Kübler, J. (2003). Ab initioestimates of the Curie temperature for magnetic compounds. *J. Phys.: Cond. Matter*. 18:9795-807(1-14).

[47] Block, T. Felser, C. Jakob, G. Ensling, J. Mühling, B. Gütlich, P. et al. 2003, Large negative magnetoresistance effects in Co2Cr0.6Fe0.4Al. *J. Solid State Chem.* 176:646-51.

[48] Inomata, K. Okamura, S. Goto, R. Tezuka, N. (2003). Large Tunneling Magnetoresistance at Room Temperature Using a Heusler Alloy with the B2 Structure. *Japan. J. Appl. Phys.* 42:L419-L22.

[49] Liu, H-x. Honda, Y. Taira, T. Matsuda, K-i. Arita, M. Uemura, T. et al. (2012). Giant tunneling magnetoresistance in epitaxial Co2MnSi/MgO/Co2MnSi magnetic tunnel junctions by half-metallicity of Co2MnSi and coherent tunneling. *Appl. Phys. Lett.* 101:132418.

[50] Jourdan, M. Minár, J. Braun, J. Kronenberg, A. Chadov, S. Balke, B. et al. 2014. Direct observation of half-metallicity in the Heusler compound Co2MnSi. *Nature Communi.* 5:3974.

[51] Mishra, S.S. Mukhopadhyay, S. Yadav, T.P. Yadav, R.M. Radhakrishnan, S. Vajtai, R. Ajayan P.M. Mukhopadhyay N.K. Singh H.K. Srivastava O.N. (2017). Effect of Fe substitution by Co on off-stoichiometric Ni–Fe–Co–Mn–Sn Heusler alloy ribbons. *Mat. Res. Express*. 4:086507(1-11).

[52] Mishra, S.S. Mukhopadhyay S. Yadav, T.P. Yadav, R.M. Romero-Aburto R, Vajtai R, et al. (2017). Structural and magnetic properties of rapidly solidified Ni45Fe5Mn40Sn10 alloy ribbon. *J. Adv. Phys.* 6:389-96.

[53] Lawn, B.R. Wilshaw, T. Rice, J. (1977). Fracture of brittle solids: Cambridge; 1977. *Cambridge solid state science series.* 517-517.

[54] Zeng, J. Sato, Y. Ohkubo, C. Hosoi, T. (2005). In vitro wear resistance of three types of composite resin denture teeth. *J. Prosthetic Dentistry*. 94:453-7.

[55] Kawaharada, Y. Kurosaki, K. Muta, H. Uno, M. Yamanaka, S. (2004). Thermophysical properties of NiZrSn1−xSbx half-Heusler compounds. *J. Alloys and Comp.* 381:9-11.

[56] Seetawan, T. Amornkitbamrung, V. Burinprakhon, T. Maensiri, S. Kurosaki, K. Muta, H. et al. (2005). Mechanical properties of Ag-doped Na1.5Co2O4. *J. Alloys and Comp.* 403:308-11.

[57] Rogl, G. Grytsiv, A. Gürth, M. Tavassoli, A. Ebner, C., Wünschek, A. et al. (2016). Mechanical properties of half-Heusler alloys. *Acta Materialia*. 107:178-95.

[58] Pugaczowa-Michalska, M. 2006. Pressure behaviour and thermal expansion of NiMnSb from first principles calculations. *Solid State Communic.* 140:251-5.

[59] Verges, M.A. Schilling, P.J. Germond, J.D. Upadhyay, P. Miller, W.K. Takas, N.J. Poudeu F.P. (2010). Indentation Testing of Bulk $Zr_{0.5}Hf_{0.5}Co_{1-x}Ir_xSb_{0.99}Sn_{0.01}$ Half-Heusler Alloys. *MRS Online Proceedings Library*. 1267.

[60] Grünberg, P. Schreiber, R. Pang, Y. Brodsky, M.B. Sowers, H. (1986). Layered Magnetic Structures: Evidence for Antiferromagnetic Coupling of Fe Layers across Cr Interlayers. *Phys. Rev. Lett.* 57:2442-5.

[61] Baibich M.N. Broto J.M. Fert, A. Van Dau F.N. Petroff F. (1988). Giant Magnetoresistance of (001)Fe/(001)Cr Magnetic Superlattices. *Phys. Rev. Lett.* 61:2472-5().

[62] Iwase, T. Sakuraba, Y. Bosu, S. Saito, K.1-4 Mitani, S. Takanashi, K. (2009). Large Interface Spin-Asymmetry and Magnetoresistance in Fully Epitaxial Co_2MnSi/Ag/ Co_2MnSi Current-Perpendicular-to-Plane Magnetoresistive Devices. *Appl. Phys. Express*. 2:063003(1-3).

[63] Sakuraba, Y. Izumi, K. Iwase, T. Bosu, S. Saito, K. Takanashi, K. (2010). Mechanism of large magnetoresistance in Co_2MnSi/Ag/Co_2MnSi devices with current perpendicular to the plane. *Phys. Rev. B*. 82:094444(1-4).

[64] Oogane, M. Kubota, T. Kota, Y. Mizukami, S. Naganuma, H. Sakuma, A. Ando Y. (2010). Gilbert magnetic damping constant of epitaxially grown Co-based Heusler alloy thin films. *Appl. Phys. Letters*. 96:252501(1-3).

[65] Mizukami, S. Watanabe, D. Oogane, M. Ando, Y. Miura, Y. Shirai, M. Miyazaki T. (2009). Low damping constant for Co_2FeAl Heusler alloy films and its correlation with density of states. *J. of Appl. Phys.* 105:07D306(1-3).

[66] Yilgin, R. Sakuraba, Y. Oogane, M. Mizukami, S. Ando, Y. Miyazaki, T. (2007). Anisotropic Intrinsic Damping Constant of Epitaxial Co_2MnSi Heusler Alloy Films. *Jap. J. Appl. Phys.* 46:L205-L8.

[67] Sakuraba, Y. Hattori aM, Oogane, M. Ando, Y. Kato, H. Sakuma, A. Miyazaki T. (2006). Giant tunneling magnetoresistance in Co_2MnSi/Al–O/Co_2MnSi magnetic tunnel junctions. *App. Phys. Lett.* 88:192508(1-3).

[68] Wolf, S., Awschalom, D., Buhrman, R., Daughton, J., von Molnár, S., Roukes, M., et al. (2001). Spintronics: a spin-based electronics vision for the future. *Science*. 294:1488-95.

[69] Datta, S., Das, B. (1990). Electronic analog of the electro-optic modulator. *App. Phys.Lett.* 56:665-7.

[70] Caballero, J., Park, Y., Childress, J., Bass, J., Chiang, W-C., Reilly, A. et al. (1998). Magnetoresistance of NiMnSb-based multilayers and spin valves. *Journal of Vacuum Science Technology A: Vacuum, Surfaces, Films*. 16:1801-5.

Chapter 8

Carbon Nanotubes: One Dimensional Carbon Nanomaterial

Chaudhary Ravi Prakash Patel[1], Amit Srivastava[2] and Thakur Prasad Yadav[1,*]

[1]Hydrogen Energy Centre, Department of Physics, Institute of Science, Banaras Hindu University, Varanasi, India

[2]Department of Physics, Tilak Dhari Post Graduate College, VBS Purvanchal University, Jaunpur, India

Abstract

Three- dimensional carbon allotropes (graphite and diamond) have been known for ages. However, since the discovery of fullerenes which added a new dimension to the knowledge of carbon materials, considerable efforts have been made to search and assess new potential allotropes. The subsequent discovery of carbon nanotubes (CNTs) have enriched the field of carbon-based materials. The CNTs are supposed to be a key component of carbon-based nanotechnology due to their outstanding inherent and tailored mechanical, electrical, thermal, and chemical properties, which make them potentially interesting candidate for applications in diverse fields such as optics, nanoelectronics, composites, optoelectronics, battery storage, sensors, etc. Further, Novoselov and Geim in 2004 have isolated the first two-dimensional material ever discovered and, christened as Graphene. Graphene is a one-atom thick material consisting of sp2-bonded carbon with a honeycomb structure. It resembles to a large polyaromatic molecule of semi-infinite size. Graphene has been regarded as a basic building block for graphitic material of all other dimensionalities. It can be wrapped up in 0D

* Corresponding Author's Email: thakurp.yadav@bhu.ac.in.

In: Materials Science: Future Aspects
Editors: Kalpana Awasthi, Arti Srivastava and Mridula Tripathi
ISBN: 978-1-68507-843-0

fullerenes, rolled into 1D as nanotube or stacked in 3D as graphite. Moreover, in the past three decades, CNTs based nanomaterials have been pivotal for materials scientists and engineers due to their fascinating properties such as high mechanical strength, high elasticity, and thermal conductivity, room-temperature quantum Hall effect, high room temperature electron mobility, and tunable bandgap. The detailed framework leading to the extensive research of the 1D carbon nanomaterials-carbon nanotubes has been manifested in the present chapter.

Keywords: nanomaterials, fullerenes, carbon nanotubes, graphene, 2D nanomaterials

1. Introduction

Carbon, one of the most abundant elements on our planet, has been an appealing candidate of immense interest for a long time as it plays a very specific role in our lives. It has its specification in the food, cloths, cosmetics and vehicle fuel for transportation. It also provides the framework for all tissues of plants and animals. Hence forth it has proved its importance as a special element in every part of life and lifecycle. Carbon exhibits three naturally occurring allotropes such as amorphous, graphite and diamond [1]. Graphite and diamond have been known from ancient days and regarded as bulk crystalline phases of pure form. Amorphous carbon is formed when a material containing carbon is burnt completely in scarcity of oxygen (Figure 1(a)). The resulting black soot is also known as lampblack, gas black, channel black or carbon black and is used in inks, paints and rubber products. Further, it can also be pressed into different shapes to form the cores of most dry cell batteries [2].

Diamond, another naturally occurring allotrope, is known as one of the hardest substances with a remarkable specific gravity of 3.51 (Figure1(b)). It may be considered as a single molecule of carbon atoms in which each carbon atom is joined to four other carbons in regular tetrahedrons, or triangular prisms. It possesses a face-centered cubic (fcc) crystal structure with high refractive index of 2.42 which signifies its optical features, particularly refraction and diffraction. Although naturally occurring diamond is generally used for jewelry, whereas most commercial quality diamonds are produced artificially. Small pieces of diamond are produced by squeezing graphite under high temperature and pressure for several days and are primarily used to make diamond tipped saw blades etc. Graphite and diamond differ not only in their crystal structure, but they display different physical properties too.

Another natural allotrope Graphite is regarded as one of the softest materials known so far and has been used primarily as a lubricant (Figure 1(c)). Although it does occur naturally, but most commercial graphites have been produced by treating petroleum coke, a black tar residue while refinement of crude oil in oxygen free oven. Natural graphite occurs in two forms- alpha and beta. These two forms possess different crystal structures but nearly identical physical properties. All artificially produced graphites are usually alpha types and are used as lubricant and in large amounts in the production of steel (graphite, in a form known as coke) are used. Coke is made by heating soft coal in an oven without allowing oxygen to mix with it. Moreover, the black material used in pencils, commonly called a lead, is actually graphite. However, in recent years carbon nanomaterials, such as fullerene, carbon nanotubes (CNTs), graphene and so forth, have garnered significant attention of researcher/scientists and led to diverse applications due to their extraordinary properties at nanoscale [3-7] since the discovery of Fullerene in 1985.

Fullerenes are zero dimensional, closed-cage, carbon clusters and are regarded as a new class of carbon allotropes (Figure 1(d)). Before the first synthesis and detection of the smaller fullerenes C_{60} and C_{70}, it was generally believed that these large spherical molecules are unstable. However, it has been confirmed experimentally that C_{60} in the gas phase is stable and has a relatively large band gap [8]. In the context of a series of numerous important scientific discoveries, fullerenes have been accidentally discovered. In 1985 Kroto and Smalley found some strange results in mass spectra of evaporated carbon samples [3] and proximate analysis confirms the presence of Fullerene. Due to this interesting result, the search for other fullerenes got started. Fullerenes are now confirmed to possess different forms such as hollow spheres, ellipsoid, or tubes. Cylindrical fullerenes are named as carbon nanotubes or buck tubes. Fullerenes are similar in structure to graphite, which is composed of a sheet of linked hexagonal rings, but they contain pentagonal (or sometimes heptagonal) rings [9].

Since the discovery of carbon nanotubes in 1991 by Iijima and coworkers [4], carbon nanotubes have been of great interest for fundamental studies and extended prospective applications [10]. CNTs can be seen as nearly one-dimensional form of fullerenes and their large length (up to several microns) and small diameter (a few nanometers) result in a large aspect ratio (Figure 1(e)). Henceforth, CNTs are expected to possess additional interesting physical and chemical properties which make them unique materials with a wide range of promising applications [11-13]. Economically achievable large-scale production is still to be achieved [14-15].

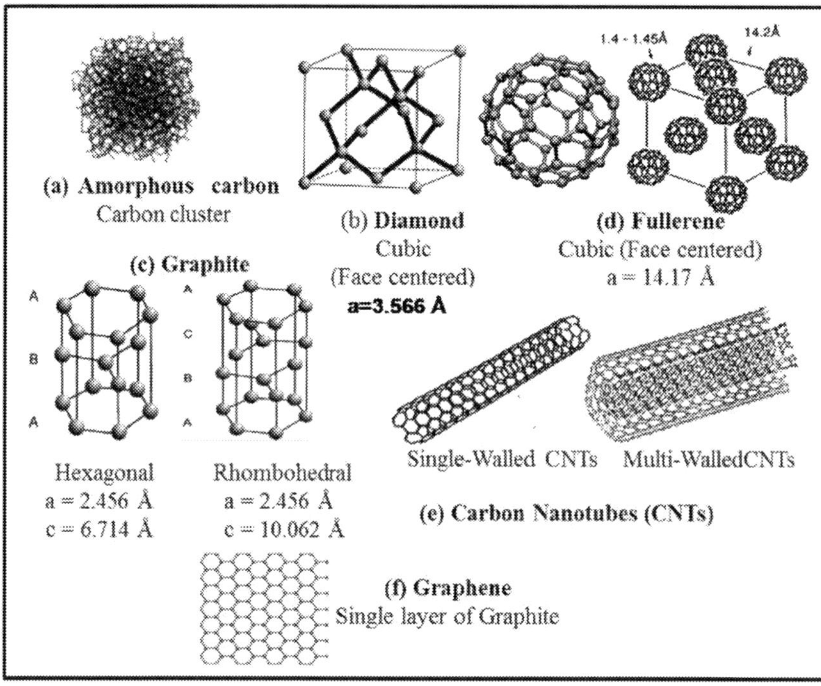

Figure 1. Schematic description of different allotropes of carbon.

Graphene is perhaps one of the most studied nanomaterial after the landmark discovery and publication about stability of a single layer graphene in 2004, by K.S. Novoselov and A.K. Geim [6-7]. After six years, they have been conferred noble prize in physics in 2010.Graphene can be thought of as one atom thick layer of carbon atoms arranged in a hexagonal lattice in two-dimension (Figure 1(f)).

2. Carbon Nanotubes

As a novel field of one-dimensional, CNTs have received substantial interest since their discovery in the year 1991 by Iijima [4]. It was first observed as elongated and concentric layered microtubules consisting of carbon atoms arranged in a graphitic structure through HRTEM studies of the synthesized filamentous carbon through electric arc discharge technique. The nanotubes that Iijima observed have been referred to as multi-walled nanotubes (MWNTs) containing at least two graphitic layers, and inner diameters of ~ 4 nm. Later after two year (1993), S. Iijima and Donald Bethune at IBM independently (IBM Almaden Research Center in California) observed single wall nanotubes (SWCNT) [16-17]. The SWCNTs have also been prepared by adopting a similar approach as used for producing MWNTs with some modification of adding metal particles as a catalyst in the carbon electrodes. The appearance of SWCNT has been quite different from that of MWNT. The SWCNTs have very small diameters (typically ~ 1nm), and are curled and looped rather than straight. The SWCNTs can either be metallic or semiconducting attributing to their chirality and diameter [18]. The most remarkable features of SWCNTs are their electronic (semiconducting or metallic), mechanical (Young modulus ~1 TPa), optical and chemical characteristics [1]. The multi-walled carbon nanotubes (MWNTs) have been assumed as a collection of concentric SWCNTs having different diameters and displaying different properties than their single-walled counterparts. The double walled carbon nanotubes (DWNTs)can be considered as a simplest example of MWNTs [19]. The diameter of CNTs varies from a few nanometres in the case of SWCNTs to several tens of nanometres in the case of MWNTs. The CNTs have length in the micrometer range usually. To understand the structure of a CNT, it can be first imagined as a rolled-up sheet with planar-hexagonal arrangement of carbon atoms in a honeycomb lattice. Therefore, since the discovery, CNTs have captured the attention of many researchers due to their outstanding properties [20-21].

3. Structural and Electronic Properties of Carbon Nanotubes

Generally, CNTs are hollow tube with sp^2-hybridized carbon atoms and can be categorized into two broad types - single wall carbon nanotubes (SWCNTs) and multi-walled carbon nanotubes (MWCNTs). Commonly a SWCNT comprises of a single layer graphene sheet (one layer of graphitic sheet) rolled over into a hollow cylinder with diameter of ~1.4 nm. The rolling of Graphene sheets leads to three peculiar types of structure i.e., armchair, zig-zag and chiral [22]. Therefore, the nanotube structure can be described by a chiral vector defined by the following equation C ($C = ma1 + na2$) (Figure 2). The chiral angles and electrical properties are dependent on the diameters of CNTs.

Generally, two integers (m, n) are commonly used to label the structures of SWCNTs. A graphene sheet is when folded into a tubular form; the beginning and end of a lattice vector (m, n) in the graphene plane join together. The chirality (m, m) tubes are called "arm-chair" tubes since the atoms around the circumference appear like an arm-chair (Figure 3) [23]. The "zigzag" tubes are (m, 0) nanotubes in view of the atomic configuration along the circumference (Figure3 b and c). The other categories of nanotubes are chiral, with the rows of spiraling hexagons along the nanotube axes (Figure. 3). An SWCNT can be considered as a metal, small-gap semiconductor or semiconductor depending on the (m, n) structural parameters.

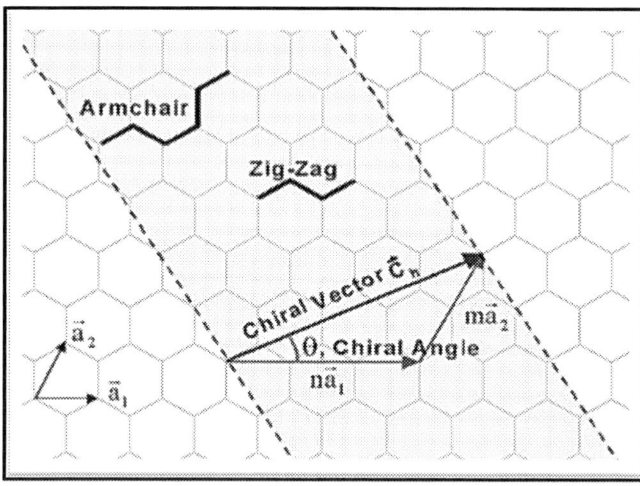

Figure 2. Schematic diagram shows chiral vector and chiral angle in a rolled graphite sheet with a periodic hexagonal structure.

For (m, m) arm-chair tubes, there have always been states crossing the corner points of the first Brillouin zone; thus arm-chair tubes always display metallic character. For (m, n) nanotubes with m–n ≠ 3 × integer, the nanotub are semiconducting. For (m, n) nanotubes with m–n = 3 × integer would be semimetal. The sensitivity of the electronic properties has always been a challenging for the chemical synthesis of CNTs in term of controlling the chirality and diameter of the nanotube. Ajayan reported that nearly two-thirds of small-diameter SWCNTs are semiconductors while one-third refers to the metallic character [24]. To generate identical electrical and optical properties, SWCNTs require being identical in diameter and chirality since SWCNTs with almost uniform diameters can have different electronic properties and different chirality. In recent years, considerable attempts have been made for the large-scale production of mono-dispersed SWCNTs with identical diameter and chirality [25].

A MWNT consists of concentric SWCNTs and closed graphite tubules with an interlayer spacing of 3.4 Å which is very proximate to the interlayer spacing in graphite. MWNTs have been generally metallic conductive but since they are composed of numerous cylinders of different helicities, therefore it complicates a simple explanation for electronic properties [24]. The walls of the tubes are different from their ends in CNTs structure and therefore attributes to local anisotropy. The side walls of CNTs are comparatively inert layer of sp^2-hybridized carbon atoms; however, the ends or "tips" of nanotubes comprising of carbon atoms bonded with oxygen to give a far more reactive species such as the edge planes of pyrolytic graphite.

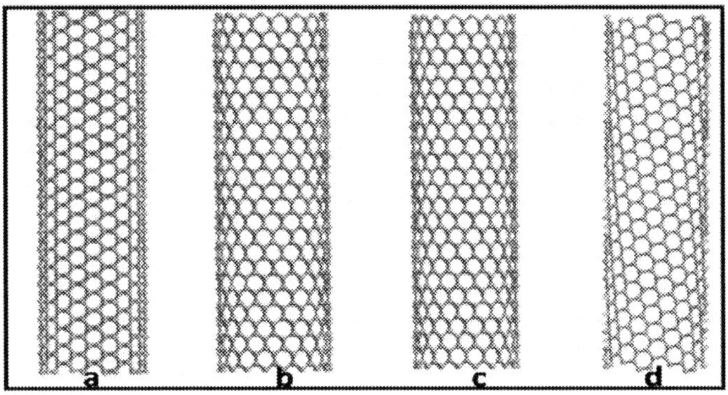

Figure 3. Schematic structures of SWCNTs (a) A (10, 10) arm-chair nanotube. (b) A (12, 0) zigzag nanotube. (c) The (14, 0) zigzag tube is semiconducting because the states on the vertical lines miss the corner points of the hexagon. (d) A (7, 16) tube is semiconducting. This Figure illustrates the extreme sensitivity of nanotube electronic structures to the diameter and chirality of nanotubes.

4. Different Synthesis Techniques for Carbon Nanotubes

There have been several approaches to synthesize CNTs (SWCNT or MWNTs such as electric arc discharge, laser vaporization and chemical vapour deposition etc. Among them, certain methods are well established to produce diverse variety of CNTs. The characteristics of the samples synthesized depend on the control and the choice of experimental parameters. The following section has been devoted to describe various synthesis methods.

4.1. Electric Arc-Discharge

The electric arc-discharge method is a high temperature process used for vaporization of carbon from solid graphite rod following which nanotubes, as well as fullerenes can be obtained. It has been used for the first time for mass production of fullerenes by the Krätschmer–Huffman method [8]. CNTs have been normally synthesized by striking an arc between graphite electrodes in an inert atmosphere (argon or helium).This process produces carbon soot containing fullerene molecules. Iijima and Ichihashi have first reported the production of SWCNT [16]. They used a Fe-graphite electrode in methane argon atmosphere for arcing. In the process, a hole has been made in the graphite anode and filled with a composite mixture of metal, and graphite powders, and pure graphite has been used as a cathode. The catalyst (transition metals such as Fe, Co, Ni and rare earth metals) and composite catalyst (Fe/Ni and Co/Ni) have been used to prepare isolated SWCNTs, and bundles of SWCNTs, respectively [1, 16, 24]. The as synthesized tubes have an average diameter as ~ 1.20 nm. To achieve the vaporization of carbon atoms into plasma, the carbon arc provides a convenient and traditional tool for generating the very high temperatures (~3000^{0}C) [24, 26]. The yield of CNTs depends on the stability of the plasma formed between the electrodes, the current density, and gas pressure and cooling of electrodes &chamber. The best result has been found with helium (He) among inert gases as it has higher ionization potential. The well-cooled electrodes and arc chamber help to facilitate the nanotube yield in the arc growth process. For the production of MWNTs, the condition required is to be optimized so that during the arc evaporation, the amount of soot production gets reduced and result in 75% of the evaporated carbon from a pure graphite anode on the graphite cathode surface. The arc deposit consists of a hard gray outer shell made of pyrolitic graphite and an interior made of a soft black powder (which contains about two thirds CNTs and one third graphitic nanoparticles). The optimized synthesis conditions have been 20-25V, 50-100Amp. d.c. (direct current) and helium pressure ~ 500 torr. The arc discharge is a very simple process to obtain structurally excellent high quality CNTs. However, the process of conventional arc discharge is discontinuous and unstable. So, it cannot lead the production of CNTs in large quantity. The CNTs are produced on the cathode surface and the electrode spacing is not constant, which leads to the non-homogeneous electric field. This problem has been resolved by generating the stable and high efficient discharge and many studies have been carried out to understand the growth mechanism of the prepared nanotube

CNTs in huge quantity by plasma rotating arc discharge method. In this process, the graphite anode has been rotated with a high velocity for the synthesis of CNTs. The anode rotation at high speed distributes the micro-discharges uniformly and generates stable plasma. Further, CNTs are collected on the graphite collector placed at the periphery of the plasma. The nanotube yield increases with the rotation speed of the anode and the collector becomes closer to the plasma. Moreover, two conditions are required to optimize this approach. First, the high density of carbon vapor is created by uniform and high temperature plasma for nucleation and second is related to the adequate temperature of collectors. The plasma rotating electrode process is stable and continuous process of the discharge and is expected to lead to the mass production of high-quality nanotubes. The CNTs have also been produced in large quantities by using plasma arc jets with optimizing quenching process in an arc between a graphite anode and a cooled copper electrode. The electric arc discharge method requires no catalyst particle for the synthesis of MWNTs (catalyst species are necessary for the growth of SWCNTs). The electric arc discharge method usually involves high-purity graphite electrodes, metal powders and high-purity inert gases (He/Ar). The crystallinity of the material is high, but there has been no control over the length and diameter of the tubes. However, some by-products have also been formed such as polyhedral graphite particles in the case of MWNTs, encapsulated metal particles in case of SWCNTs during the process.

4.2. Laser Ablation

The laser ablation method involves the production of carbon vapor species from graphite by employing a high-energy laser beam target followed by condensation. The laser ablation approach usually has the advantage in ease of operation and production of high-quality products. Furthermore, it permits better control over the processing parameters. However, it has several disadvantages particularly the high cost of the laser source and low yield of nanotubes.

Laser beams are very intense and coherent with the capability of achieving a very fast rate of vaporization of target materials. In the process, graphite target has been placed inside a quartz tube surrounded by a furnace (operated at 1200^{0}C as reaction temperature), in the inert atmosphere. Synthesis at lower temperature (<1200^{0}C) leads the presence of many defects in the CNTs. The target exposed with a laser beam forms hot carbon vapor species (e.g., C3, C2 and C). Further, the carbon species have been swept by flowing inert gas from the high-temperature zone to a conical copper collector located at the exit end of the furnace [27].

The laser ablation method has originally been used as a source for the production of clusters and ultrafine particles. This approach has been developed for the production of fullerene and CNT by Smalley's group [28]. The laser furnace consists of a quartz tube with a window, a target carbon composite doped with catalytic metals, a water-cooled trap, and flow systems for the buffer gas to maintain constant pressure and flow rate. In 1996, a dual pulsed laser was used to optimize the laser oven (for producing SWCNT with yields of 70%) [28]. Samples were prepared by laser vaporization of graphite rods (with 1.2 at% of a 50:50 mixture of Co and Ni powder) at 1200^{0}C in argon ambience (at 500 Torr). Moreover, the soot of carbon was minimized by employing two successive laser pulses: first to ablate the carbon-metal mixture and second to break the larger ablated particles and in-turn feed into the growing nanotube structures. The as received SWCNTs have been almost uniform in diameter and self-assemble into ropes (bundles) consisting of 100 to 500 tubes in a two-dimensional triangular lattice. It has also been possible to synthesis SWCNTs using a CO_2-laser focused on a graphite-metal target without an oven [29]. Further studies reveal Ar and N_2 as the best atmospheres to produce SWCNT bundles, whereas, the ambience facilitates small amounts of CNTs. Additionally, Dillon and coworkers found that the diameter of the nanotubes depends precisely on the laser power [30].

4.3. Chemical Vapor Deposition Technique

Chemical vapor deposition (CVD) is one of the most simple and popular methods to produce high yield CNTs [31]. CVD is different from the two above-discussed methods used for CNT production. Arc discharge and laser vaporization have been classified as very high temperature (>3000 ºC) and short time reaction (micro to millisecond) techniques, while CVD expedites a medium temperature (700–1200 °C) and longtime reaction (typically minutes to hours) technique. The main technological short comings associated with arc discharge and laser vaporization techniques are that the CNTs get produced and stand on their own. Substantial and rapid progress has been made in the development of CVD to establish it as a highly controlled technology for the production of CNTs [32]. CVD can be classified as thermal, plasma-enhanced, laser-assisted and spray pyrolysis CVD processes depending upon the heating sources used to start the chemical reactions [32]. The heating sources used to decompose the molecules exist in gas or liquid reactants (*e.g.*, methane, ethylene, acetylene, benzene or hexane) into reactive atomic carbon. The carbon diffuses on catalyst particles of nanometer scale. The catalyst can be synthesized by sputtering or evaporating metal such as Fe, Ni or Co on the top of a substrate. This is trailed by either thermal annealing or chemical etching to prompt a high density of catalyst particle nucleation on a substrate. Metal nanoparticles can be nucleated and dispersed more uniformly on a substrate by using a spin-coating method [33]. The size and nature of the catalyst perform a very vital role on the growth rate, diameter, morphology and structure of the synthesized nanotubes. It can be added that the diameter and length of CNTs can be controlled by the size of metal nanoparticles and deposition conditions. The growth rate and density of nanotubes directly depend on the size of particles [15]. As the size of particles reduces, the diffusion period becomes shorter for carbon atoms at the nucleation sites, and thus the growth rate and density of nanotubes increases. Thus the yield of nanotubes depends significantly on the properties of the substrate material. Zeolite is generally known as a good catalyst support as it favors the production of high yield CNTs: Alumina is also an excellent catalyst support which pertains to strong metal-support interaction and prevents agglomeration of metal particles and thus offers very dense catalytic sites for CNTs growth [34].

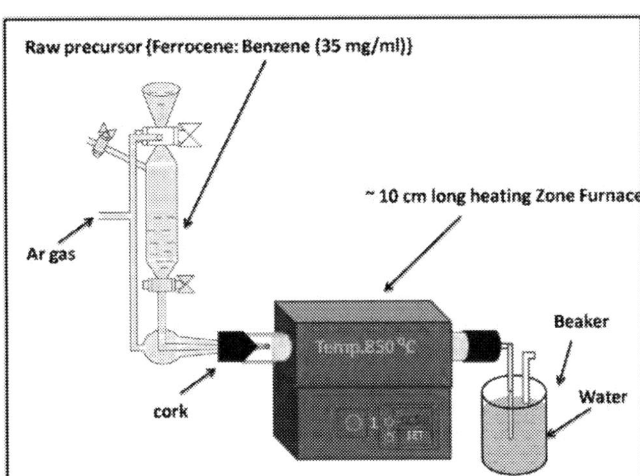

Figure 4. Schematic diagram of the Chemical vapor deposition set up.

CVD is basically a thermal dehydrogenation reaction because of a transition metal catalyst that lowers the temperature as required to "crack" a gaseous hydrocarbon feed, into carbon and hydrogen. Figure 4illustratesthe apparatus employed in the CVD process. All CVD techniques are capable of producing free-standing CNTs. The CVD method produces both single-wall and multi-wall nanotubes. Generally, the CVD method uses hydrocarbons as carbon source and flow in the quartz tube heated to an optimized temperature (700–1200 °C). At high temperature, the carbon precursor decomposes into carbon and forms carbon nanotubes.

Table 1. Comparisons of the established techniques for CNT synthesis

Method	Arc discharge	Laser ablation	Chemical vapor deposition
Description	arc evaporation of graphite in the presence of inert gas	vaporization of graphite target by laser in the presence of inert gas	decomposition of hydrocarbons over transition metal catalyst to form CNT
Temperature	>3000 °C	>3000 °C	<1200 °C
Pressure	50-760 Torr under vacuum	generally, 200-750 Torr under vacuum	May occur at low pressure or at atmospheric pressure
Advantages	good quality CNTs	good quality CNTs (SWNTs)	easy scale up; synthesis possible on the substrate
Disadvantages	difficult to scale up	difficult to scale up expensive	quality of CNT not as good

The growth of SWCNTs needs high temperature than for MWNTs. Most of the thermal CVD methods use acetylene (C_2H_2) or ethylene (C_2H_4) gas as the carbon precursor and Fe, Ni, or Co nanoparticles as catalyst for the synthesis of MWNTs. The growth temperature used is typically in 600-900 °C range. The carbon precursor also affects the growth of CNTs. Thus, methane (CH_4) and carbon mono oxide (CO) are commonly used as a source of hydrocarbons for SWCNT growth, while C_2H_2, C_2H_4, xylene, benzene, hexane, toluene and acetone, which are unsaturated and have high carbon content, are typically used as a hydrocarbons source for MWNT. Organometallic compounds (*e.g.*, metallocenes, iron pentacarbonyl, and iron (II) phthalocyanine are regularly used as precursors for the catalyst [35]. These precursors on heating usually get sublimed and catalyst nanoparticles are formed in situ when the compound is decomposed/ reduced. Aligned CNTs have also been prepared by the spray pyrolysis of ferrocene-benzene solution

[36]. The main advantages of the CVD technique have been the requirement of low temperature (600-900 °C) with low power input, comparatively high purity of CNTs and probability for scale up in production [37]. A comparative study of the established techniques for CNT synthesis has been presented in Table 1.

5. Properties of Carbon Nanotubes

CNT shows metallic type electrical conductivity due to π-e^- donated by each atom, which can move freely in the entire network of the structure. It, also exhibits high intrinsic thermal conductivity as a consequence of high frequency C-C bond vibrations. CNTs possess extraordinary heat conductivity, mechanical property and thermal conductivity due to possible electron configuration, owing to nitration of C atom with neighbor atoms or the several bonding arrangements [38]. It also exhibits remarkably low electrical resistance [39]. It is known fact that the resistance is produced when electron get deflected from the path with collision of defect present in the crystal structure of the material. The probability of electron scattering is very high in case of 3D conductor as it may happen at any angle. This in turn gives rise to the electrical resistance. However, in 1 D material, the only possible scattering is backscattering as the possibility of electron motion is either forward or backward. So there are rare chances for backscattering to take place, as it requires strong collision between the electrons. This low scattering provides CNTs with very good electrical conductance or very low resistance. In comparison to other known materials like CNTs, possesses high current density as ~ 109 A/cm^2. Theoretically, its value is much higher (1013 A/cm2) [39]. To understand the conductivity mechanism on individual crystalline tube of MWCNTs and SWCNTs, direct 4-probe (or 2- probe) measurements have been performed which suggest each individual MWCNT to display both metallic and semiconducting characteristics (resistivity 1.2×10^{-4} to 5.1×10^{-6} ohm cm at 300K). Metallic nature has been demonstrated by the bundle of SWCNTs with resistivity range (0.34×10^{-4} to 1.0×10^{-4}ohm cm) [39].

The elastic modulus and tensile strength suggest that CNTs are the strongest and stiffest material, owing to covalent sp^{2}bonds between the individual carbon atoms. On average, the highest and lowest values of the elastic modulus as 4.15 TPa and 0.40 TPa, respectively for MWCNTs and the respective tensile strength as 0.63GPa [40]. This is much higher than for steel (1 to 2 GPa). Table 2 illustrates some of the basic properties of CNTs.

Table 2. Basic properties of CNTs

Property	Tensile strength	Young's modulus	Resistivity	Maximum current density	Thermal conductivity
SWCNTs	75 GPa	900-1700 GPa	10^{-6} Ω.m	10^7-10^9 A.m^{-2}	1750-5800 Wm^{-1}K^{-1}
MWCNTs	150 GPa	690-1870 GPa			3000 Wm^{-1}K^{-1}

6. Applications of Carbon Nanotubes

CNTs exhibit outstanding novel properties, such as very high current density, high thermal conductivity, high resilience, high tensile strength and electrical property ranging from semiconducting to metallic regions. These proprieties render CNTs favorable for many applications in diverse fields. Some of these applications have been presented and described briefly in the following section:

6.1. Water Filtration

Nanostructures that possess membrane-like pores finds attractive application in water filtration. Recently, instead of porous polymer membrane, native porous carbon nanostructures, particularly carbon nanotubes (CNTs), have drawn significant attention for filtration of water [41]. Polymeric membranes have flexible chains and thus they often lack the presence of well-defined pores necessary for filtration. Due to the unique porous structure, use of CNT as a supplementary entity on to conventional filters, or as filters on their own, has been increased. Moreover, the pore sizes and configurations, of CNTs can be altered with tailoring the growth parameters. CNT based filters have also been used to remove bacteria from water and heavy hydrocarbon from petroleum [36]. Based on extensive calculations, it has been anticipated that CNTs membranes can provide an efficient means of water filtration. The narrow pores of CNTs have been capable of filtering water while rejecting ions (Na^+ / Cl^-) [42].

6.2. Electromagnetic Interference (EMI) Shielding

In the host matrix, intrinsic conductivity and formation of conducting paths by CNTs play a very crucial role in tailoring the conductivity of the composite. The conductive pathway provided by CNT offers it to be used as an efficient EMI shielding material. The percolation threshold of CNTs is achieved at very low loading due to the above-mentioned properties, so that CNTs help in enhancing certain aspects of the resultant composite for EMI shielding without disturbing the properties of the host matrices. There have been few reports on CNT in different matrices, in which CNTs present along with catalyst. A thin and flexible composite film of 60 to 165 μm thickness with different weight ratio of MWCNTs in PMMA {poly (methylmethacrylate)} matrix was prepared for EMI shielding [43]. The electrical conductivity and EMI percolation phenomenon of matrix have been improved due to the presence of Fe catalyst in MWCNTs. This shows overall EMI shielding as 27 dB and the dominating contribution is due to the absorption rather than reflection. This composite matrix was found as an efficient EMI shielding material in RF and microwave frequency ranges. Similarly, Liu et al. also reported dispersion of layered composite of Fe_3O_4 and Fe fillers in MWCNT in the epoxy matrix. EMI shielding through absorption was achieved by Fe-filled CNTs in the epoxy matrix (tri-layer composite) as well as Fe_3O_4 filled CNT in the frequency range of 3.22–40 GHz. Another composite (MWCNT with nickel impurity in wax and epoxy matrix) has been used in the frequency range of 3–18 GHz for EMI shielding. For researchers, CNTs dispersion in polymer matrix has been a major issue. For the preparation of MWCNT/PMMA composite, two methods (In-situ and ex-situ) have also been explored. In the in-situ method, the preparation of composite of PMMA bonds with MWCNT due to free radical reaction reports that the shielding effectiveness (SE) of composite depends on the fabrication and processing of composites. It exemplifies that ten layers stacked MWCNT/PMMA (each layer of thickness ~ 0.1 mm, i.e., total thickness of layered as1mm) illustrates better shielding efficiency than a single 1 mm thick MWCNT/PMMA composite layer (when thickness of both films was same). It achieved maximum shielding of 58.73 dB from layered composite. In the ex-situ method, the composite of functionalized MWCNTs with PMMA, MWCNT has been functionalized via Friedel–Crafts acylation with maleic anhydride. For better dispersion properties of the composite, the solvent casting method has been used. It has been observed that interfacial polarization of PMMA through CNTs result in the difference between SE through reflection and absorption increases with increasing CNT concentration [43].

6.3. Electronic Devices as Field-Emission Sources

Due to their unique property, CNTs can be employed in electric devices application in the form of field-emission sources [44]. It can be worked as a field- emission source by applying a potential between the anode and CNTs surface. Electrons are easily emitted through CNTs tips due to the curvature present in the CNTs in the form of pentagons or due to the presence of oxidized on their tips. Due to this principle, CNTs can be used for the production of numerous electronic devices, such as X-ray sources, intense light sources, flat-panel displays, and bright lamps [45]. CNTs are noble emitters too and its nano-composites have proved to be excellent electron-emission surfaces. CNT as an electron emitters have several advantages such as high current densities, stable field-emission over long time periods, long lifetimes of the components, the absence of the need for an ultra-high vacuum and low emission threshold potentials [46].

6.4. Sensors

Sensors have very crucial detecting devices to be used in various fields these days. The efficiency of the biosensors and molecular sensors can be enhanced with the attachment of CNTs. It was demonstrated that functional chemical groups attached to the end of CNTs is sensible [46]. Therefore, it has been feasible to construct several kinds of sensors containing nanotube composite pellets. These are highly sensitive to gases and can be used as an observer of leaks in chemical plants [47]. The SWCNT samples are very sensitive to air and vacuum through observing huge variations in the electrical resistance levels of SWCNTs [48]. They report that MWNTs can also be used as efficient sensors for CO_2, H_2O, CO and NH_3 [47]. Certain changes have been found in the capacitance and resistance levels of CNTs, with the considerable modification in the environment. Fast-responsive and very sensitive sensors of microwave-resonant have been prepared for sensing of NH_3 through either SWCNTs or MWNTs [49]. CNTs and their composites can also be used to make sensitive sensors for environmental pressure. Some studies reveal that the CNTs have to be extremely sensitive for liquid immersion or polymer-embedding processes. Carbon nanotubes can slightly be deformed in the existence of different liquid media [50].

6.5. Gas and Hydrogen Storage

Due to the tube (hollow cylinder) nature of CNTs, they can be used as an efficient gas and metal containers. CNTs have also been found as source for numerous important chemical species in form of oxides, metal carbides and metals. These can directly be encapsulated into CNT core by numerous techniques such as arc-discharge methods, chemical treatments, electrochemical and solid-state reaction techniques. However, the main problems associated to these methods are that CNTs cannot encapsulate gaseous materials. However, it has been revealed that H_2 and Ar can be stored in CNTs [51]. The capacity of hydrogen storage in the CNTs is high while contaminations present during the CNT synthesis may be responsible for 7 wt% hydrogen storages [51]. A theoretical calculation has also been done through density functional theory to find out the H_2 storage capacity in CNTs [52]. Possibly adsorb hydrogen molecules in the CNTs were proposed through different mechanisms such as adsorption at interstitial sites, chemisorption and the swelling of the nanotube array. CNTs may not be the best material for storage of hydrogen due to the above mentioned conditions, but further investigations and additional calculations must be carried out to augment the findings. Moreover, CNTs as well as carbon foams have been checked as adsorption material for other gases [53].

6.6. Hydrogen Production

CNTs as efficient electrocatalyst have also been studied due to their unique electronic properties [54]. Some studies report a high electrocatalytic activity of nitrogen doped CNTs for oxygen reduction [55, 56]. They have concluded that the inclusion of electron accepting nitrogen atoms facilitates a relatively high positive charge density on the adjacent carbon atoms and provide more electron pathway for oxygen reduction. Moreover, theoretical calculations have shown thermal dissociation of water at lower temperatures (~ 1000 K) in the presence of CNT whereas the temperature ~2000 K has been required with graphite as catalyst. The dissociation of water on charged defective SWCNT has also been studied and demonstrates that the charged carbon nanotubes with

C dopant exhibit an activation barrier of 0.167 eV between the transition state and the reactant. The low activation barrier has been found due to electrostatic interaction between the delocalized π electrons of the nanotubes and water molecules which promote adsorption and dissociation of water molecules. The Graphene-CNT composite as an anode for the hydrogen production through water electrolysis has also been studied and shows remarkably good results [54].

Conclusion

The discovery of carbon nanotubes has opened an entirely new field in the domain of physical, chemical and material science. It has drawn much attention due to several distinctive properties of CNTs and thereby potential applications in electronic devices, field emission displays, batteries, nanotechnology tools e.g., atomic force microscopy tips, transistors etc. The architecture of CNTs has been built from the aligned arrangement and is of particular interest as it can lead to the building of integrated systems. The high yield CNTs can be prepared by spray pyrolysis under Ar atmosphere. The morphology of the as-grown carbon nanotubes can be verified by electron microscopy. Functionalization of nanotubes can improve the interfacial bonding between the nanotubes and other suitable elements/ materials. But electrical conductivity has been decreased due to attachment of functional group onto CNT surface. Henceforth, many challenges associated with CNTs have yet to be addressed and further research is needed to alter/enhance the properties of the material to meet the prospective applications.

Acknowledgement

The authors offer their thankfulness to Dr. Kalpana Awasthi (KNPG College, Bhadohi) for her critical comments and constructive suggestions related to the outcome/findings of the useful experiments executed in the area of 1D nanomaterial.

References

[1] Awasthi, K. Srivastava, A. and Srivastava, O.N. (2005). Synthesis of carbon nanotubes. *J. Nanosci. Nanotech.* 5 (10): 1616-1636.

[2] Ajayan, P. and Zhou, O. (2001). Applications of carbon nanotubes. *Carbon Nanotubes* 391–425.

[3] Kroto, H.W., Heath, J.R., O'Brian, S.C., Curl, R.F. and Smalley, R.E. (1985). C60: Buckminsterfullerene. *Nature* 381: 162–163.

[4] Iijima, S. (1991). Helical microtubules of graphitic carbon. *Nature* 354: 56-58.

[5] Ebbesen, T.W. and Ajayan P.M. (1992). Large-scale synthesis of carbon nanotubes. *Nature* 358: 220-222

[6] Novoselov, K.S., Geim, A.K., Morozov, S.V., Jiang, D., Zhang, Y., Dubonos, S.V., Grigorieva, I.V. and Firsov, A.A. (2004). Electric Field Effect in Atomically Thin Carbon Films. *Science* 306: 666-669.

[7] Novoselov, K.S., Geim, A.K., Morozov, S.V., Jiang, D., Katsnelson, M.I., Grigorieva, I.V., Dubonos, S.V. and Firsov, A.A. (2005). Two-dimensional gas of massless Dirac fermions in graphene. *Nature* 438: 197-2000.

[8] Krätschmer, W., Lamb, L.D., Fostiropoulos, K. and Huffman, D.R. (1990). Solid C60: A new form of carbon. *Nature* 347: 354–358.

[9] Wragg, J.L., Chamberlain, J.E., White, H.W., Krätschmer W. and Huffman, D.R. (1990). Scanning tunnelling microscopy of solid C_{60}/C_{70}. *Nature* 348: 623–624.

[10] Awasthi, K., Kumar, R., Raghubanshi, H., Awasthi, S., Pandey, R., Singh, D., Yadav, T.P. and Srivastava, O.N. (2011). Synthesis of nano-carbon (nanotubes, nanofibres, graphene) materials. *Bulletin of Mat. Sci.* 34 (4): 607-614.

[11] Qu, L., Zhu, G., Ji, J., Yadav, T.P., Chen, Y., Yang, G., Xu, H. and Li, H. (2018). Recyclable visible light-driven O-g-C_3N_4/Graphene oxide/N-carbon nanotube membrane for efficient removal of organic pollutants. *ACS applied materials & interfaces* 10 (49): 42427-42435.

[12] Yadav, T.P., Palma, J.L., Singh, D.P., Michea, S. (2020). Facile synthesis and magnetic behavior of 1D g-C_3N_4, *J. Solid State Chem.* 290: 121539.

[13] Prajapati, V.K., Awasthi, K., Gautam, S., Yadav, T.P., Rai, M., Srivastava, O.N. and Sundar, S. (2011).Targeted killing of *Leishmania donovani* in vivo and in vitro with amphotericin B attached to functionalized carbon nanotubes. *J. Antimicrobial Chemotherapy* 66: 874-879.

[14] Awasthi, K., Yadav, T.P. and Srivastava, O.N. (2018). Growth of carbon nanotubes on icosahedral quasicrystalline Al–Cu–Fe Alloy. *Adv. Sci. Engg. and Medicine* 10 (7-8): 773-777.

[15] Kajiwara, K., Matsui, Y., Yadav, T.P., Mukhopadhyay, M.K. and Srivastava, O.N. (2016). Quasicrystal as a catalyst for the synthesis of carbon nanotubes. *J. of Nanosci. Nanotech.* 16 (3): 3084-3089.

[16] Iijima, S. and Ichihashi, T. (1993). Single-shell carbon nanotubes of 1-nm diameter. *Nature* 363: 603- 605.

[17] Bethune, D.S., Kiang, C.H., de Vries, M.S., Gorman, G., Savoy, R., Vazquez, J. and Beyers, R.(1993). Cobalt-catalysed growth of carbon nanotubes with single-atomic-layer walls. *Nature* 363: 605-607.

[18] Guo, T., Nikolaev, P., Rinzler, A.G., Tomanek, D., Colbert, D.T. and Smalley, R.E. (1995). Self-Assembly of Tubular Fullerenes. *J. Phys. Chem.* 99: 10694-10697.

[19] Shen, C., Brozena, A.H. and Wang, Y. (2011). Double-walled carbon nanotubes: Challenges and opportunities. *Nanoscale*3: 503-518.

[20] Yadav, T.P., Singh, D., Tiwari, R.S. and Srivastava, O.N. (2012). Enhanced microhardness of mechanically activated carbon-quasicrystal composite. *Mat. Letts.* 80: 5-8.

[21] Awasthi, K., Yadav, T.P., Mishra, P.K., Awasthi, S. and Srivastava, O.N. (2008). Investigation on carbon nanomaterials: Coaxial CNT-cylinders and CNT-polymer composite. *Bulletin of Materials Science* 31(3): 313-318.

[22] Ivchenko, E.L. and Spivak, B. (2002). Chirality effects in carbon nanotubes. *Phys. Rev. B* 66: 155404.

[23] Amelinckx, S., Bernaerts, D., Zhang, X.B., Van Tendeloo, G. and Landuyt, J. Van (1995). A Structure Model and Growth Mechanism for Multishell Carbon Nanotubes. *Science*267: 1334-1338.

[24] Ajayan, P.M. in *Handbook of Nanostructured Materials and Nanotechnology* edited by H.S. Nalwa, Academic Press, San Diego (2000). Vol.5, p.375.

[25] Graf, A., Zakharko, Y., Schießl, S.P., Backes, C., Pfohl, M., Flavel, B.S., and Zaumseil, J. (2016). Large scale, selective dispersion of long single-walled carbon nanotubes with high photoluminescence quantum yield by shear force mixing. *Carbon* 105: 593-599.

[26] Ajayan, P.M., Ebbesen, T.W., Ichihashi, T., Iijima, S., Hiura, H. and Tanigaki, K. (1993). Opening carbon nanotubes with oxygen and implications for filling. *Nature* 362; 522.

[27] Kuo, T.F., Chi, C.C. and Lin, I.N. (2001). Synthesis of carbon nanotubes by laser ablation of graphites at room temperature. *Jap. J. Appl. Phys.* 40: 12.

[28] Thess, A., Lee, R., Nikolaev, P., Dai, H.J., Petit, P., Robert, J., Xu, C.H., Lee, Y.H., Kim, S.G., Rinzler, A.G., Colbert, D.T., Scuseria, G.E., Tomanek, D., Fischer, J.E. and Smalley, R.E. (1996). Crystalline ropes of metallic carbon nanotubes. *Science* 273: 483.

[29] Maser,W.K., Munoz, E., Benito, A.M., Martinez, M.T., de la Fuente, G.F., Maniette, Y., Anglaret, E., Sauvajol, J.L. (1998). Production of high-density single-walled nanotube material by a simple laser ablation method. *Chem. Phys. Lett,* 292: 587-593.

[30] Munoz, E., Maser, W.K., Benito, A.M., Martinez, M.T. Fuente, de la Fuente, G.F., Maniette, Y., Righi, A., Anglaret, E., Sauvajol J.L.(2000). Gas and pressure effects on the production of single-walled carbon nanotubes by laser ablation. *Carbon*, 38, 1445-1451.

[31] Jung, M., Eun, K.Y., Lee,J.-K, Baik, Y.-J. Lee, K.-R. and Park,J.W. (2001). Growth of carbon nanotubes by chemical vapor deposition. *Dia. Relat. Mat.*10: 1235-1240.

[32] Terranova, M.L., Sessa,V. and Rossi, M. (2006). The world of carbon nanotubes: An overview of CVD growth methodologies. *Chem. Vap. Dep.*12: 315–325.

[33] Chen, Y. and Zhang, J. (2014). Chemical vapor deposition growth of single-walled carbon nanotubes with controlled structures for nano device applications, *Acc. Chem. Res.* 47: 2273–2281.

[34] Satio, Y., Okuda, M., Fujimoto, N., Yoshikawa, T., Tomita M. and Hayashi, T.(1994). Single-wall carbon nanotubes growing radially from Ni fine particles formed by arc evaporation. *Jpn. J. Appl. Phys.* 33: L526.

[35] Iijima, S.(2002). Carbon nanotubes: past, present, and future. *Physica B: Condensed Matter*323 (1–4).

[36] Srivastava, A., Srivastava, O.N., Talapatra, S., Vajtai, R. and Ajayan, P.M. (2004). Carbon nanotube filters. *Nature Materials* 3: 610-613.

[37] Qin, L.C. (1997). CVD synthesis of carbon nanotubes. *J. Mat. Sci. Lett.* 16: 457-459.

[38] Ruoff, R.S. and Lorents, D.C. (1995). Mechanical and thermal properties of carbon nanotubes. *Carbon* 33 (7):925-930.

[39] Dresselhaus, G., Saito, R., Dresselhaus, M.S. (1998). *Physical Properties of Carbon Nanotubes.* Singapore: World Scientific Publishing Company.

[40] Yakobson B.I., Avouris P. (2001). Mechanical properties of carbon nanotubes. In: Dresselhaus M.S., Dresselhaus G., Avouris P. (*eds*) Carbon Nanotubes. *Topics in Applied Physics,* vol 80. Springer, Berlin, Heidelberg.

[41] Huang, Y., Zhu, G., Zou, K., Tian, F., Yadav, T.P., Xu, H., Yang, G., Li, H. and Qu, L. (2021). Highly efficient removal of organic pollutants from wastewater using a recyclable graphene oxide membrane intercalated with g-C_3N_4@ TiO_2-nanowires. *J. Molecular Liquids* 337; 116461.

[42] Patel, C.R.P., Tripathi, P., Srivastava, O.N. and Yadav, T.P. (2014). Filtration of sodium chloride from seawater using carbon hollow tube composed of carbon nanotubes. *Int. J. Smart and Nano Materials* 5 (3): 194-206.

[43] Tripathi, P., Gupta, B.K., Bhatnagar, A., Patel, C.R.P., Banker, P.K., Late, D.J., More, M.A., Lalla, N.P., Phase, D.M., Choudhary, R.J., Shaz, M.A., Ajayan, P.M. and Srivastava, O.N. (2018). Highly efficient field emission properties of radially aligned carbon nanotubes. *J. Mat. Chem. C* 6 (24); 6584-6590.

[44] Milne, W.I., Teo, K.B.K., Amaratunga, G.A. J., Legagneux, P., Gangloff, L., Schnell, J.-P., Semet, V., Thien Binh V. and Groening, O. (2004). Carbon nanotubes as field emission sources. *J. Mater. Chem.*14: 933-943.

[45] Terrones, M. (2004) Carbon nanotubes: synthesis and properties, electronic devices and other emerging applications, *International Materials Reviews*, 49: 325-377.

[46] Zhu, W., Bower, C., Zhou, O., Kochanski, G., Jin, S. (1999). Large current density from carbon nanotube field emitters. *Appl. Phys. Lett.* 75:873-875.

[47] Gao, C., Guo, Z. Liu, J.-H. and Huang, X.-J. (2012). The new age of carbon nanotubes: An updated review of functionalized carbon nanotubes in electrochemical sensors. *Nanoscale,* 4: 1948-1963.

[48] Kumar, U., Sikarwar, S., Sonker, R.K., Yadav, B.C. (2016). Carbon Nanotube: Synthesis and Application in Solar Cell, *Journal of Inorganic and Organometallic Polymers and Materials* 26: 1231–1242.

[49] Mao, S., Lu, G. and Chen, J. (2014). Nanocarbon-based gas sensors: progress and challenges. *J. Mater. Chem. A*2: 5573-5579.

[50] Sinha, N. Ma, J., Yeow, John T.W. (2006). Carbon nanotube-based sensors. *J. Nanosci. Nanotech.*6: 573-590.

[51] Ibrahim, K.S. (2013). Carbon nanotubes-properties and applications: a review, *Carbon letters* 14:131-144.
[52] Ming, H., Hong, C.Q., Liu, Y.C. (2001). Hydrogen storage in carbon nanotubes. *Carbon* 39:1447-1454.
[53] Zollo, G. and Gala, F. (2012). Atomistic modeling of gas adsorption in nanocarbons. *J. Nanomat.* Article ID 152489, 32 pages.
[54] Askari, S., Jafari, A. (2020).A novel model for natural gas storage on carbon nanotubes. *Appl. Nanosci.* 10:1115-1129.
[55] Patel, C.R.P., Tripathi, P., Vishwakarma, A.K., Talat, M., Soni, P.K., Yadav, T.P. and Srivastava, O.N. (2018). Enhanced hydrogen generation by water electrolysis employing carbon nano-structure composites. *Int. J. Hydrogen Energy* 43: 3180-3189.
[56] Zhang, C., Xie, Y., Deng, H., Zhang, C., Su, J.-W. and Lin, J. (2019). Nitrogen doped coal with high electrocatalytic activity for oxygen reduction reaction. *Engineered Science* 8: 39-45.

Chapter 9

Graphene: A Green, Metal Free and Sustainable Catalyst

Chandani Singh, Arti Srivastava, Bhaskar Sharma and S. K. Singh*

Department of Chemistry, Guru Ghasidas Central University, Bilaspur, Chhattisgarh

Abstract

The environmental concerns of the present times have compelled the scientists to adopt sustainable protocols. Nowadays, greener protocols such as performing reaction in solvent-free and metal-free condition, performing microwave assisted reaction, use of green catalysts and solvents are in practice. By using such a greener protocol, pollution can be easily reduced. One such attempt is using green catalysts such as Graphene, an allotrope of carbon having a honeycomb-like structure with unique properties. It is now an established green catalyst for numerous organic transformations. Today, the wide applicability of this material has evolved many sustainable approaches for its synthesis. Apart from catalysis, Graphene and its different forms (monolayer graphene, grapheme superlattice, grapheme nanoribbons, bilayer graphene and Graphene oxide) find wide application in electronics and transistors. It finds applications in a capacitor, sensor, solar cells, and is used in biomedical applications too. Above all, graphene also catalyzed various organic reactions in good yield. Its outstanding performance as a catalyst for various reactions with excellent yield, good recyclability, and high catalytic activity, addressing all environmental concerns, is now well documented. It also performs reactions in neat conditions and also as a

* Corresponding Author's Email: singh.skumar@gmail.com.

In: Materials Science: Future Aspects
Editors: Kalpana Awasthi, Arti Srivastava and Mridula Tripathi
ISBN: 978-1-68507-843-0

metal-free system. It shows fabrication of metal on it which increases its catalytic performances; it also works as a phase transfer catalyst. It is also tunable by doping different elements for requisite properties. It is a convenient carbo-catalyst. It catalyzes many known reactions like Michael addition, condensation, cyclo-condensation, multicomponent reaction, and trans-esterification with good yield and excellent recyclability in an eco-friendly manner. Some of them are discussed in this chapter.

Keywords: metal free, carbo-catalyst, sustainable, trans-esterification, tunable, recyclable

1. Introduction

Carbon has many allotropes due to its variable valancy. Graphene is an allotrope of carbon which is classified under crystalline type (Figure1). It has a long lasting history and was first successfully prepared by Andren Geim and Konstantin Novoselov in 2004 [1] and for their groundbreaking experiment they were awarded the Nobel Prize in Physics in 2010. Graphene films are found to be two-dimensional, with a single layer of carbon atom closely packed in a hexagonal structure like a honeycomb. It shows sp^2 hybridization s, p_x, p_y, hybrid orbital involved in forming sigma bond and p_z orbital which are perpendicular to hexagonal structure forming π bond like benzene.

Graphene is found to be semimetal as its conduction and valence band meets dirac point. It has different forms. It has a unique optical property and also shows a strong magnetic field. During its production, it was found that graphene showed some lattice defects [2-4]. Methodologies of reduction of defects in graphene are also employed. For preparing grapheme, various methods are employed like vacuum-assisted synthesis of graphene from thermal exfoliation and reduction of graphite oxide [5], high–yield production of graphene by liquid-phase exfoliation of graphite [6], large-area synthesis of high-quality and uniform graphene films on copper foils [7] etc. Many green synthetic methods are also employed for the preparation of graphene like synthesis of graphene from biomass (dead camphor leaves) [8], synthesis of chemically converted grapheme nano sheets based on reducing sugar, using exfoliated graphene oxide [9], and successfully synthesized graphene nano-sheets in the presences of pomegranate juice [10]. Graphene oxide is an oxidized form of graphene (Figure 2); it was first prepared by Benjamin Brodie in 1859 by exposing graphite to strong acid [11]. It has a two-dimensional single layer of carbon atom like graphene but it shows sp^3 hybridization, with a carbon atom attached to oxygen atom forming hydroxyl group, carboxylic acid and epoxide ring.

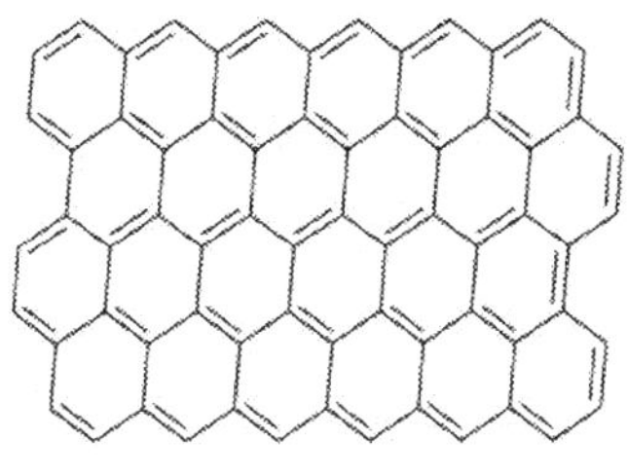

Figure 1. Structure of graphene.

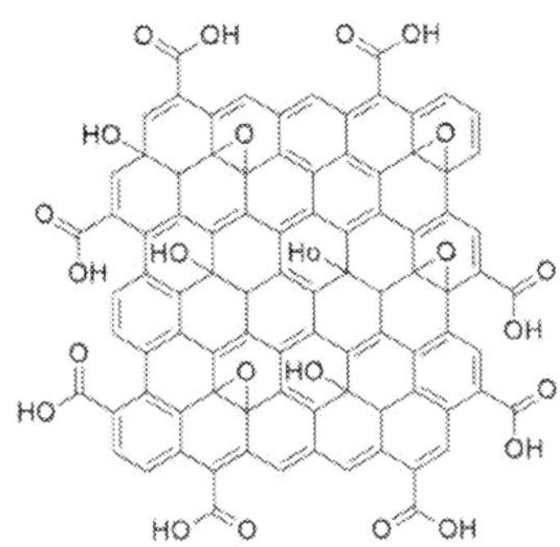

Figure 2. Structure of graphene oxide.

Graphene oxide behaves as an insulator due to a wide band gap. Various synthetic methods for preparing graphene oxide are employed like (Hummers [12], Hofmann [13], Brodie [14], Staudenmaier [15] etc.) Green approaches are also employed for preparing graphene oxide. For example, widely soluble graphene oxide sheets decorated by silver nano-prisms were prepared through green synthesis [16]. Hummers method for improved eco-friendly synthesis of graphene oxide is also reported [17].

2. Reaction Catalyzed by Graphene

Graphene is a carbon material that has unique properties, due to which it shows outstanding performances in various fields. As it is eco-friendly, it can be used in an environmentally benign manner. Here we mention some reactions which are catalyzed by graphene and its forms.

2.1. Oxidation and Hydration Reactions Catalyzed by Graphene Oxide

Daniel R. Dreyer et al. reported graphene oxide as a convenient carbocatalyst for facilitating oxidation and hydration reactions. In this experiment, graphene oxide is a carbocatalyst which performs oxidation and hydration reaction. In a preliminary experiment, oxidation of neat benzyl alcohol was performed using 20 wt% GO for 14 hours under ambient conditions, heating at a temperature of 100°C [18].

Further study was done by using GO as a catalyst for the oxidation of benzyl alcohol into benzaldehyde and benzoic acid under various conditions. From this, we found that heating at 100°C $PhCH_2OH$ for 24 hrs using 200 wt% GO as a catalyst gives high yield (>98%) and shows good selectivity for the aldehyde versus acid (92:7). After that, various alcohols are oxidized using 200 wt% GO under the same condition, showing results from moderate to high yield. GO oxidizes various alcohols, so its ability is also checked for unsaturated hydrocarbons under the same condition [18].

Further, GO is applied as a catalyst for the hydration of various alkynes under the same condition and it also shows excellent yield. From the above experiments, we conclude that GO shows broad chemoselectivity and functional group tolerance, and makes possible the recovery of catalysts by simple filtration methods. GO is an eco-friendly and convenient carbocatalyst [18].

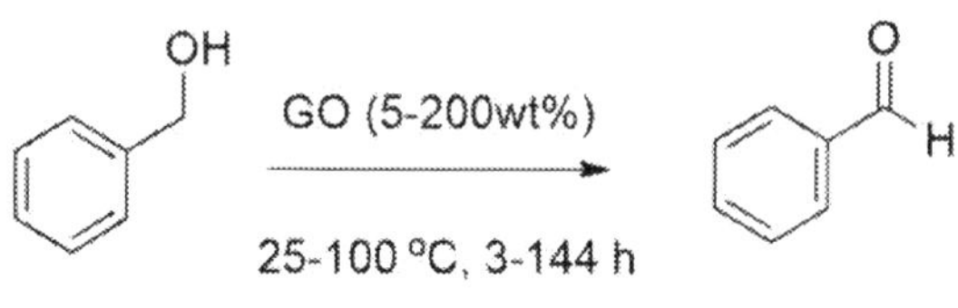

Scheme 1. Oxidation of Alcohol catalyzed by GO.

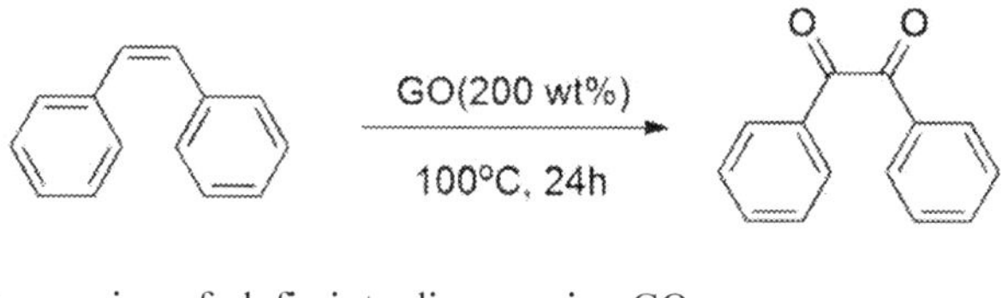

Scheme 2. Conversion of olefin into diones using GO.

2.2. Graphene Oxide Mediated Solvent-Free Three Component Reaction

Gupta et al. reported three solvent-free component reactions catalyzed by GO. In this report, graphene catalyzed a solvent-free multicomponent reaction. Here,1-amidoalkyl-2-naphthols and 1,2-dihydro-1-arylnaphth[1,2-e] [1, 3] oxazin-3-ones were synthesized from 2-naphthol, aldehyde and amide/urea using graphene oxide as heterogeneous catalyst in solvent free synthesis [19]. 1-amidoalkyl-2-naphthols were synthesized from (1mmol) 2-naphthol, anisaldehyde (1 mmol) and (1.2 mmol) benzamide using 35 wt% GO, and it gave 35% yield.

To improve it, various conditions were studied; it was found that the yield increases from 35% to 95% at 120°C temperature using 35 wt% GO as catalyst. For 1, 2-dihydro-1-arylnaphth [1, 2-e] [1, 3] oxazin-3-ones thiophene 2, carboxaldehyde was used which gave 85% yield [19].

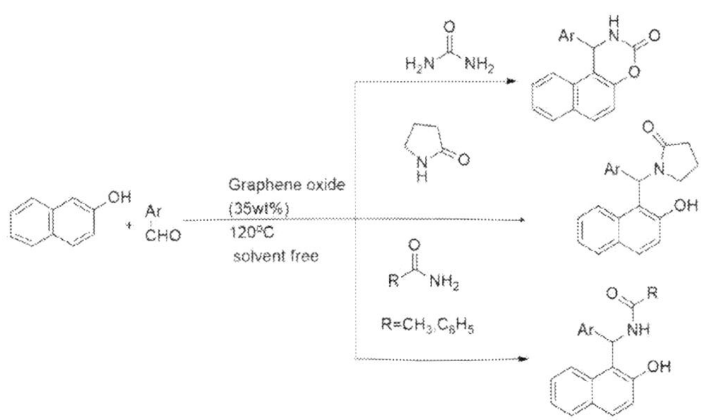

Scheme 3. Solvent free three component reaction catalyzed by GO.

This protocol shows high yield and environmentally-benign solvent free synthesis of multicomponent reaction and catalyst show good recyclability [19].

2.3. Metal-Free Oxidation of Biomass-Derived 5-Hydroxymethylfurfural

Due to the continuous use of fossil fuels, the sources of fossil fuels are getting depleted as they are non-renewable sources. So, we have to find renewable sources for our requirements. Yongshen Ren et al. reported that nitrogen-doped carbon materials performed as a catalysts for the production of 2,5-diformylfuran (DFF) from 5-hydroxymethylfurfural (HMF). HMF are biomass derived [20]. Nitrogen doped carbon material (NC-T) is prepared from the pyrolysis of chitosan. It was observed that on the gradual increase in the pyrolysis temperature (650^0C to 850^0C). the pore size and surface area of NC-T catalyst are affected. On increasing the pyrolysis temperature, they show a decrease in nitrogen content, showing that the nitrogen atoms are destroyed in the NCT catalyst. This was also observed by other researchers [21]. The investigation of oxidation of HMF using NC-T at different pyrolysis temperatures showed that using NC-950 catalyst, high conversion of 95.1% was obtained in the nitric-acid mediated oxidation of HMF to DFF with molecular oxygen.

This experiment was performed with several common solvents from which acetonitrile was fitted to be the best solvent for the oxidation. Its catalytic performance is observed using different parameters like different temperatures, different loading of HNO_3, and reusability.

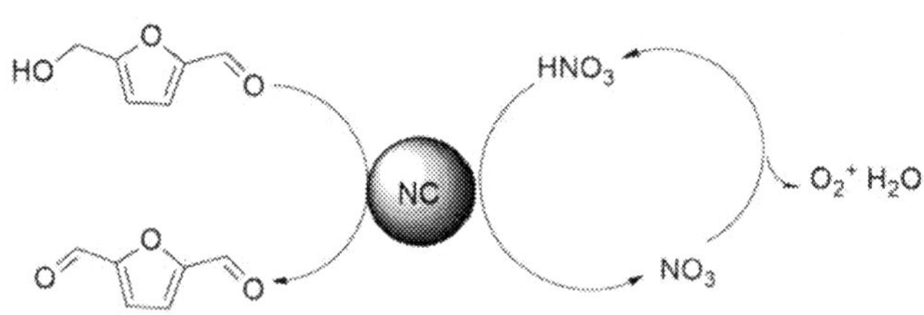

Scheme 4. Mechanism of the oxidation of HMF to produce DFF.

This is a new method for the oxidation of biomass derived HMF to DFF using nitrogen doped carbon material as catalyst mediated by nitric acid which is renewable and ecofriendly. NC-950 shows high catalytic activity and good reusability [20].

3. Michael Addition Reaction Catalyzed by Graphene and Their Forms

Graphene and its different forms performed various reactions in an eco-friendly manner. It also catalyzed Michael addition reaction. We included a few reactions from this category.

3.1. Aza-Michael Addition of Amines to Activated Alkene

Sanny Verma et al. reported graphene oxide as a carbocatalyst for the synthesis of β amino compounds via Michael addition [22]. In this experiment, aza-Michael addition of (1 mmol) amines to α, β unsaturated compounds (1.2 mmol) using graphene oxide (0.5 mmol) as the catalyst was performed at room temperature for a shorter time. From the experiment, it was found that in general, aromatic amines are less reactive than aliphatic amines. All observed reactants are efficiently converted into the corresponding β amino compounds in excellent yield with no evidence of by-product.

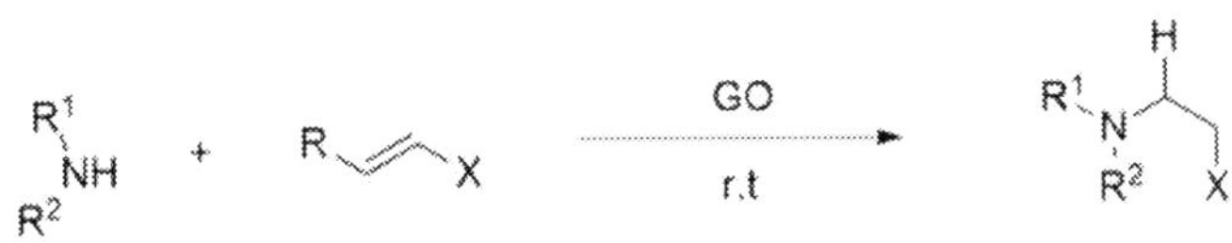

R= H, CH_3, R^1= Et, n- Bu, n-Pr, Bz
R^2= Et, n- Bu, n- Pr, Bz
X= CN, $CONH_2$, $COOCH_3$

Scheme 5. Aza-Michael addition using GO.

GO was found to be a simple and efficient catalyst for the synthesis of various amino compounds via aza-Michael addition of amines and α, β unsaturated compounds which was performed in mild conditions and it gave excellent yield within a shorter time. Catalyst was easily renewable and it is the first report of GO work as catalyst for coupling type reaction [22].

3.2. Thiol-Michael Addition Click Chemistry

Lie Nie et al. synthesized, micro patterned gold nanoparticles on GO nanosheet (AuNPs@GO) via thiol Michael addition click chemistry [23]. Here, Au NP was functionalized by acrylamide and GO nanosheet was functionalized by dithiothreitol.

Then AuNP was fabricated on GO nanosheets through thiol-Michael addition click reaction under U. V. light. It showed potentials for future cancer diagnosis and its therapy [23].

3.3. Graphene Oxide as a Recyclable Phase Transfer Catalyst for Michael Addition

Youngmin Kim et al. reported an experiment in which GO works as a recyclable phase transfer catalyst for the Michael addition. Michael addition reaction of trans β nitro styrene and 2, 4 pentanedione takes place using GO as phase transfer catalyst, KOH as base, and water-dichloromethane as solvent [24]. It provides a novel method for the synthesis of c-c bonds, showing high yield and, simple and green methods. GO show excellent recyclability [24].

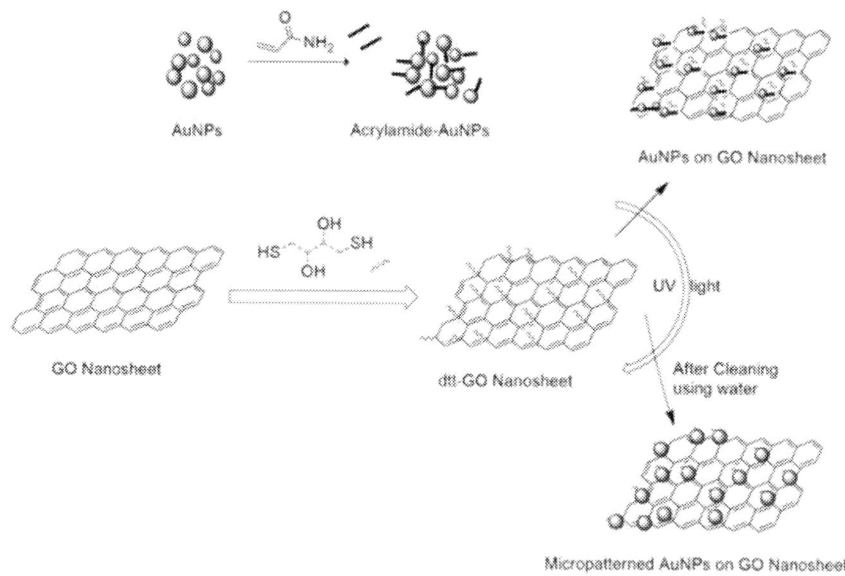

Scheme 6. Systematic representation of fabrication of micropatterned gold nanoparticles on graphene oxide nanosheet via thiol-Michael addition click chemistry.

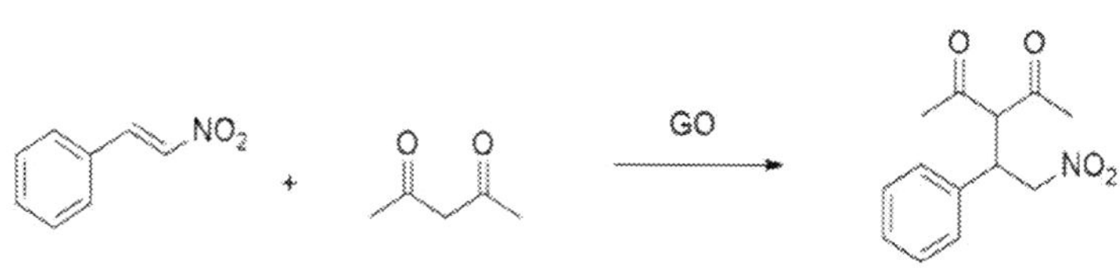

Scheme 7. Michael addition reaction using GO as phase transfer catalyst.

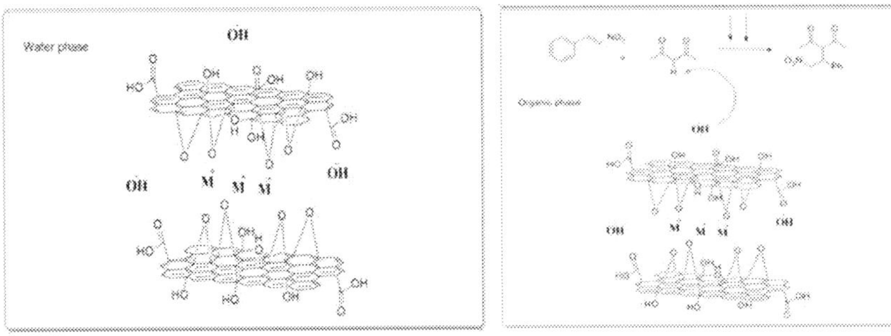

Scheme 8. GO works as a recyclable phase transfer catalyst for the Michael Addition.

4. Condensation Reactions Performed by Graphene and Their Forms

4.1. Graphene Oxide with Ethylenediamine as a Solid Base Catalyst for Knoevenagel Condensation Reaction

Bing Xue et al. reported the work in which Knoevenagel condensation reaction of benzaldehyde and malononitrile was performed, using ethylenediamine modified GO as a catalyst. This GO is grafted by ethylenediamine which enhances its catalytic activity for the Knoevenagel reaction [25].

The grafting of amine on the surface of GO was mainly due to the reaction between the epoxide ring and carboxylic acid of GO with ethylenediamine.

This report shows a novel solid base catalyst of GO grafting with ethylenediamine, which shows excellent catalytic performance for the Knoevenagel condensation reaction [25].

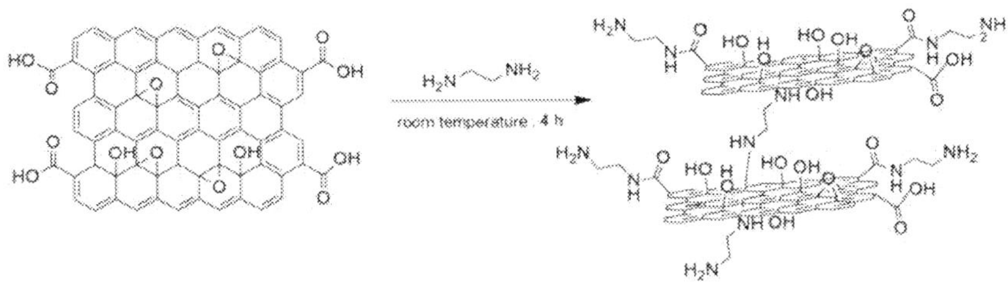

Scheme 9. Knoevenagel condensation reaction of benzaldehyde and malononitrile was performed, using ethylenediamine modified GO as catalyst.

4.2. Oxidative Condensation of Toluene and Hydrazine/Aniline Catalyzed by Copper Complex Immobilized on Functionalized Graphene Oxide

H. R. E. Zand et al. synthesized GO-IL-CO_2H-Cu catalyst and studied its catalytic activity on oxidative condensation of toluene and hydrazine/aniline [26].

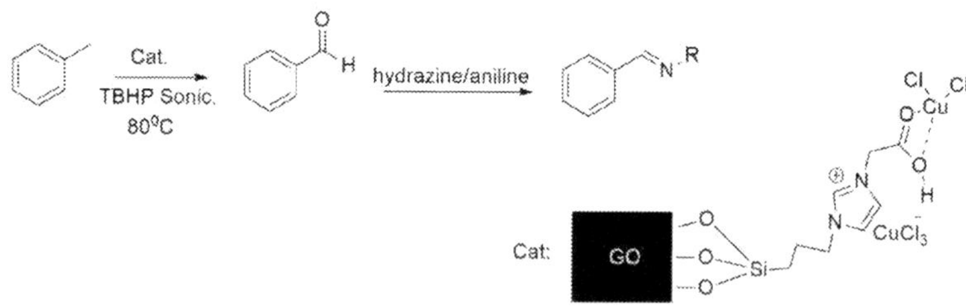

Scheme 10. Oxidative condensation of toluene and hydrazine/aniline catalyzed by copper complex immobilized on functionalized GO.

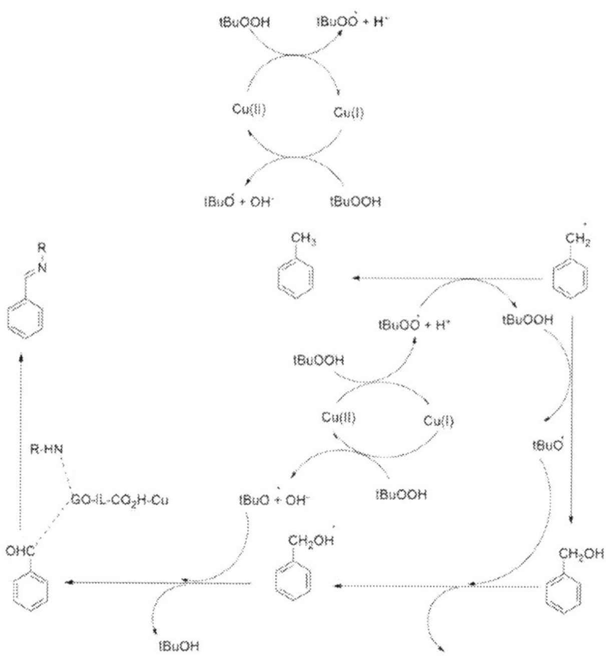

Scheme 11. Mechanism of the reaction.

Cu-functionalized imidazolium Ionic liquid was firstly synthesized and then immobilized on solid substrates like graphene oxide-imidazolium ionic liquid (GO-IL-Cu), graphene oxide-imidazolium ionic liquid-CO_2H-Cu (GO-IL-CO_2H), MCM- 41- imidazolium ionic liquid–Cu(MCM-41-IL-Cu), MCM-41-imidazolium ionic liquid- CO_2H-Cu(MCM-41-IL-CO_2H-Cu), chitosan-imidazolium ionic liquid CO_2H-Cu (chit-IL-CO_2H-Cu), and g-C_3N_4-imidazolium ionic liquid- CO_2H-Cu (g-C_3N_4-IL-CO_2H-Cu).

To find out the best catalyst for the oxidative condensation, the synthesized catalysts were compared and from the observation it can be concluded that GO with CO_2H is the best catalyst for performing the reaction of oxidative condensation.

4.3. Knoevenagel Condensation of Aldehydes with Malononitrile

N. Lolak et al. synthesized heterogeneous nanocatalysts GO supported PdNi nanoparticles and used as a catalyst for the Knoevenagel reaction between malononitrile and aromatic aldehyde under mild reaction conditions [27].

From the observation, it can be concluded that 2 mg of catalyst (GO-supported PdNi) is used for the Knovenagel condensation reaction with Et-H_2O as solvent at room temperature.

This is an environmentally friendly, clean, and efficient method of Knoevenagel condensation reaction using uniformly dispersed GO-supported PdNi nanoparticles as a catalyst which gives excellent yield and shows good reusability.

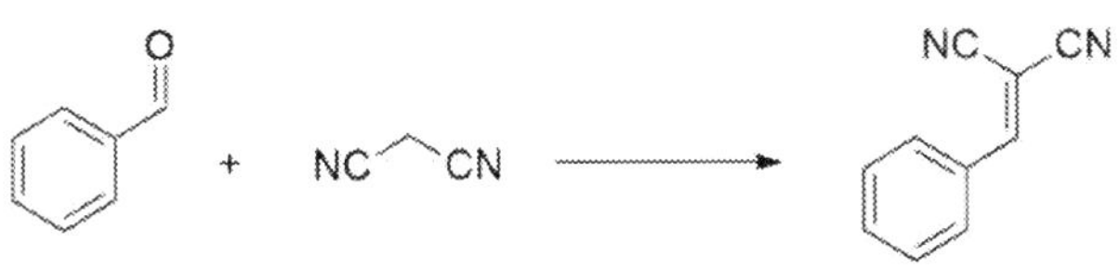

Scheme 12. Catalyst (GO-supported PdNi) is used for the Knovenagel condensation reaction.

5. Cyclocondensation Reaction

S. Abdolmohammadi et al. synthesized of [1]-benzopyranopyrido[d] pyrimidines using Porous Graphene/MoO_3. In this experiment, ultrasound assisted cyclocondensation of reaction of 4-hydroxycoumarin, aromatic aldehyde, and 6-amino-1,3-dimethyluracil was performed for the synthesis of [1] -benzopyranopyrido[d]pyrimidines, which is catalyzed by porous graphene/MoO_3 in water as solvent [28].

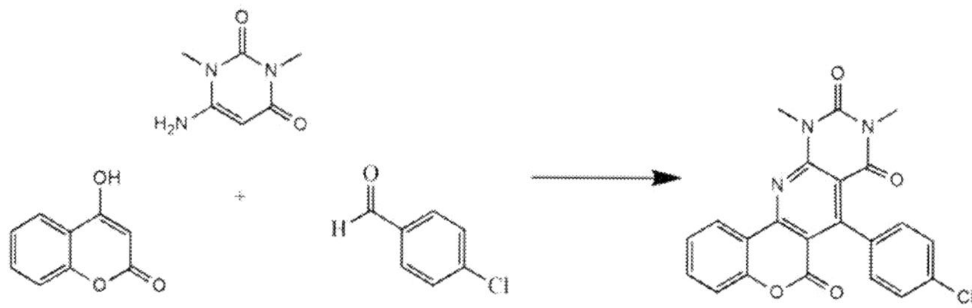

Scheme 13. Ultrasound assisted cyclocondensation reaction, catalyzed by porous graphene/MoO_3.

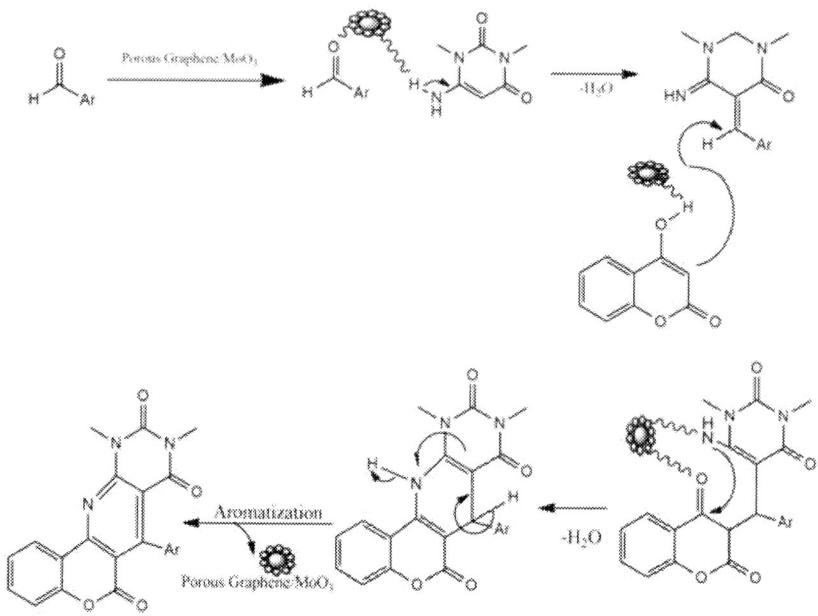

Scheme 14. Mechanism of the reaction.

This is a novel procedure that shows high yield in short time with the, use of green solvent (water) and ultrasound irradiation. The catalyst also shows good reusability. It shows notable features, such as in the absence of ultrasound waves, the final cyclized product was not formed [28].

6. Trans-Esterification

B. Xue et al. synthesized heterogeneous catalyst for trans-esterification. In this work, for the first time basic ionic liquid (1-trimethoxysilyl) propyl-3-methylimidazolium hydroxide grafted on GO in the water phase was synthesized at room temperature. Then its catalytic performance was investigated in trans-esterification [29].

This is a simple, efficient, and green work that shows excellent performances and maximum dispersibility of GO in water.

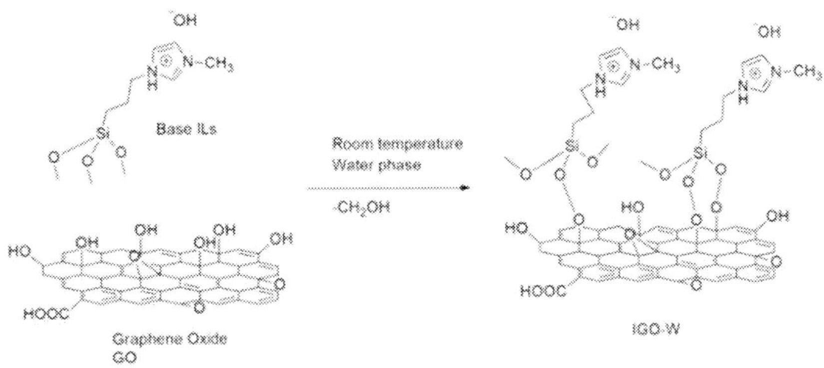

Scheme 15. GO was grafted with basic Ils.

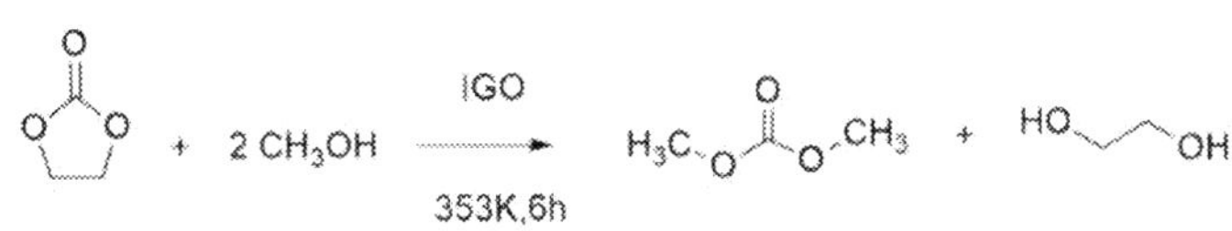

Scheme 16. Transesterification reaction using IGO.

Conclusion

Graphene, an allotrope of carbon with unique properties, gives a wide platform to researchers as it is eco-friendly, tunable in different forms, and can be fabricated with other materials, which enhances its catalytic properties. Various reactions were performed with it in an environmentally benign manner, furnishing good product yield, shorter time, and good recyclability.

References

[1] Novoselov, K. S., Geim, A. K., Morozov, S. V., Jiang, D., Zhang, Y., Dubonos, S. V., Grigorieva, I. V. and Firsov, A. A. (2004). Electric field effect in atomically thin carbon films. *Science* 306: 666-669.

[2] Tian, W., Li, W., Yu, W. and Liu, X. (2017). A review on lattice defects in graphene: Types generation effects and regulation. *Micromachines* 8:163(1-15).

[3] Liu, L., Qing, M., Wang, Y. and Chen, S. (2015). Defects in Graphene: Generation, Healing, and Their effects on the Properties of Graphene. J. *Mater. Sci. Technol.* 31, 599-606.

[4] Banhart, F., Kotakoski, J. and Krasheninnikov, A. V. (2011). Structural defects in graphene. *ACS Nano*5: 26-41.

[5] Zhang, Hao-Bin., Wang, Ji-Wen., Yan, Q., Zheng, Wen-Ge., Chen, C., and Yu, Zhong-Zhen (2011). Vacuum-assisted synthesis of graphene from thermal exfoliation and reduction of graphite oxide. *J. Mater. Chem.* 21: 5392-5397.

[6] Hernandez, Y., Nicolosi, V., Lotya, M. et al. (2008). High-yield production of graphene by liquid- phase exfoliation of graphite. *Nature Nanotechn.* 3: 563-568.

[7] Li, X. et al. (2009). Large-Area Synthesis of High-Quality and Uniform Graphene Films on Copper Foils. *Science* 324, 1312.

[8] Shams, S. S., Zhang, Li S., Hu, R., Zhang, R. and Zhu, J. (2015). Synthesis of graphene from biomass: A green chemistry approach. *Mat. Lett*. 161:476-479.

[9] Zhu, C., Guo, S., Fang Y., and Dong S. (2010). Reducing Sugar: New Functional Molecules for the Green Synthesis of Graphene Nanosheets. *ACSNANO* 4: 2429-2437.

[10] Tavakoli, F., Salavati-Niasari M., Badiei A. and Mohandes, F. (2015). Green synthesis and characterization of graphene nanosheets. *Mat. Res. Bullt*. 63: 51-57.

[11] Brodie, B. C. 1859. On the atomic Weight of Graphite. *Philos Trans R Soc London* 14, 249-259.

[12] Hummers Jr, W. S. and Offeman, R. E. (1958). Preparation of Graphitic Oxide. *J. Am. Chem. Soc.* 80: 1339.

[13] Hofmann, U., and König, E. (1937). Untersuchungenüber Graphitoxyd [Investigations on graphite oxide]. Z. Anorg. *Allg. Chem.* 234: 311-336.

[14] Brodie, B. C. (1859). On the atomic weight of graphite. *Philos. Trans. R. Soc. London*149: 249-259.

[15] Staudenmaier, L. (1898). Verfahrenzur Darstellung der Graphitsäure [Process for preparing graphitic acid] *Ber. Dtsch. Chem. Ges*. 31: 1481-1487.

[16] Yuan, W., Gu, Y. and Li, Li. (2012). Green synthesis of graphene/Ag nanocomposite. *Appl. Surf. Sci.* 261: 753-758.

[17] Chen, Ji., Yao B., Li, C. and Shi, G. (2013). An improved Hummers method for eco-friendly synthesis of graphene oxide. *Carbon* 64: 225-229.

[18] Dreyer, D. R., Jia, Hong-Peng. and Bielawski, C. W. (2010). Graphene Oxide: A Convenient Carbocatalyst for Facilitating Oxidation and Hydration Reactions. *Angew. Chem.* 122: 6965-6968.

[19] Gupta, A., Kour, D., Gupta, V. K. and Kapoor, K. K. (2016). Graphene oxide mediated solvent-free three component reaction for the synthesis of 1-amidoalkyl-

2-naphthols and 1, 2-dihydro-1-arylnaphth[1,2-e][1,3]oxazin-3-ones. *Tetrahedron Letters* 57: 4869-4872.

[20] Ren, Y., Yuan, Z., Lv, K., Sun, J., Zhang Z. and Chi, Q. (2018). Selective and metal-free oxidation of biomass-derived 5-hydroxymethylfurfural to 2,5-diformylfuran over nitrogen-doped carbon materials. *Green Chem.* 20: 4946-4956.

[21] Nguyen, C. V., Liao, Y. T., Kang, T. C., Chen, J. E., Yoshikawa, T., Nakasaka, Y., Masuda T. and Wu, K. C. W. (2016). A metal-free, high nitrogen-doped nanoporous graphitic carbon catalyst for an effective aerobic HMF-to-FDCA conversion. *Green Chem.* 18, 5957-5961.

[22] Verma, S., Mungse, H. P., Kumar, N., Choudhary, S., Jain, S. L., Sain B., and Khatri Om P. (2011). Graphene oxide: an efficient and reusable carbocatalyst for aza-Michael addition of amines to activated alkenes. *Chem. Commun.* 47: 12673-12675.

[23] Nie, L., Zhang, J., Wu, Q., Fei, G., Hu, K., Fang, L. and Yang S. (2020). Fabrication of micropatterned gold nanoparticles on graphene oxide nanosheet via thiol-Michael addition click chemistry. *Mat. Lett.* 261, 127014.

[24] Kim, Y., Some, S. and Lee, H. (2013). Graphene oxide as a recyclable phase transfer catalyst. *Chem. Commun.* 49: 5702-5704.

[25] Xue, B., Zhu, J., Liu, Na. and Li Y. (2015). Facile functionalization of graphene oxide with ethylenediamine as a solid base catalyst for Knoevenagel condensation reaction. *Catalysis Communications* 64: 105-109.

[26] Zand, H. R. E., Ghafuri, H., Rashidizadeh, A. and Khoushab, Z.(2019). Study on oxidative condensation of toluene and hydrazine/aniline catalyzed by Copper complex immobilized on functionalized grapheneoxide. *Ind. Eng. Chem. Res.* 58:5379-5387.

[27] Lolak, N., Kuyuldar, E., Burhan, H., Goksu, H., Akocak, S. and Sen, F. (2019). Composites of Palladium-Nickel Alloy Nanoparticles and Graphene Oxide for the Knoevenagel Condensation of Aldehydes with Malononitrile. *ACS Omega* 4: 6848-6853.

[28] Abdolmohammadi, S. andAfsharpour, M., (2020). An ultrasound assisted cyclocondensation reaction for the efficient synthesis of [1]benzopyranopyrido[d] pyrimidines using porous graphene/ MoO_3. *Appl Organomet Chem.* 35: 1.

[29] Xue, B., Wu, J., Liu, Na., Zhu, X. and Li, Y. (2017). Facile immobilization of basic ionic liquids onto graphene oxide in water at room temperature as heterogeneous catalysts for transesterification. *J. Mol. Catal.* 428: 1-8.

Chapter 10

Graphene Based 2-D Nanocomposites for Heavy Metal Ions Detection

Upasana Choudhari[1], Shweta Jagtap[1,*], Niranjan Ramgir[2,3,†], A. K. Debnath[2,3], K. P. Muthe[2] and D. K. Aswal[2,3]

[1]Department of Electronics and Instrumentation Science, Savitribai Phule Pune University, Pune, India

[2]Technical Physics Division, Bhabha Atomic Research Centre, Mumbai, India

[3]Homi Bhabha National Institute, Anushaktinagar, Mumbai, India

Abstract

Water is the most important and essential component for living beings on Earth. In the last decades, anthropogenic water has gradually increased owing to the drastically increased global population. Anthropogenic activities mainly mining, smelting operations, agricultural and industrial production and manufacturing industry are the major source of heavy metal ions into water. As a result, millions of people still lack improved drinking water sources. Also, accumulation of heavy metals especially Hg (II), Pb (II), Cr (VI), Cd (II) and As (III)/(V) has adverse effects on human health like neurological and mental disorders. Numerous conventional analytical methods have been employed for detection of these heavy metals. They however, suffer from drawbacks like tedious or lengthy sample preparation, complex operational procedures, long detection times and importantly, are expensive. Alternatively, electrochemical sensors are looked upon as a promising method for heavy metal ion detection. This has been attributed mainly to advantages

* Corresponding Author's Email: shweta.jagtap@gmail.com.

† Corresponding Author's Email: niranjanpr@yahoo.com.

In: Materials Science: Future Aspects
Editors: Kalpana Awasthi, Arti Srivastava and Mridula Tripathi
ISBN: 978-1-68507-843-0

like ease of operation, high sensitivity, rapid detection, portability, suitability for field monitoring and overall cost effectiveness. Recently, research has tempted great attention on graphene-based sensing materials as an efficient and low-cost adsorbent for detection of heavy metals. They show outstanding physical and chemical features like large surface area, presence of functional groups, high intrinsic mobility, good electrical conductivity and ease of surface modification. The present chapter focuses on the recent development of electrochemical sensors using graphene-based materials for heavy metal ions detection. Recent advancement is discussed elaborately taking help of literature and some of our recent findings. The challenges that need to be overcome for possible commercial deployments are also discussed.

Keywords: electrochemical sensors, heavy metals, graphene, sensitivity, selectivity

1. Introduction

Heavy metal ions (HMIs) are mainly responsible for environmental pollution. They get accumulated in the biosphere and through the alimentary chain enter into living organisms thereby deteriorating human as well as aqueous animal's health [1]. Some of these metals like lead, cadmium, arsenic, mercury and chromium are considered as "Environmental health hazards" compounds even at lower concentrations and ranked in the top 10 according to "Agency for Toxic Substances and Disease Registry Priority List of Hazardous Substances" [2]. The accumulation of these HMIs causes strange diseases like Minamata disease and sometimes could be fatal. In order to assess toxicity of these HMI, several international agencies like Centre for Disease Control (CDC) [3], World Health Organization (WHO) [4] and International Agency for Research on Cancer (IARC) [5] are established.

Detection of HMIs is possible using various conventional methods such as atomic fluorescence spectrometry (AFS) [6], atomic absorption spectroscopy (AAS) [7], X-ray fluorescence spectrum (XRF) [8], flameless atomic absorption spectrophotometry (FAAS), ion-coupled plasma-mass spectroscopy (ICP-MS) and inductively coupled plasma (ICP) [9] technique. Although, these traditional methods have high sensitivity and selectivity, but being bulky are not suitable for rapid and in-situ detection. Also, they suffer from drawbacks like requirement of biomolecule reagents, complicated sample pretreatment procedures, skilled operators, and expensive and time consuming. For evaluation of heavy metal ion, these techniques are often coupled with other chromatographic techniques and hence, pose the risk of sample change during storage and handling. Optical techniques like spectrophotometric measurements involve complex and expensive equipment's, requires high power for operation and hence are not considered to be suitable for in-field applications. Electrochemical techniques (EC) offer numerous and notable advantages viz. economic reliability, rapid, simple procedures, user friendly and suitable for fabrication of compact portable devices and in-field applications. Besides, these techniques can be used for on-line monitoring of real samples. As compared to other techniques, these techniques are highly sensitive and have lower limit of detection (LoD).

Recently, 2D carbon-based materials like fullerenes, carbon nanotubes and graphene nanoparticles have exhibited improved sensing response towards HMIs detection and accordingly, have been investigated in detail [10]. Among these, graphene exhibits many outstanding and unique properties including, excellent thermal conductivities (5000 $W\ m^{-1}\ K^{-1}$), high specific surface area, low electronic noise, high-speed electron mobility of 200000 $cm^2\ V^{-1}\ s^{-1}$, great electrical conductivity (3,189 $S\ cm^{-1}$) and electrocatalytic activity [11-14]. The International Union of Pure and Applied Chemistry (IUPAC) defines graphene as '2D carbon atoms having a single layer of graphite and hexagonal pattern structure, analogous to a polycyclic aromatic hydrocarbon of quasi-infinite size' [15]. Novoselov and co-workers in 2004, produced graphene through mechanical exfoliation [16] and were awarded Nobel Prize in Physics for the same in 2010. Graphene family consist of three derivatives namely pristine graphene, graphene oxide (GO), and reduced graphene oxide (rGO).

GO consists of a single-layer of graphite oxide and usually produced through oxidation of graphite by the chemical treatment, with subsequent dispersion and exfoliation in suitable organic solvents or water. GO sheets can also be produced by a modified Hummers' method [17]. The oxidized graphene oxide has various oxygen-containing functional groups in the form of epoxy and hydroxyl groups on the basal plane, with less quantity of carbonyl, carboxyl, lactone and phenol at the sheet edges. Such presence of wide range of oxygen functional groups on GO strongly affect its mechanical, electronic and electrochemical properties. This makes GO different than the pristine graphene. It can be deposited onto different solid as well as flexible substrates as a thin conductive film. Reduction of GO to form rGO nanocomposites has been studied extensively so as to enhance mechanical, electrical and thermal properties. rGO is often synthesized using chemical/electrochemical reduction method that involves removal of some of the oxygen-containing functional groups. This enables the tuning of chemical and electrical properties (improved conductivity) of rGO. The control over number of oxygen functional groups and defects further helps to tailor the charge transport and analyte adsorption properties. [18].

Accordingly, they have been used for the detection of various HMIs (Pb(II), Cu(II), Co(II), Cd(II), Cr(VI), Hg (II), etc.). The strong inter-functional bond between graphene sheets may lead to agglomeration and poor dispersion in aqueous solutions. The presence of inactive surface chemical properties further restricts the detection performance. To overcome this drawback GO is often doped with hetero-atom and/or a composite is formed using metal oxide, noble metals and organic compounds including polymers

1.1. Heavy Metal Ions

Heavy metal pollution has become a severe public health and environmental problem. Although their presence is necessary to carry out several cellular functions, but if their concentration exceeds the permissible limit, they pose a great threat to human health causing adverse cytological and physiological effects. The main sources of these heavy metals are automobile exhausts, cosmetic products and different kinds of radiations. Table 1 lists the various sources, toxicity range as per the world health organization (WHO) and adverse health effects of different heavy metals.

2. Heavy Metal Ions Detection Techniques

Monitoring of heavy metals in the environment, food, drinking water and body tissues is often realized using optical and electrochemical sensors. Of these, the optical sensors include colorimetric, fluorescent, surface-enhanced Raman scattering (SERS) and surface plasmon resonance (SPR) sensors. Although, these sensors exhibit good sensitivity, selectivity and LoD but they suffer from numerous imitations. For example, in case of colorimetric and fluorescent sensors, it is a challenge to apply sensor to real-world sample detection [29]. SERS provides a fingerprint signatures of analytes but is unable to directly detect HMIs. Use of microfluidic SERS devices may lead to reduction of number of active sites and accessibility of analytes to the SERS active sites [30].

Table 1. Heavy metal ions sources, effects and WHO limits

Metal	Sources	Effects	WHO limits (µg/L)	Ref.
Arsenic (As)	Oxidation of pyrite (FeS) and arseno pyrite (FeAsS), extensive use of pesticides and fertilizers, wood preservation for electricity poles, industry, light-emitting diodes, lasers and release of untreated effluents	Hypertension, spontaneous abortion, damage of nervous systems, hepatic damage affected on peripheral nervous system (PNS) and central nervous system (CNS), genitourinary (GU), skin cancer, dermatologic, localized edema and teratogenic diseases	10	19, 20
Cadmium (Cd)	Anthropogenic sources, electroplated parts (protective coatings), stabilizers for photovoltaic cells, preparation of Cd–Ni batteries, photographic process, fertilizers, pigments, paints, synthetic rubber and plastics	Hypertension, microcytic hypochromic anaemia, weight loss, renal toxicity, pulmonary fibrosis, fatigue, lymphocytosis, atherosclerosis, lung cancer and Kidney damage, peripheral neuropathy, osteomalacia and hyperuricemia	3	21, 22
Chromium (Cr)	Naturally occurred by burning of oil and coal, Leather and chrome plating industries, chromium steel, pigment oxidants and catalyst, metal plating tanneries and oil well drilling, petroleum from ferro chromate refractory material	Lung cancer, Reproductive toxicity, skin ulcers, embryo toxicity, perforation of septum, carcinogenicity, teratogenicity, mutagenicity, irritant contact dermatitis and dermatitis and DNA damages due to oxidative stress in cell	50	23, 24
Lead (Pb)	PVC pipes in sanitation, plumbing pipes, house paints, lead batteries, agriculture, jewelry, toys, lunch boxes, lead bullets, Industrial processes and gasoline	Nervous system damage learning disabilities, behavior issues, attention deficit disorder, kidney damage, decreased bone growth and muscle, Penetrates through protective Blood Brain Barrier (BBB)causing risk factor for Senile Dementia and Alzheimer's disease	10	25, 26
Mercury (Hg)	Incinerators, combustion of coal, thermometers, fossil fuels, municipal, volcanic emissions, dental amalgams, old latex paint	Effects on eyes, skin, digestive system, immune system and lungs, memory loss, vomiting, headaches, thyroid, impaired neurologic development, emotional liability, minamata, acrodynia, Tremors, increases salivation, hypotonia, insomnia, hypertension and abdominal pain	1	27, 28

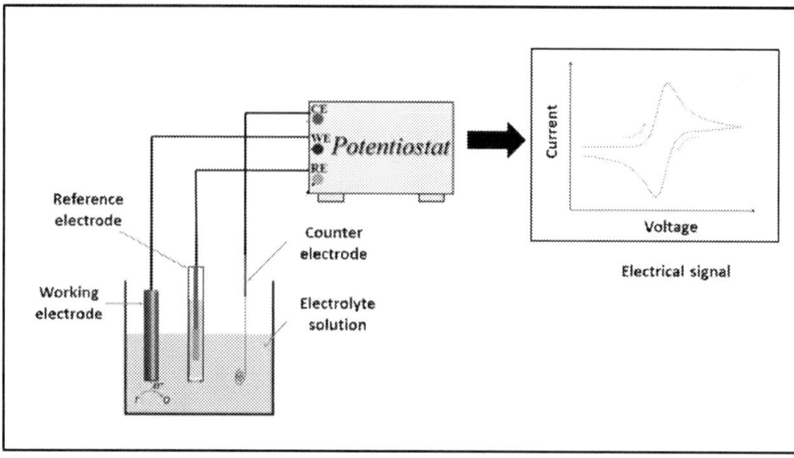

Figure 1. Setup of electrochemical workstation.

Electrochemical sensor has several advantages over optical sensor like it can be easily packed into a compact system and suitable for real time sample detection. In addition, defined redox potential of bare electrodes can achieve selectivity toward specific HMIs without the use of molecular recognition probes [31]. The details of various HMIs detection methods and their examples are shown in Table 2. The simple, compact and portable electrochemical setup renders this technique effective for HMI detection [32]. The technique generates measurable and useful electrical signal that is equivalent to the electrochemical reactions occurring in the solution in presence of analyte. The set-up usually has three electrodes, counter electrode (CE), working electrode (WE) and reference electrode (RE). The WE play a vital role in selective and sensitive determination of HMI and accordingly, is modified with different materials. Initially, carbon and mercury were most frequently employed as interface material for construction of WE. The toxicity and mechanical instability restrict the use of mercury as electrode. To achieve improved sensing properties the use of chemically modified electrodes has been studied. The electrodes are modified by incorporation of various materials like metal oxides, polymers, and carbon nanotubes. Figure 1 shows a general electrochemical setup for detection of HMIs in aqueous solution is shown in. The current generally flows through WE and CE, while the potential is measured between WE and RE. The material of CE is chosen such that it doesn't affect the WE. As a precautionary measure to prevent current being drawn from RE, a high input impedance device is connected to RE [33].

Table 2. Various techniques for HMIs detection

S.N.	Materials	HMIs	Range	LOD	Ref
Atomic adsorption spectroscopy					
1	Activated charcoal as bio sorbent	As(III)	5 – 100 ppb	1 ppt	7
2	Pre-concentration using ion exchange media	Cr(III)	1 - 50 ppm	0.4 ppm	35
3	Pyrolysis and atomization using rhodium permanent modified	Cr(III) Cd(II) Pb(II)	-	0.2 ppb 0.1 ppb 0.5 ppb	36
4	Chromium(III)-imprinted silica gel	Cr(III)	-	0.004ppt	37
Inductive coupled plasma spectroscopy					
5	Silica-coated magnetic graphene oxide (Fe_3O_4-GO@SiO_2) composites	Cr(III) Pb(II) Cd(II)	0.05 - 60 ppb	3.77 ppb 3.64 ppb 3.79 ppb	38
6	Graphene oxide–silica (GO–silica) composite	Cd(II) Pb(II)	0.03 - 50 ppb 0.1 - 50 ppb	6.7 ppt 28 ppt	39
Fourier transform Raman spectroscopy					
7	Glutathione (GSH)/4-mercaptopyridine (4-MPY)-modified silver nanoparticles	As(III)	4 - 300ppb	0.76ppb	40
8	Gold nanoparticles	Hg (II)	0.1–2.0 ppb	100ppt	41
Optical methods					
9	Peroxidase-mimicking DNAzyme using Chemoluminescense	Hg (II)	12 – 600 nM	12 nM	42
10	DNAzyme and molecular bracon (MB)-based Pb^{2+} sensor using colorimetry	Pb(II)	0.05 – 5 nM	20 pM	43
11	Urease and Glucose Oxidase based biosensor using photoluminescence	Cd(II)	10 nM - 1 mM	10nM	44
12	Dimethylglyoxime, bathocuproine, 5-diphenylcarbazide	Cr(VI)	0.5 – 10ppm	18ppb	45
X-ray fluorescence spectroscopy					
13	Alumina particles	Pb(II)	10 - 100nM	10nM	46
14	MoS_2-rGO nanocomposite	Cr(III) Pb(II)	1 – 200 ppb	110ppb 90ppb	57
15	Tricaprylmethylammonium chloride ($CH_3N[(CH_2)_7CH_3]_3Cl$; Aliquat 336)	Cr(VI)	1 - 250 ppm	-	48
Screen printed electrodes					
16	L-cysteine conjugated to lipoic acid capped gold nanoparticles	As(III)	3 - 25ppb	3ppb	49
17	GR-IL	Pb(II) Cd(II)	110 - 500 ppb	3.30ppb 4.43 ppb	50
18	Graphene/ polyaniline/ polystyrene nanoporous fibers	Pb(II) Cd(II)	6.9 - 100.9 ppb 3.7 - 100.3 ppb	2.1 ppb 1.1 ppb	51
19	SPCE/ PtNPs	As(III)	0.16 to 1.3 mM	0.8 ppb	52
20	Ag@AuNPs	Hg(II)	0.01 - 0.16 nM	6.0 pM	53
21	Cr_2O_3NPs	Cd(II) Pb(II)	80 – 800 ppb 10 - 800ppb	25.0 ppb 3.0 ppb	54

Table 2. (Continued)

S.N.	Materials	HMIs	Range	LOD	Ref
Microelectrode and Nanoelectrode Arrays					
22	Au-microelectrode array	As(III)	0.125 - 5 ppb	42.3 ppt	55
23	Au- gel integrated microelectrode	As(III)	2 – 10 nM	1nM	57
24	Bi-band silver microelectrode (b- BAgmE)	Pb(II)	-	0.4 nM	58
25	Au microelectrode/MB-tagged DNA	Hg(II)	0.2 – 100 nM	0.1nM	59
26	Bismuth film microelectrodes	Pb(II)	0.2 nM – 12 μM	0.22 nM	60
Microfluidic based sensor					
27	Carbon	Pb(II) Cd(II)	10 – 100 ppb	2 ppb 2.3 ppb	61
28	ZnSe quantum dots	Cd(II) Pb(II)	1-70 ppb 1-60 ppb	0.245 ppb 0.335 ppb	62
29	Graphene/PANI nanocomposite	Cd(II) Pb(II)	1 - 300 ppb	0.1 ppb	63
30	Silver nanoparticles	As (VI)	-	1 ppb	64

Electrochemical techniques are differentiated according to the generation of different electrical signals (voltage, current, charge, electrochemical impedance, and electroluminescence) and their type in the electrolytic solution used for determining the analyte concentration. The techniques are classified into voltametric, amperometry, impedance measurement, potentiometric, electro chemiluminescent and coulometric techniques. Most techniques do not allow simultaneous control of current and potential, instead either a current or a potential is controlled to measure the variation in the other parameter. Potentiometric techniques are used to measure potential for determining the type of HMI. The impedance measurement technique is used on the basis of solution resistance, change in double layer capacitance and charge transfer resistance. In electrochemiluminescence technique the luminescence effect is generated in presence of HMI concentration. [34].

3. Go-Based Nanomaterials for Detection of HMIs in Water

The surface of nanoparticles plays a crucial role in the interaction with other molecules. Use of metal nanoparticles (NPs) is one of the most promising approaches to improve the selectivity of composite materials towards heavy metals. The improved sensing response is assigned to the high conductivity and catalytic activity via the size effect. The metal oxide is used as substitute to metal NPs because of their high electrocatalytic activities and large surface areas [65]. Conducting polymers are also used for HMIs detection due to various advantageous such as high mechanical strength, good film forming properties, flexibility, and transparency. Among these nanomaterials, carbon-based nanomaterials such as graphene and its derivatives have the advantages of wide electrochemical window, high electrical conductivity and large specific surface area. This material is often utilized to fabricate electrochemical sensors for HMIs detection. Additionally, in graphene-based materials the large surface coverage can easily be achieved via either direct absorption or binding with other functional groups on graphene derivatives [66]. Accordingly, tuning of the surface area of graphene materials is considered to be the most favorable approach to adjust graphene materials for different purposes. Table 3 summarizes and tabulates the graphene composites used for HMI detection using electrochemical methods.

3.1. Heteroatom Doped GO

Heteroatom (N, S, F, etc.) doped graphene is a fascinating approach to achieve improved optical, electromagnetic, structural and physicochemical properties. For example, Xing et al. modified electrode with N-doped graphene (NG) using differential pulse stripping voltammetry (DPSV) and demonstrated its use for the simultaneous determination of Cd^{2+}, Pb^{2+}, Hg^{2+} and Cu^{2+}. Herein, N-doped graphene was synthesized using a one-step electrochemical strategy The incorporation of pyridine and pyrrole into graphene results in the formation of a nitrogen group that enhances the electrochemical performance towards presence of Pb^{2+}, Hg^{2+}, Cd^{2+}, and Cu^{2+}compared with rGO. The detection limit of 0.05 μM for Cd^{2+} and Hg^{2+} and 0.005 μM for Pb^{2+} and Cu^{2+}, was achieved. Also, the NG modified electrode has exhibited high selectivity and a long-term usage possibility [67]. The presence of N atoms in graphene increases the number of catalytically active sites thereby enhancing the electron transfer ability of the modified electrode. Similarly, Liu et al. reported a nano-carbon paste electrode modified with N-doped graphene for tracing of Pb and Cd using square wave anodic stripping voltammetry [68]. Cheng et al. assembled nitrogen/sulfur doping on the rGO (SN-rGO) which act as "hooks" for electrochemical determination of Hg (II) using square wave anodic stripping voltammetry (SWASV). The fabricated sensor has excellent sensitivity (20.48 μA/μM), minimal interference for other ions and LOD (8.93 nM) for Hg (II) detection. The presence of multifunctional groups on the graphene nanosheets can also be used for routine water quality monitoring [69].

3.2. GO/Metal Nanoparticles Composite

Graphene modified with noble metal nanoparticles (NPs) is one of the approaches to improve the selectivity of composite materials towards heavy metals. The high conductivity and catalytic activity as a result of the size effect has been attributed for the improved sensing response. Besides, graphene provides a conduction path for electrons obtained from catalytic process of metal NPs to be transferred to electrodes. The different metal modifiers used for detection of HMI are Au, Pt, Ag, Pd, Sn, Bi, etc.

Au NPs is commonly used owing to its high chemical stability and ease of preparation. Au NPs have been predominantly used for mercury and arsenic detection. For example, Liu et al. reported the determination of As^{3+} using Au NPs-electro-reduced graphene oxide (ERGO) composite film. The high sensitivity with LoD of 2.7 nM, obtained was attributed to the formation of Au–As intermetallic compounds, that further improves the efficiency of cathodic pre-concentration of As (0) [70]. Wang et al. electrochemically deposited AuNPs onto the surface of rGO. The sensor thus realized was found highly sensitive towards Hg^{2+} in the range of 10 ng/L–1.0 μg/L and also applicable for real sample analysis [71]. In some cases, Ag NPs decorated on graphene has also been used. For example, Sang et al. reported Ag NPs/rGO via an in-situ method for simultaneous HMIs detection [72].

Another approach of employing metal-organic frameworks (MOFs). has shown greater advantages, such as well-defined crystalline nature, structural diversity, tunable pore size, high surface area, and easy fabrication process. Particularly, self-assembled porous structure of MOFs with their organic ligands can help to improve sensitivity for electrochemical sensing of HMIs [73]. Zou et al. fabricated Bi-MOFs for Pb^{2+} detection. The multilayer composite structure increased the surface area of reactive contact and active sites that eventually exhibited wide linear range from 0.062 to 20.72 μg/L. A LoD of 0.021 μg/L for lead ion sensing and identification of presence of Pb^{2+} in industrial wastewater with recoveries of 99.2–104% was achieved [74]. Antimony (Sb) film electrodes are considered to be a beneficial and complementary alternative to Bi film electrodes for measurements in acidic or oxidative media. Bismuth hydroxide forms at pH 4, while Sb films can be synthesized in a wider pH range and could contribute towards irreproducible measurements. For example, Martin et. al. [75] used rGO-SbNPs nanocomposite films and stripping analysis for successful determination of Cd^{2+}, Pb^{2+}, Cu^{2+} and Hg^{2+} in natural water samples. Under optimal conditions, the electrode showed a good linear relationship with the concentration of Cd^{2+},

Pb^{2+}, Cu^{2+} and Hg^{2+} ranging from 0.1 to 4.5 μmol L^{-1} with LoD of 45.3, 16.5, 34.1 and 4.8 nmol L^{-1}, respectively for simultaneous determination. Ruengpirasiri et al. [82] has also shown simultaneous determination of Cd^{2+}, Pb^{2+}, Cu^{2+}, and Hg^{2} using GO-Sb film-modified electrode. However, the toxicity of Sb metal ions, particularly Sb^{3+}, cannot be completely ignored and thus, cheaper and less toxic tin is also used for HMIs detection [76].

3.3. GO/Metal Oxide Nanocomposites

Use of metal oxide as a substitute to metal NPs have also been frequently employed in HMI detection and is attributed to high electrocatalytic activities and improved surface areas. In case of metal oxide, the detection is done via strong electrocatalytic activity or adsorption ability or both simultaneously. However, most metal oxides have minor stabilities and conductivity, which decreases the long-term stability of the sensor and electron transfer during the detection process. The composite of metal oxide with graphene provides a new electrochemical platform. Up to now, MnO_2, ZnO, Cu_2O, TiO_2, SnO_2 and Co_3O_4-based graphene nanocomposites have been successfully used to detect HMI in water. The morphology and average size of the metal oxide is known to have a strong influence on the performance of the sensor. Sun et al. used a one-step in-situ co-precipitation method to create rGO decorated with three different shapes of Fe_3O_4. For the three morphologies, the sensitivity was found to varying as band > spherical > rod [77]. Wei et al. reported SnO_2/rGO nanocomposite modified GCE to detect ultra-trace amounts of Cd^{2+}, Pb^{2+}, Cu^{2+} and Hg^{2+} ions in drinking water [78]. The simultaneous analysis of Cd^{2+}, Pb^{2+}, Cu^{2+} and Hg^{2+} was done by SWASV. Individual peaks for Cd^{2+}, Cu^{2+}, Pb^{2+} and Hg^{2+}ions are located at approximately -0.8, -0.1, -0.6 and 0.24 V, respectively. For all these metals corresponding linear calibration curves were valid from 0.3 to 1.2 μM. The LoD of 0.1015, 0.2269, 0.1839 and 0.2789 nM, respectively for Cd^{2+}, Cu^{2+}, Pb^{2+} and Hg^{2+} electrodes were obtained. The LoD, were significantly lower than the WHO recommended values. Vajedi et al. studied various mass ratios of TiO_2/rGO nanocomposites modified GCE for the detection of Cd (II), Pb (II) and Cu (II). The TiO_2/rGO synthesized using the ratio of 1:1) (TG3) has shown excellent electrochemical activity and high sensitivity for the simultaneous determination of Pb^{2+} and Cu^{2+}, attributed due to well-dispersion of TiO_2 nanoparticles in rGO matrix, high surface-specific area and high electron transfer ability of the nanocomposites [79].

Bimetallic oxides have higher electrochemical activity than monometallic oxides due to metal electron hopping between different valence states in oxygen sites. Huang et al. have shown that $NiCo_2O_4$ exhibited better electrochemical performance for detecting Pb^{2+} and Cu^{2+} than Co_3O_4 and NiO [80]. Zhou et. al. developed GO with mesoporous $MnFe_2O_4$ that increased the specific surface area thereby improving the electrochemical activity toward Pb^{2+} detection [81].

3.4. GO/Organic Materials Composite

The addition of organic molecules to graphene improves its sensing performance. GO and organic compounds both contain functional groups, and consequently, can interact with HMIs via ion exchange, chelation and surface complexation. GO/organic composite also has a large specific surface area, that increases the adsorption capacity of composites by providing more adsorption sites. The surface grafted organic compound has large oxygen-containing functional groups and more active sites, which forms the surface complexes with metal ions. Another advantage of such composites is that the organic compounds can prevent GO aggregation in aqueous solutions thereby improving HMI sensing performance. Accordingly, various organic molecules containing small molecules (electron-rich groups like –SH, – NH_2 and –OH) and caged molecules have been explored [5]. For example, Muralikrishna et al. created L-cysteine functionalized GO by reacting the carboxyl groups in graphene with the amino group in L-cysteine and used it for the simultaneous detection of Cd^{2+}, Pb^{2+}, Cu^{2+}, and Hg^{2+}. Herein, HMI adsorption was aided by the oxygen-containing groups of GO and the electron donor group in L-cysteine [82]. Priya et al. described penetrable nature of rGO (PrGO)/L-cysteine by decorated Au NPs and shown its efficacy for simultaneous determination of Cd^{2+} and Pb^{2+} [83]. The modified electrodes exhibited sensitivity of 2.2 μA/nM for Cd^{2+} and 3.2 μA/nM for Pb^{2+}along with LoD of 0.06 (Cd^{2+}) and 0.04 nM (Pb^{2+}), respectively. The PrGO/AuNPs/Sal-Cys/GCE showed excellent performance than L-cys/GR–CS/GCE and L-cys/rGO [84, 85] for toxic metal analysis in water samples. Zhiwei et al. fabricated flexible electrochemical sensors by converting commercial polyimide (PI) films into a laser-engraved graphene electrode (LEGE) via a direct laser engraving process. The activation procedure was essential for significantly improving electrochemical performance.

Using in situ electrochemical polymerization of l-cysteine the micro-patterned LEGE with porous 3D morphology can form poly-(l-cysteine) (PLC)-modified LEGE (PLC/LEGE). The synergetic effect of newly designed IL/PLC/LESE with a large active surface area, abundant active sites, and fast electron-transfer ability is provided by the combination of the enlarged specific surface area of LEGE, good conductivity of IL, and specific complexion ability of PLC. The ultralow detection limit of 0.17 g/L (S/N = 3) for Pb^{2+} was obtained with a wide linear response of 1–180 μg L^{-1}. The developed sensor also exhibited outstanding stability, dependability, reproducibility and repeatability. [86].

3.5. GO/Polymer Composite

The most well-known polymers used for the synthesis of membranes include: polysulfone, polyethersulfone, polyphenylene oxide, polyphenylene sulfate, polyamide, polycarbonate and polyethylene terephthalat. These polymers have excellent chemical and thermal stability, high mechanical strength, flexibility, transparency, excellent physical (thermal resistance) and chemical properties (soluble in a wide range of aprotic polar solvents, high versatility in electrophilic aromatic substitution reactions for functionalization of the aromatic ring) as well as a high glass transition temperature and good film forming properties.

Zou et al. reported an electrochemical detection of Hg^{2+} at trace level based on poly(3,4-ethylenedioxythiophene) nanorods/GO nanocomposite modified GCE (PEDOT/GO/GCE) using simple liquid – liquid interfacial polymerization [87]. The DPSV was used for the detection and the sensor exhibited a linear range (10 nM - 30 μM) of Hg^{2+} with LoD of 2.78 nM. A good response and selectivity were observed for real water samples. Similarly, Yasri et al. reported a composite containing poly(3,4-ethylenedioxythiophene)/poly(styrenesulfonate) (PEDOT: PSS) on a partially oxidized graphene flake (Po-Gr). The Po-Gr was obtained by electrochemical exfoliation of graphite sheet and composite was formed by dispersing Po-Gr with mixture of PEDOT: PSS. The fabricated films exhibited 0.19 μM of lower detection limit along with good selectivity and sensitivity for Hg^{2+} in presence of real sample [88]. Muralikrishna et al. reported another most promising polymer composite namely a novel hydrogel of 1D conducting polymer namely polyaniline (PANI) nanofibers with GO via polymerization of aniline onto the surface of 2D graphene oxide (GO). The electrode exhibited better selectivity towards lead along with good reproducibility and repeatability. Additionally, a wide range detection of 0.2– 300 nM with LoD of 0.04 nM was achieved. Here PANI has dual role, it provides the necessary electrical conductivity and also nitrogen functional group such as amine (-NH-) and imine (=N-) that adsorb Pb^{2+}ions onto its surface [89]. On the other side, Dai et al. developed simple, rapid and high performing electrochemical sensor by functionalizing polypyrrole (PPy)/GO electrode with phytic acid. In situ chemical oxidation polymerization was used to synthesize PPy/GO nanocomposites and the modified electrode was functionalized with phytic acid (PA) molecules by electrostatic attraction. The PA/PPy/GO modified electrode exhibited linear working range of 5–150 μg/L for Cd(II) and Pb(II) detection. Moreover, the sensor also showed good selectivity, repeatability and reproducibility compared with the PA/GO and PPy/GO modified electrodes. These improved sensing responses were observed due to (i) high mobility and surface area of GO, (ii) excellent conductivity of PPy and (iii) presence of large number of functional groups on PA and GO, which improves the adsorption capacity of HMIs [90].

3.6. 3D Graphene Based Materials

The use of low-dimensional nanomaterials like 0-D nanoparticles, 1-D nanowires / nanotubes and 2D nanosheets has been limited due to:

i. agglomeration in aqueous medium that reduces the surface area of material;
ii. limited productivity of low-dimensional nanomaterials, current research progress mainly remains at laboratory scale;
iii. it is difficult to completely recover exhausted nanomaterials and may be harmful to human beings and ecosystem.

To overcome these drawbacks researchers have attempted and investigated effect of transformation of two-dimension (2D) graphene-materials to three-dimension (3D). It was found that the 3D graphene-based adsorbents (3D GBAs) retain intrinsic graphene properties even while exhibiting new physiochemical properties such as low density, high porosity, strong mechanical strength, large specific surface area, and unique electro-chemical performance [91]. Recently, a set of 3D graphene-based adsorbents (3D GBAs) with various frameworks (e.g., aerogels, foams, sponges and hydrogels) have been created. 3D GBAs get internal connection framework and multi-level pores structures (micropore, macropore and mesopore), endowing them with better mass transport and improved surface accessibility. For example; Pan et al. reported 3D porous graphene aerogels (GAs) by ozone treatment, solvothermal assembling and high temperature annealing of GO which acts as a conductive adsorbent model to investigate its electrochemical regeneration mechanism [92]. The other approach was double hydrothermal reduction approach that is used to synthesize graphene hydrogel (GH). 3D graphene foams (GFs) have been grown on a nickel foam template by microwave plasma CVD [93] or plasma-enhanced CVD. Seashell-derived 3D GFs as shown in Figure 2, have demonstrated high porosity, ultralow density, high purity, and exceptional mechanical flexibility [94]. Another most effective approach is 3D printing that can easily design and build 3D graphene with tailored macrostructures in a wide range of sizes [95].

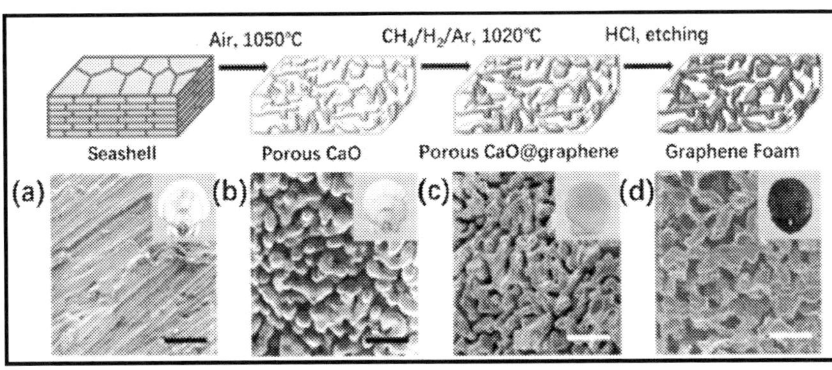

Figure 2. Fabrication of 3D GFs derived from scallops; D: pristine scallop micrographs (a), scallop calcined in air for 30 minutes at 1050°C (b); graphene-coated calcined scallop after 30 minutes of CVD growth at 1020 °C (c), and 3D GF after removal of the CaO framework (d) [94]. (Reprinted with permission from L. Shi et. al., J. Am. Chem. Soc. 138 (2016) 6360–6363.)

Thus, the skeletons of pure 3D graphene have been functionalized with various nanocomposites to enhance sensing properties of 3D GBAs towards different pollutants. For instance, L. Shi et al. [96] fabricated a novel 3D graphene framework/Bi nanoparticle (GF/BiNP) film using a facile preparation route. Porous structured 3D graphene framework (3DGF) was electrochemically reduced and in situ assembled on the electrode. Further, reduced 3DGF were tuned with BiNPs using chemical reduction with highly dispersion of NPs on the framework. This newly-designed 3D GF/BiNP films exhibited fast electron transfer ability, large active area, high mass transfer efficiency, excellent binding strength and structure stability on electrode.

Table 3. Different graphene-based nanocomposite for HMI detection using electrochemical studies

S.N.	Materials	HMIs	Range	LoD	Ref.
Metal nanoparticles doped graphene oxide					
1	AuNPs/graphene/GCE	Hg(II)	0.01 to 5 μM	-	71
2	Au NPs/rGO/GCE	As(III)	4-26.69 pM	2.7nM	70
3	SnNPs/rGO/glassy carbon sheets	Cd(II) Pb(II)	10-100 nM	0.63 nM 0.6 nM	98
4	Sb film/GO/SPE	Cd(II) Hg(II) Pb(II)	0.3-1.5 μM 0.1-1.3 μM 0.1-1.3 μM	0.054 μM 0.066 μM 0.026 μM	76
5	Bi film/graphene nanosheets/GCE	Cd(II) Pb(II)	0.00445-0.89 μM 0.48-480 nM	3.11nM 0.22 nM	99
Heteroatom doped graphene					
7	Nitrogen-doped laser engraved graphene (N@LEG)	Cd (II) Pb(II)	5 to 10 and 10 to 380 ppb	1.08 g/ L 0.16 g/ L	100
8	ETBD functionalized nitrogen-doped graphene/AuNPs and Fe_3O_4@TiO_2 core–shell nanoparticles	Pb(II)	75 pM	40 pM - 0.2 nM	101
9	Porous three-dimensional sulfur-doped reduced graphene oxide (*p*S-rGO)	Hg(II)	0.1 – 100 ppb	0.1 ppb	102
10	Nitrogen/sulfur doping on the rGO	Hg(II)	-	8.93nM	69
Metal oxide /GO nanocomposite					
15	Fe_3O_4/rGO/GCE	As(III)	0.00013-67.4 nM	0.0016 nM	103
16	SnO_2/rGO/GCE	Cd(II) Pb(II)	0.3-1.2 μM	0.1015 nM 0.1839 nM	78
17	MnO_2/rGO/GCE	As(III)	1.33-667.37 nM	0.67 nM	104
18	PbO/rGO/GCE	As(III)	-	10 nM	105
19	$MnFe_2O_4$/GO/GCE	Pb(II)	0.2-1.1 μM	0.0883 μM	81
20	Co-dope ZnO/rGO/GCE	Cd(II) Pb(II)	0.089-0.8 μM 0.048-0.43 μM	8.36nM 4nM	106
Organically modified graphene					
21	Cysteamine-GO/Au electrode	Hg(II)	5-40 nM	3 nM	107
22	p-aminophenyl-GO/GCE	Cd(II)	0.01-0.5 nM	3.3 pM	108
23	Trithiocyanuric acid/rGO/Au electrode	As(III)	2.67-133.5 nM	0.72 nM	109
24	N-[(1-pyrenyl-sulfonamido)-heptyl]-gluconamide/rGO/Au electrode	Hg(II)	0.1-4 nM	0.1 nM	110
25	poly-(l-cysteine) (PLC)-modified LEGE (PLC/LEGE)	Pb(II)	1 – 180 ppb	0.17ppb	86

S.N.	Materials	HMIs	Range	LoD	Ref.
26	penetrable nature of rGO (PrGO)/L-cysteine by decorated Au NPs	Pb(II) Cd(II)	1-10 nM	0.06 nM 0.04 nM	83
Polymer modified graphene					
27	Partially oxidized graphene flakes (po-Gr)/ PEDOT:PSS	Hg (II)	0.2–14 μM	0.19 μM	88
28	1D PANI/GO	Pb (II)	0.2– 300nM	0.04nM	89
29	PPy-graphene/ß-CDs/SPCE	Hg(II)	1-300 nM	0.47 nM	111
3D graphene based composite					
30	3D graphene-framework/Bi	Cd (II)	40 - 300 μg/L	0.02μg/L	96
		Pb(II)		0.05 μg/L	
31	Pyrrole doped graphene aerogels	Cd (II)	1 - 100μg/L	3.6 μg/L	97
32	Ag nanoparticles-modified 3D GF	Hg(II)	0.11 - 2 μM	0.11 μM	112
33	3D honeycomb-like N-doped carbon nanosheet framework embellished with bismuth nanoparticles (3D Bi-NCNF)	Cd (II) Pb(II)	1–120 μg/L	0.02μg/L 0.03μg/L	113

The superior sensing ability of materials were demonstrated by conducting simultaneous ultralow detection of Pb^{2+}(0.02 μg L^{-1}) and Cd^{2+}(0.05 μg L^{-1}) with linear range from 1 to 120 μg L^{-1}. Meanwhile, a separate optimum response of Zn^{2+} was performed with LoD of 4.0 μg $L^{-1}$and linear range of 40 - 300 μg L^{-1}. Moreover, excellent reliability, reproducibility, stability and repeatability were observed in real water samples. Due to its convenient preparation method, favorable structures and excellent sensing properties, the fabricated sensor holds promise for its application in environment, biomedicine and energy systems.

X. Guo et al. [97], reported effect of 3-D graphene aerogels (3D-GAL) on Cd (II). The properties and structure of prepared 3D-GAL were regulated by incorporation of pyrrole into 3D-GAL and pyrrole was also used as electrode materials for sensing metal ions. Using different characterization techniques, the role of doped pyrrole was studied. It indicated that the pyrrole provides N source as active binding sites for adsorption of metal ions as well as serves as a regulator and reductant to adjust the structure and composition of graphene aerogels. This greatly influences the electrochemical performance of Cd (II) sensing. Additionally, the reaction ratio of pyrrole and GO plays a key role towards optimizing the effect of pyrrole. The square wave anodic stripping voltammetry (SWASV) technique was used to fabricate 3D GO-Py10 electrode and successfully applied to determine Cd (II) in aqueous solutions.

4. Various Advanced Approaches for Detection of HMI from Aqueous Medium

4.1. Screen Printed Electrodes

Recently, one of the methods that has gained widespread attention is disposable screen-printed electrodes (SPEs). SPE incorporates three electrodes on an insulated substrate (such as polyvinyl chloride, ceramic and polyethylene terephthalate), making it portable, inexpensive and user friendly. The thickness of SPE can be controlled between few and 100 μm by adjusting the templates and the amount of printed ink. Concurrently, a small sample volume is required in the case of a small electrode size that helps to improve the mass transmission rate and signal-to-noise ratio. In addition, SPE, avoid convoluted pretreatment steps like electrode polishing enabling rapid detection. SPEs are typically made from commercial conductive inks such as silver, carbon, and gold inks. However, the presence of several insulating substances in the traditional inks results in poor electrochemical performance and conductivity, as well as limited sensitivity. As a result, the incorporation of nanomaterials with many distinguishing characteristics like high conductivity, large surface area and electrochemical window, and high stability, received a great attention [114].

SPE based sensors have two distinct methods for incorporating other materials. The first option is to make SPEs by directly incorporating materials such as reduced graphene (GR) into printing ink. For example, Teng et al. developed a disposable screen-printed graphene electrode (SPGE) that is directly printed with GR paste and can be used for rapid cadmium analysis. Shuai et al. simultaneously determined Cd^{2+} and Pb^{2} using SPGE [50]. In their studies, they thermally reduced GO to GR. Further, the viscosity and conductivity of GR was improved by adding IL (n-butylpyridinium hexafluorophosphate) into the GR ink. The developed sensor showed a wide linear range (1.0 to 300.0 mg L^{-1}) with LoD of 0.1 mg $L^{-1}$for both $Pb^{2+}$and Cd^{2+}. Additionally, the sensor is applicable for real sample analysis (river water). The second method is to modify GR on the surface of SPEs. For example, Wu et al. reported simultaneously detection of Pb^{2+} and Cd^{2+} using GR sheets-enabled electrochemical sensor [115]. The GR sheets were obtained by stirring the well dispersed GO with hydrazine solution for 24 h at 80°C. Then drop-casted on to the surface of screen-printed carbon electrode (SPCE/GR). The solution was prepared by dispersing the GR sheets in a

Nafion solution. Further SPCE/GR/AuNPs was formed by electrodeposition of AuNPs onto the surface of SPCE/GR. The developed sensor indicated significant enhancement in the linear range for Pb^{2+} (0.5 - 60.0 μg L^{-1}) and Cd^{2+} (0.8 - 50.0 μg L^{-1}) along with LOD to be around 0.23 mg $L^{-1}$and 0.35 mg L^{-1} respectively for Pb^{2+} and Cd^{2+}. Also, it was successfully applied to detect HMIs present in pond and river waters.

The preparation of GR using chemical reduction method is sometimes toxic and can easily pollute the environment. This eventually restrict the application of graphene-based sensor. Recently, one-step electrode deposition process was used to prepare SPCE/GR, in which the SPCE was dipped into the GO suspension and under application of deposition potential of -0.8 V for 600 s causes assimilation of reduction as well as deposition. This green fabrication method avoids chemical reducing reagents and most crucially, an additional deposition step required for deposition of material. For comparison, J. F. Ping et al. coated the SPCE/GR with chemically reduced GR and then drop-casted the GR suspension on the SPCE surface. They observed that the electrochemically reduced GR enabled sensor has a sharper and higher current response than the chemical reduced GR sensor. However, in this process, optimization of different types of electrolytes, deposition time and deposition potential are critical [116]. Further, to improve the sensing response the SPCE is modified with other nanomaterials. For example, Promphet et al. fabricated modified SPCE composite nano-fibers (GR-PANI-PS) by directly electrospinning the mixture (GR, PANI, and PS) of solution on the SPCE surface. Electrochemical tests revealed that the current response of the as prepared SPCE was three times greater than that of the unmodified SPCE [51]. Arcos-Martínez's group modified SPCE (SPCE/ Pt NPs) by direct electrodeposition of platinum NPs (using K_2PtCl_6 solution) and used it for determination of As^{3+} [52]. They obtained broad linear range from 0.16 to 1.3 mM with a LoD of as low as 77 μM. In addition, selective detection of As^{3+} in tap water samples was evaluated. Similarly, Lu et al. modified a SPCE with AuNPs (SPCE/AuNPs) using chronoamperometry to achieve simultaneous detection of Zn^{2+}, Pb^{2+}, and Cu^{2+} [117]. The performance of SPCE/ AuNPs was enhanced and attributed to the formation of an in-situ plated BiF during the deposition of Pb^{2+}, Zn^{2+} and Cu^{2+}. The stripping voltammogram exhibited four distinct peaks for Zn^{2+}, Bi^{2+}, Cu^{2+}, and Pb^{2+}over the linear range of 1.0 - 150.0 μg L^{-1} with LoD of 55, 27, and 28 ngL^{-1} for Zn^{2+}, Pb^{2+} and Cu^{2+}, respectively. Besides, they also implemented real sample (lake water) analysis for Pb^{2+}, $Zn^{2+}$and Cu^{2+}.

4.2. Microelectrode and Nanoelectrode Arrays

Microelectrode arrays (MEAs) and nanoelectrode arrays (NEAs) are used to enhance sensitivity and lower detection limits. Microelectrodes based sensing is a very powerful and versatile technique owning to their fast mass transport rates, low interfacial capacitance and steady-state diffusion current and very small ohmic drop. The small dimension of microelectrodes is highly beneficial to study the concentrations in meager region which macroelectrodes are unable to access. Moreover, microelectrodes have high spatial resolution to monitor chemicals without disrupting the system. In general, the active area of the electrode is one of the important parameters, as the signal is based on the total geometric area of the diffusion field. This possess large signal-to-noise ratio rendering micro- and nano-electrodes superior than conventional macro-electrodes. Interestingly, small current is required for stripping analysis. This eliminates use of supporting analytes which reduces the interference in the analysis. Hence this technique is used to detect HMIs [118].

In case of microelectrodes, Hg-based WEs have been widely used as they exhibit high reproducibility and sensitivity for HMI detection. Another frequently used WE is Au electrode as it has higher oxygen over potential than Pt, while other electrode material namely silver suffers from inferring metals present in water. Consequently, Au was discovered to be a superior material for WEs [118]. M.-L. Tercier-Waeber et al. [56] reported development and field validation of nanostructured Au-plated gel integrated MEAs (Au-GIMEs) for As (III) detection in marine aquatic systems. The microelectrode is rendered of renewable Au nanofilaments (AuNF) or Au nanoparticles (AuNP) that have been electroplated on interconnected iridium (Ir)-based micro disk arrays before being covered with a gel. The gel shields the sensor's surface from contamination. It also ensures that analyte mass transport toward the sensor surface occurs through diffusion and is independent of the media's uncontrolled hydrodynamic conditions. SWASV was used to quantify the responses of these sensors for As (III) in 0.1 M $NaNO_3$ (pH 7.6) and marine water samples. The small size of single electrodes reduces the electroactive area, lowering overall current densities. Thus, to increase electroactive surface area, arrays of interconnected microelectrodes were developed. The analytical responses were found to be analogous to the number of interconnected microelectrodes and the morphology of the Au electrode is found independent of media composition. Hence, AuNF-GIME could detect As (III) at sub-nanomolar levels in fresh and salt water systems. Further, remote high-resolution monitoring was also performed by incorporating AuNF-GIME in a submersible multi-channel trace metal sensing probe.

Figure 3. (a) Overview of Sensor in presence of As (III) [118]. Reprinted with permission from [118] P. Podesva et al. Anal. Chem. 90, 2 (2018) 1161–1167.

Electrochemical methods involving noble metal electrodes that have large specific surface area are suitable for monitoring of HMIs, owing to their improved signal-to-noise ratio, sensitivity and lower LoD. P. Podesva et al. modified a large surface area (by a factor of ≈ 1440) nanostructured Au microelectrode array by selective electrodeposition [118]. The morphology of the fabricated sensor as shown in Figure 3 resembled a tree trunk, growing from lithographically defined circular areas and individual branches of the main trunk, with forking. The pore diameters and thickness of interconnected Au leaves are ~50 – 300 nm and ~ 100 nm, respectively. This array was utilized for ultrasensitive detection of the As^{3+} ions and exhibited LoD of ~ 0.0212 ppb.

4.3. Microfluidic Electrochemical (EC) Devices

Microfluidics is a technology which enables manipulation of liquids in volumes lower than 10^{-8} to 10^{-18} litres. Conversely, electrochemical detectors completely overcome the problem of downscaling detectors within a 10 mm dimensional device without any loss of performance. As a result of the incorporation of EC sensors, the lab-on-chip microfluidic channel is formed. The advantages of microfluidic electrochemical devices include the need for a small analyte volume, miniaturization, on-chip sample preparation and pre-concentration of analytes, single chip multiplexed detection, shortening assay time and the absence of complicated instruments. Besides, it is lightweight, flexible, portable and disposable. It also has excellent accuracy, sensitivity, and reproducibility. Moreover, microfluidic paper-based analytical devices (μ-PADs) are more user-friendly and widely available than silicon, glass, and super polymer. The μ-PADs provide opportunities for the development of low-cost, portable, and disposable diagnostic tools for healthcare applications and environmental monitoring. The hydrophobic or hydrophilic micro-channel networks, in conjunction with the analytical devices, allow μ-PADs to handle quantitative analysis and fluid performance. The μ-PADs can be manufactured using a variety of techniques, including flexographic printing, wax printing, photolithography, screen printing laser treatment, inkjet printing, plasma treatment, wax screen-printing, and wet etching. Similarly, effective detection methods include electrochemistry, chemiluminescence (CL), photoelectro-chemistry (PEC), electrochemiluninescence (ECL), fluorescence, and colorimetry [119].

For example, Zhihong et al. developed a low-cost microfluidic paper-based electrochemical sensing (μ-PEDs) device for environmental monitoring [120]. They used photolithography and wax printing to construct paper-based microfluidic channels, and screen-printed electrodes were formed using conducting inks (e.g., carbon or Ag/AgCl). Using SWASV, the fabricated μ-PEDs demonstrated the selective measurement of Pb (II) in an aqueous mixture of Zn(II) and Pb(II). The target Pb^{2+} ions show linear behavior over the concentration range from 0 to 100 μg L^{-1}. Furthermore, hydrodynamic μ-PEDs have exhibited five times higher sensitivity (0.17 μA ppb^{-1}) than stagnant systems. This result was obtained due to the use of a large volume (~800 mL) of sample flows across the electrode surface and the accumulation of metals on the electrodes. The accumulation of metals on the electrode is caused by convection of flowing fluids in the porous matrix of paper over the electrodes. Similarly, Jianjun et al. reported a microfluidic device for detecting Pb (II) and Cd (II) in aqueous samples by assembling the filter paper strips with SPCE [121]. The portable microfluidic device greatly enhanced analytical performance in contaminated aqueous samples (ground water and gas dissolved salty soda water) without pretreatment and detected Cd(II) and

Pb(II) over the concentration range of 0 to 100 ppb with LoD of 2.0 ppb (Pb^{2+}) and 2.3 ppb (Cd^{2+}), respectively using SWASV signal.

Zhiwei et al. presented in situ and online monitoring of environmentally friendly disposable sensor for HMIs detection. [122]. Miniaturized, non-toxic and micro-fabricated bismuth electrode was used as WE in place of conventional mercury electrodes. The fabricated Bi working electrode detected Pb^{2+}(25–400 ppb; $R^2 = 0.991$; LoD - 8 ppb) and Cd^{2+} (28–280 ppb; $R^2 = 0.986$; LoD – 9.3 ppb) using anodic stripping voltammetry. Interestingly, the sensor exhibited *in situ* measurement of Cd^{2+} concentration in ground water and importantly, an online direct measurement of Cd^{2+} concentration in cell culture media in its native environment was performed using LabVIEW.

5. Challenges, Issues and Opportunities

Although the development of GO-based nanocomposites has made remarkable progress as a hot research topic in the field of the environment monitoring, there are still some major issues to overcome. For example, effective synthesis needs to be improved by designing nanocomposite materials with selective functional groups for in-depth study and achieve best selectivity towards HMIs under stringent conditions. Importantly, improvement is also needed in case of stability of the nanocomposites at extreme conditions. These issues can be resolved using techniques such as electrochemical nucleic acid sensor, electrochemical enzyme sensor, the ion imprinted sensor, and novel nanomaterial modified electrochemical sensors [123-126]. Electrochemically immobilized enzymes overcome the lack of recycling process after enzyme reaction and determine the selectivity, sensitivity and stability of enzyme sensors. Similarly, ion imprinted polymers have attracted great attention due to low preparation cost, predetermined recognition ability, good stability, effective quantification and identification of metal ions from water environment [123].

Recently, the electrochemical sensors based on functional nucleic acid exhibited great potential towards HMIs detection. For example, Tang et al. [127] developed a specific electrochemical biosensor by coupling synergistic catalysis of hemin/G-quadruplex-based DNAzyme with porous Au-Pd bimetallic nanoparticles for Pb (II) detection. Functional nucleic acid – DNAzymes, exhibits high binding affinity and specificity towards selective HMIs detection. DNAzyme-based biosensor indicates best selectivity towards targeted Pb (II) with linear concentration range from 1.0 pM to 100 nM and

LoD of 0.34 pM. Similarly, Oroval et al. [128] developed a novel sensor for selective As (III) detection using nanocomposite of mesoporous silica NPs with aptamers (capping agent). Aptamers are one of the easier and cheaper nucleotides and have high binding affinity and specificity towards HMIs.

Although, the nanotechnology has enabled the fabrication of highly sensitive platforms, but still there are some big challenges when sensor operates in low concentrations. This includes homogeneity of nanostructures on large scales, difficulties in controlling the size distribution, signal stability and reproducibility. Also, owing to the fact that the limited research study is available to achieve selective and sensitive detection of HMIs with good stability it paves way for future study.

Another challenge is to create a sensor system for commercialization or on-site monitoring of HMIs based on a variety of criteria such as affordability, efficiency, effectiveness, sensitivity, selectivity, field applicability and user friendliness. As a result, this new system could include substrates, sensing materials, potentiostats, control circuits, sampling methods, a wireless communication module, a sensor array for selective detection of HMIs, and LabVIEW control.

Over the last few years, smart-phones have gained great attention in research domain for in-situ or on field HMIs detection. To study water quality parameters such as pH, total dissolved solids, oxidation-reduction potential and temperature several mobile applications have been developed and linked to a portable sampling device. Additionally, several auto-recognized devices such as smartphone- testing kits have been developed that analyze samples based on Android/IOS application and provide cloud data storage. They also display location information and analyzed water quality data on Google Maps for easy comprehension and decision-making. The Android/IOS application also has several advantages, such as quick on-site usage, warnings via SMS texts and email, and the ability to share data [129-131]. For example, A. Nemiroski et al. [129] developed a low-cost, hand-held device that connects the most common types of electrochemical analysis to "the cloud" via any mobile phone. Universal-mobile-electrochemical-detector (μMED) was referred as a device as it is suitable for a wide range of electrochemical techniques; universal compatibility with mobile technology and paper-based electrodes and different commercially available electrodes. Here, HMIs (Zn (II), Cd (II), and Pb (II)) were detected on SPE using the SWASV method and the sensing element was carbon ink modified with CNT. Also, they incorporated a small vibration motor (the one used in mobile phones (3.3V; 530 Hz)) into the μMED to mix the droplet of analyte directly on the SPE and to reduce the cost and complexity of fabricated hand-held device. Interestingly, greater compatibility of the fabricated sensor with other commercial sensor was observed. Similarly, in 2017, M. Biyani et al. described affordable (<1 USD) and simple disposable electrode printed (DEP)-On-Go, for early warning and rapid (within 5 min) on-site monitoring of HMIs present in environment (Figure 4 a, b) [130]. Here they developed USB charged plasma sized potentiostat and disposable GO based SPE chips for simultaneous detection of zinc, cadmium, lead, copper, mercury and arsenic. The utility of this device is observed by simultaneous detection of HMIs in ground water, tap water, house dust and soil, in the northern parts of India and Japan.

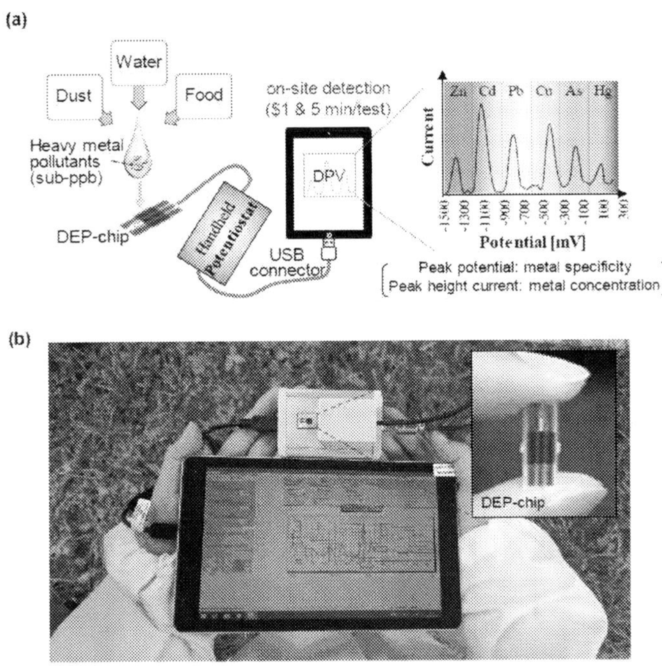

Figure 4. (a) Working flow of "DEP-On-Go" sensor and (b) photograph of on field HMIs detection [130]. Reprinted with permission from Madhu Biyani et al. Sensors, 17(1) (2017) 45.

Although various commercial kits have been developed for HMIs detection but very few literature is available on electrochemical sensors based commercial systems. Also, real-time water monitoring still remains a daunting challenge as these sensor systems have complex fabrication procedures, less stability and expensive. Additionally, there is still high demand of other novel material synthesis for selective detection of HMIs along with fully-automated portable device for real-time monitoring. This clearly indicates that nanomaterials-based electrochemical smart sensors have a promising future in environmental monitoring and needs to be studied extensively for possible commercial availability.

Conclusion

Non-biodegradable HMIs contaminate most of the natural resources found in the environment including aqueous media. Even at low concentrations, heavy metals such as arsenic (As), mercury (Hg), chromium (Cr), lead (Pb), and cadmium (Cd) are highly toxic and dangerous to humans. This paves the way for the development of sensitive, selective, fast and user-friendly techniques to acquire real-time data on water borne pollutants. In view of this, a variety of user-friendly and low-cost electrochemical detection techniques for on-site and real-time monitoring of HMIs that are superior to other techniques are extensively discussed in the present chapter. The classification of various electrochemical techniques is further elaborated in terms of the various types of signals generated as a result of the presence of HMIs in the solution, such as potential, current, electrochemical impedance, conductivity, and electrochemiluminescence. Also, the evaluation of most efficient adsorbents that can operate at various environmental circumstances like solution temperature, pH, contact time and initial pollutant concentration is essentially studied. The recent trend in electrochemical detection with various graphene oxide nanocomposite sensing platforms includes heteroatom, metals, metal oxides, organic, polymers and 3D graphene. It has been discovered that the conventional detection systems lag in terms of rapid screening and field portability. Now, it is essential to focus the research on the development of low-cost portable detection technologies. Therefore, in this chapter along with electrochemical techniques some other most effective, miniaturized, cheap, calibration-less and portable techniques such as screen-printed electrodes (SPEs), microelectrode and nanoelectrode arrays and microfluidic electrochemical devices are elaborately discussed. Recently research is focused on the portable and smart devices for HMIs detections. The portable devices show good sensitivity and selectivity attributed to the limited integrated components and smaller size. The combination of nanomaterials and portable devices cover the shortages of the portable devices and are used to detect the various targets. Although various portable devices or commercial devices are available for HMIs detection, still there are some challenges in the development of nano-devices such as stability, in-bulk preparation and high cost. We believe that in future, the advancement in nanotechnology ensures the development of portable and smart devices and surely are within reach.

Acknowledgement

The authors would like to thank "*Board of Research in Nuclear Science* (*BRNS*)" for funding under DAE YSRA scheme.

References

[1] Maria, L. *Biosensors for Environmental Applications* (2011). 1–16, http://www.intechopen.com/books/environmental-biosensors/biosensor-forenvironmental-applications.

[2] Chen, Y., Wu, F. Liu, M., Parvez, F., Slavkovich, V. and Eunus, M. A. (2013). Prospective study of arsenic exposure, arsenic methylation capacity, and risk of cardiovascular disease in Bangladesh. *Environ. Health Perspect.* 121: 832–838.

[3] Yantasee, W., Lin, Y., Hongsirikarn, K., Fryxell, G. E., Addleman, R. and Timchalk, C. (2007). Electrochemical sensors for the detection of lead and other toxic heavy metals: the next generation of personal expo-sure biomonitors, *Environ. Health Perspect.* 115: 1683–1690.

[4] Sharma, P., Bihari, V., Agarwal, S. K., Verma, V., Kesavachandran, C. N. and Pangtey, B. S. (2012). Groundwater contaminated with hexavalent chromium [Cr(VI)]: a health survey and clinical examination of community inhabitants (Kanpur, India), *PLoS ONE* 7: 1–7.

[5] Aragay, G., Pons, J. and Merkoc¸ A. (2011). Recent trends in macro-, micro-, and nanomaterial-based tools and strategies for heavy-metal detection, *Chem. Rev.* 111: 3433–3458.

[6] Sotolongo, A. C., Messina, M. M., Ibañez, F. J. and Wuilloud, R. G. (2020). Hybrid ionic liquid-3D graphene-Ni foam for on-line preconcentration and separation of Hg species in water with atomic fluorescence spectrometry detection, *Talanta* 210: 120614.

[7] Shahlaei, M. and Pourhossein, A. (2014). Determination of arsenic in drinking water samples by electrothermal atomic absorption spectrometry after preconcentration using the biomass of Aspergillus Niger loaded on activated charcoal, *Chemistry* 912619.

[8] Kocot, K. and Sitko, R. (2014). Trace and ultratrace determination of heavy metal ions by energy-dispersive X-ray fluorescence spectrometry using graphene as solid

sorbent in dispersive micro solid-phase extraction, *Spectrochem. Acta B* 94 – 95: 7-13.

[9] Shabani A., Dadfarnia S., Moosavinejad T. and Ahmadi S. H., (2009). On-line preconcentration system using a microcolumn packed with Alizarin Red S-modified alumina for zinc determination by flame atomic absorption spectrometry, *Quimica Nova,* 32: 1202–1205.

[10] Pujol, L., Evrard, D., Serrano, K. G., Freyssinier, M., Cizsak, A. R. and Gros, P., (2014). Electrochemical sensors and devices for electrochemical assay in water: the French groups contribution. *Front. Chem., Anal. Chem.* 19: 1–24.

[11] Rassaei, L., Marken, F., Sillanpää, M., Amiri, M., Cirtiu, C. M. and Sillanpää, M. (2011). Nanoparticles in electrochemical sensors for environmental monitoring. *Trend Anal. Chem.* 30(11): 1704–1715.

[12] Bhangre, B., Ramgir, N. S., Sinju, K., Pathak, A., Jagtap, S., Debnath, A., Muthe, K. and Gosavi, S. (2020). *Reduced Graphene Oxide (rGO) – Based Nanohybrids as Gas sensors: State of the art*, Functional Nanomaterials, 189 – 217.

[13] Pumera, M. (2009) Electrochemistry of graphene: new horizons for sensing and energy storage. *Chem. Rec.* 9(4): 211–223.

[14] Aragay, G., Pons, J. and Merkoc¸ A. (2011) Recent trends in macro-, micro-, and nanomaterial-based tools and strategies for heavy-metal detection, *Chem. Rev.* 111: 3433–3458.

[15] Fang, Y. and Wang, E. (2013). Electrochemical biosensors on platforms of graphene. *Chem. Comm.* 49(83): 9526–9539.

[16] Novoselov, K. S., Geim, A. K., Morozov, S. V., Jiang, D., Zhang, Y., Dubonos, S. V., Grigorieva, I. V. and Firsov A. A., (2004). Electric field effect in atomically thin carbon films, *Science* 306 (5696): 666–669.

[17] Hummers, W. S. and Offeman, R. E. (1958). Preparation of Graphitic Oxide, *J. Am. Chem. Soc.* 80: 1339.

[18] Naveen, M. H., Gurudatt, N. G. and Shim, Y. B. (2017). Applications of conducting polymer composites to electrochemical sensors: a review, *Applied Mater. Today* 9: 419–433.

[19] Chouhan, B., Meena, P. and Poonar, N. (2016). Effect of heavy metal ions in water on human health, *Int. J. Sci. Eng. Res.* 4: 2015–2017.

[20] ATSDR, *Toxicological profile for arsenic TP-92/09*. Center for Disease Control, Agency for Toxic Substances and Disease Registry, Atlanta, GA.

[21] Waalkes, M. P. (1995). Cadmium and carcinogenesis. In: Berthan G. (Ed) *Handbook on metal ligand interactions of biological fluids*, Dekker, New York, NY, 2 471–482.

[22] Wittman, R. and Hu, H. (2002). Cadmium exposure and nephropathy in a 28-year old female metals worker, *Environ. Health Perspect.* 110: 1261–1266.

[23] Cohen, M. D., Kargacin, B., Klein, C. B. and Costa, M. (1993). Mechanisms of chromium carcinogenicity and toxicity. *Crit. Rev. Toxicol.* 23: 255–281.

[24] ATSDR (2008) *Toxicological profile for chromium.* U.S. Department of Health and Human Services, Public Health Service, Agency for Toxic Substances and Disease Registry, Atlanta, GA.

[25] ATSDR (1999) *Toxicological profile for Lead. Public Health Service*, U.S. Department of Health and Human Services, Agency for Toxic Substances and Disease Registry, Atlanta, GA.

[26] Apostoli, P., Kiss, P., Stefano, P., Bonde, J. P. and Vanhoorne, M. (1998). Male reproduction toxicity of lead in animals and humans, *Occup. Environ. Med.* 55: 364–374.

[27] Sarkar, B. A. (2005). Mercury in the environment: effects on health and reproduction. *Rev. Environ. Health* 20: 39–56.

[28] Tchounwou, P. B., Ayensu, W. K., Ninashvilli, N. and Sutton, D. (2003). Environmental exposures to mercury and its toxicopathologic implications for public health, *Environ. Toxicol.* 18: 149–175.

[29] Li, M., Cushing, S. K., Zhou, X., Guo, S. and Wu, N. Q. (2012). Fingerprinting photoluminescence of functional groups in graphene oxide. *J. Mater. Chem.*, 22: 23374−23379.

[30] Yang, X., Shi, C., Wheeler, D., Newhouse, R., Chen, B., Zhang, J. Z. and Gu, C. (2010). High-sensitivity molecular sensing using hollow-corephotonic crystal fiber and surface-enhanced Raman scattering. *J. Opt. Soc. Am. A*, 27: 977−984.

[31] Kimmel, D. W., LeBlanc, G., Meschievitz, M. E. and Cliffel, D. E. (2012). Electrochemical sensors and biosensors. *Anal. Chem.*, 84: 685−707.

[32] Kudr, J., Nguyen, H. V., Gumulec, J., Nejdl, L., Blazkova, I., Ruttkay- Nedecky, B., Hynek, D., Jindrich, K., Vojtech, A. and Rene, K. (2015). Simultaneous automatic electrochemical detection of zinc, cadmium, copper and lead ions in environmental samples using a thin-film mercury electrode and an artificial neural network, *Sensors* 15: 592–610.

[33] Cui, L., Wu, J. and Ju, H., (2015). Electrochemical sensing of heavy metal ions with inorganic, organic and bio-materials, *Biosens. Bioelectron.* 63: 276–286.

[34] Fan, L., Chen, J., Zhu, S., Wang, M. and Xu, G. (2009). Determination of Cd2+ and Pb2+ on GCE modified by electrochemical reduction of aromatic diazonium salts, *Electrochem. Comm.* 11: 1823–1825.

[35] Men′endez-Alonso, E., Hill, S. J., Foulkes, M. E. and Crighton, J. S., (1999). Speciation and preconcentration of Cr III and Cr VI in waters by retention on ion exchange media and determination by EDXRF, *Anal. Atomic Spect.* 14: 187–192.

[36] de Almeida Pereira, L., de Amorim, I. G. and da Silva, J. B. B. (2004). Development of methodologies to determine aluminum, cadmium, chromium and lead in drinking water by ET AAS using permanent modifiers, *Talanta*, 64: 395–400.

[37] Zhang, N., Suleiman, J. S., He, M. and Hu, B. (2008). Chromium (III)-imprinted silica gel for speciation analysis of chromium in environmental water samples with ICP-MS detection, *Talanta*, 75 5: 36–543.

[38] Suo, L., Dong, X., Gao, X., Xu, J., Huang, Z., Ye, J. and Zhao, L. (2019). Silica-coated magnetic graphene oxide nanocomposite based magnetic solid phase extraction of trace amounts of heavy metals in water samples prior to determination by inductively coupled plasma mass spectrometry, *Microchemical Journal*, 149: 104039.

[39] Su, S., Chen, B., He, M. and Hu, B. (2014). Graphene oxide–silica composite coating hollow fiber solid phase microextraction online coupled with inductively coupled plasma mass spectrometry for the determination of trace heavy metals in environmental water samples, *Talanta,* 123: 1-9.

[40] Li, J., Chen, L., Lou, T. and Wang, Y. (2011). Highly Sensitive SERS Detection of As^{3+} Ions in Aqueous Media using Glutathione Functionalized Silver Nanoparticles, *ACS Appl. Mater. Interfaces* 3, 10: 3936–3941.

[41] Wang, G., Lim, C., Chen, L., Chon, H., Choo, J., Hong, J. and deMello, A. J. (2009). Surface-enhanced Raman scattering in nanoliter droplets: towards high-sensitivity detection of mercury (II) ions, *Anal. Bioanal. Chem.* 394: 827–1832.

[42] Gribas, A. V. and Sakharov, I. Y. (2018). Homogeneous chemiluminescent determination of mercury (II) using a peroxidase-mimicking DNAzyme assay, *Anal. Lett.* 51: 1280–1290.

[43] Tang, D., Xia, B., Tang, Y., Zhang, J. and Zhou, Q. (2019). Metal-ion-induced DNAzyme on magnetic beads for detection of lead (II) by using rolling circle amplification, glucose oxidase, and readout of pH changes, *Microchimica Acta,* 186: 318.

[44] Sc, Skryshevsky, V. A., Soldatkin, O. O. and Soldatkin A. P. (2015). Enzyme biosensor systems based on porous silicon photoluminescence for detection of glucose, urea and heavy metals. *Biosens. Bioelec.* 66: 89–94.

[45] Sun, X., Li, B., Qi, A., Tian, C., Han, J., Shi, Y., Lin, B. and Chen, L. (2018). Improved assessment of accuracy and performance using a rotational paper-based device for multiplexed detection of heavy metals, *Talanta* 178: 426-431.

[46] Hutton, L. A., O'Neil, G. D., Read, T. L., Ayres, Z. J., Newton, M. E. and Macpherson, J. V. (2014). Electrochemical X-ray Fluorescence Spectroscopy for Trace Heavy Metal Analysis: Enhancing X-ray Fluorescence Detection Capabilities by Four Orders of Magnitude, *Anal. Chem.* 86, 9: 4566–4572.

[47] Aranda, P. R., Moyano, S., Martinez, L. D. and De Vito, I. E. (2010). Determination of trace chromium(VI) in drinking water using X-ray fluorescence spectrometry after solid-phase extraction, *Anal. Bioanal. Chem.* 398: 1043–1048.

[48] Chekin, F., Boukherroub, R. and Szunerits, S. (2017). MoS_2/reduced graphene oxide nanocomposite for sensitive sensing of cysteamine in presence of uric acid in human plasma, *Mater. Sci. Engineer*. C 73: 627 – 632.

[49] Jijana, A. N., Mphuthi, N., Shumbula, P. Vilakazi, S. and Sikhwivhilu, L. (2021). The Ultra-sensitive Electrochemical Detection of As(III) in Ground Water Using Disposable L-cysteine/Lipoic Acid Functionalised Gold Nanoparticle Modified Screen-Printed Electrodes, *Electrocatalysis*, 12: 310–325.

[50] Shuai, H. and Lei, Y. J. (2016). Graphene ink fabricated screen printed electrode for Cd^{2+} and Pb^{2+} determination in Xiangjiang river, *Int. J. Electrochem. Sci.* 11: 7430-7439.

[51] Promphet, N., Rattanarat, P., Rangkupan, R., Chailapakul, O. and Rodthongkum, N. (2015). An electrochemical sensor based on graphene/polyaniline/polystyrene nanoporous fibers modified electrode for simultaneous determination of lead and cadmium, *Sens. Actuators B* 207: 526-534.

[52] Domínguez-Renedo, O. and Arcos-Martínez, M. J. (2007). A novel method for the anodic stripping voltammetry determination of Sb(III) using silver nanoparticle modified screen-printed electrodes, *Electrochem. Comm.* 9: 820-826.

[53] Vilian, A. T. E., Shahzad, A., Chung, J., Choe, S. R., Kim, W. S., Huh, Y. S., Yu, T. and Han, Y. K. (2017). Square voltammetric sensing of mercury at very low working potential by using oligomer-functionalized Ag@Au core-shell nanoparticles, *Microchim. Acta* 184: 3547-3556.

[54] Lim, J., Mubarak, N., Khalid, M., Abdullah, E. and Arshid, N. (2018b). Novel fabrication of functionalized graphene oxide via magnetite and 1-butyl-3-methylimidazolium tetra fluoroborate. *Nano-Struct. Nano-Objects* 16: 403-411.

[55] Podešva, P., Gablech, I. and Neužil, P. (2018). Nanostructured gold microelectrode array for ultrasensitive detection of heavy metal contamination, *Anal. Chem.* 90 (2): 1161-167.

[56] Fighera, M., Abdou, M., Bakker, E. and Van Der Wal, P. (2021). Newly designed gel-integrated nanostructured gold-based interconnected microelectrode arrays for direct arsenite monitoring in aquatic systems, *Sens. Actuators B* 328: 128996.

[57] Jedlińska, K., Porada, R., Lipińska, J. and Baś, B. (2020). The bi-band silver microelectrode: Fabrication, characterization and analytical study, *Sens. Actuators B* 302: 127152.

[58] Tortolini, C., Bollella, P., Antonelli, M. L., Antiochia, R., Mazzei, F. and Favero, G. (2015). DNA based biosensors for Hg2+ determination by polythymine-methylene blue modified electrodes, *Biosens. Bioelectron.* 67 (15): 524-531.

[59] Rehacek, V., Hotovy, I. and Vojs, M. (2007). Bismuth-coated diamond-like carbon microelectrodes for heavy metals determination, *Sens. Actuators B* 127: 193-197.

[60] Zou, Z., Jang, A., MacKnight, E., Wu, P.-M., Do, J., Bishop, P. L. and Ahn, C. H. (2008). Environmentally friendly disposable sensors with microfabricated on-chip planar bismuth electrode for in situ heavy metal ions measurement, *Sens. Actuators B* 134: 18–24.

[61] Shi, J., Tang, F., Xing, H., Zheng, H., Bi, L. and Wang, W. (2012). Electrochemical Detection of Pb and Cd in Paper-Based Microfluidic Devices, *J. Braz. Chem. Soc.*, 23: 1124-1130.

[62] Zhou, J., Li, B., Qi, A., Shi, Y., Qi, J., Xu, H. and Chen, L. (2020). ZnSe quantum dot based ion imprinting technology for fluorescence detecting cadmium and lead ions on a three-dimensional rotary paper-based microfluidic chip, *Sens. Actuators B* 305: 127462.

[63] Ruecha, N., Rodthongkum, N., Cate, D. M., Volckens, J., Chailapakul, O., Henry, C. S. (2015). Sensitive electrochemical sensor using a graphene-polyaniline nanocomposite for simultaneous detection of Zn(II), Cd(II), and Pb(II), *Anal. Chim. Acta*, 874: 40-48.

[64] Kim, U., Vandergiessen, J., Demaree, B., Reynolds, M. and Perricone, K. (2013). In Development of low-cost plastic microfluidic sensors toward rapid and point-of-use detection of arsenic in drinking water for global health, *Biomed. Circuits Sys. Conf.*, 113-117.

[65] Molina, J., Cases, F. and Moretto, L. M. (2016). Graphene-based materials for the electrochemical determination of hazardous ions. *Anal. Chim. Acta*, 946: 9–39.

[66] Zhang, L., Li, D. W., Song, W., Shi, L., Li, Y. and Long, Y. T. (2010). High Sensitive On-Site Cadmium Sensor Based on AuNPs Amalgam Modified Screen-Printed Carbon Electrodes, *IEEE Sensors* 10 (10): 1583-1588.

[67] Xing, H., Xu, J., Zhu, X., Duan, X., Lu, L., Wang, W., Zhang, Y. and Yang, T. (2016). Highly sensitive simultaneous determination of cadmium (II), lead (II), copper (II), and mercury (II) ions on N-doped graphene modified electrode. *J. Electroanal. Chem.* 760: 52–58.

[68] Liu, X. S., Li, Z. J., Ding, R. M., Ren, B. B. and Li, Y. H. (2016). A nanocarbon paste electrode modified with nitrogen-doped graphene for square wave anodic stripping voltammetric determination of trace lead and cadmium. *Microchim. Acta* 183(2): 709–714.

[69] Bi, C.-C., Ke, X.-X., Chen, X., Weerasooriya, R., Hong, Z.-Y., Wang, L.-C. and Wu, Y.-C. (2020). Assembling Reduced Graphene Oxide with Sulfur/Nitrogen- 1 "Hooks" for Electrochemical Determination of Hg(II), *Anal. Chem. Acta* 1100: 31-39.

[70] Liu, Y., Huang, Z., Xie, Q., Sun, L., Gu, T., Li, Z., Bu, L., Yao, S., Tu, X., Luo, X. and Luo, S. (2013). Electrodeposition of electro reduced graphene oxide-Au nanoparticles composite film at glassy carbon electrode for anodic stripping voltammetric analysis of trace arsenic(III), *Sens. Actuators B* 188: 894–901.

[71] Wang, N., Lin, M., Dai, H. and Houyi, M. (2016) Functionalized gold nanoparticles/reduced graphene oxide nanocomposites for ultrasensitive electrochemical sensing of mercury ions based on thymine-mercury-thymine structure, *Biosens. Bioelectron.* 15 79: 320-326.

[72] Sang, S., Li, D., Zhang, H., Sun, Y., Jian, A., Zhang, Q. and Zhang, W. (2017). Facile synthesis of AgNPs on reduced graphene oxide for highly sensitive simultaneous detection of heavy metal ions. *RSC Adv.* 7(35): 21618–21624.

[73] Stassen, I., Styles, M. J., Van Assche, T. R., Campagnol, N., Fransaer, J., Denayer, J. F., Tan, J. C., Falcaro, P., DED, V. and Ameloot, R. (2015). Electrochemical film deposition of the zirconium metal-organic framework UIO-66 and application in a miniaturized sorbent trap. *Chem. Mater.* 27(5): 1801–1807.

[74] Zou, J., Zhong, W., Gao, F., Tu, X., Chen, S., Huang, X., Wang, X., Lu, L. and Yu, Y. (2020). Sensitive electrochemical platform for trace determination of Pb2+ based on multilayer Bi-MOFs/reduced graphene oxide films modified electrode, *Microchimica Acta* 187: 603.

[75] Silva, Martin K. L. and Cesarino, I. (2020). Electrochemical sensor based on Sb nanoparticles/reduced graphene oxide for heavy metal determination, *Environ. Anal. Chem.* 1029: 0397.

[76] Ruengpirasiri, P., Punrat, E., Chailapakul, O. and Chuanuwatanakul, S. (2017). Graphene oxide-modified electrode coated with in-situ antimony film for the simultaneous determination of heavy metals by sequential injection-anodic stripping voltammetry, *Electroanal.* 29(4): 1022–1030.

[77] Sun, Y., Zhang, W., Yu, H., Hou, C., Li, D., Zhang, Y. and Liu, Y. (2015). Controlled synthesis various shapes Fe3O4 decorated reduced graphene oxide applied in the electrochemical detection. *J. Alloy Comp.* 638: 182–187.

[78] World Health Organization, *Guidelines for drinking-water quality*, 3rd Ed. incorporating the first and second addenda, Vol. 1, Recommendations. Silver, (2008) 434–435. Geneva, Switzerland.

[79] Vajedi, F. and Dehghani, H. (2019). The characterization of TiO_2-reduced graphene oxide nanocomposites and their performance in electrochemical determination for removing heavy metals ions of cadmium(II), lead(II) and copper(II), *Mater. Sci. Eng. B* 243: 189–198.

[80] Sako, R., Tsuchiya, H. and Ogawa, M. (2011). Influence of band-gap opening on ballistic electron transport in bilayer graphene and graphene nanoribbon FETs, *IEEE Trans Electron Dev.* 58(10): 3300–3306.

[81] Zhou, S. F., Han, X. J., Fan, H. L., Huang, J. and Liu, Y. Q. (2018). Enhanced electrochemical performance for sensing Pb(II) based on graphene oxide incorporated mesoporous $MnFe_2O_4$ nanocomposites. *J. Alloy Comp.* 747: 447–454.

[82] Muralikrishna, S., Suresh kumar, K., Varley, T. S., Nagaraju, D. H. and Ramakrishnappa, T. (2014). In situ reduction and functionalization of graphene oxide with l-cysteine for simultaneous electrochemical determination of cadmium (ii), lead (ii), copper (ii), and mercury (ii) ions, *Anal. Methods* 6(21): 8698–8705.

[83] Priya, T., Dhanalakshmi, N., Thennarasu, S., Karthikeyan, V. and Thinakaran N., (2019). Ultra sensitive electrochemical detection of Cd^{2+} and Pb^{2+} using penetrable nature of graphene/gold nanoparticles/modified L-cysteine nanocomposite, *Chem. Phys. Lett.* 731: 136621.

[84] Zhou, W., Li, C., Sun, C. and Yang, X. (2016). Simultaneously determination of trace Cd^{2+} and Pb^{2+} based on l-cysteine/graphene modified glassy carbon electrode, *Food Chem.,* 192: 351-357.

[85] Muralikrishna, S., Sureshkumar, K., Varley, T. S., Nagaraju, D. H. and Ramakrishnappa, T. (2014). In situ reduction and functionalization of graphene oxide with L-cysteine for simultaneous electrochemical determination of cadmium (II), lead (II), copper (II), and mercury (II) ions, *Anal. Methods,* 6: 8698-8705.

[86] Lu, Z., Lin, X., Zhang, J., Dai, W., Liu, B., Mo, G., Ye, J. and Ye, J. (2019). Ionic liquid/poly-L-cysteine composite deposited on flexible and hierarchical porous laser-engraved graphene electrode for high-performance electrochemical analysis of lead ion, *Electrochem. Acta* 295: 514-523.

[87] Zuo, Y., Xu, J., Zhu, X., Duan, X., Lu, L., Gao, Y., Xing, H., Yang, T., Ye, G. and Yu, Y. (2016). Poly(3,4-ethylenedioxythiophene) nanorods/graphene oxide nanocomposite as a new electrode material for the selective electrochemical detection of mercury (II), *Syn. Metals* 220: 14-19.

[88] Yasri, N. G., Sundramoorthy, A. K., Chang, W.-J. and Gunasekaran, S. (2014). Highly selective mercury detection at partially oxidized graphene/poly (3,4 ethylenedioxythiophene): poly(styrenesulfonate) nanocomposite film - modified electrode, *Front. Mater.* 1: 33.

[89] Muralikrishna, S., Nagaraju, D. H., Geetha Balakrishna, R., Surareungchai, W., Ramakrishnappa, T. and Shivanandareddy, A. B. (2017). Hydrogels of polyaniline with graphene oxide for highly sensitive electrochemical determination of lead ions, *Anal. Chem. Acta*, 990: 67 – 77.

[90] Dai, H., Wang, N., Wang, D., Ma, H. and Lin, M. (2016). An electrochemical sensor based on phytic acid functionalized polypyrrole/graphene oxide nanocomposites for simultaneous determination of Cd(II) and Pb(II), *Chem. Eng.* 299: 150-155.

[91] Shehzad, K., Xu, Y., Gao, C. and Duan, X. F. (2016). Three-dimensional macro-structures of two dimensional nanomaterials, *Chem. Soc. Rev.* 45: 5541–5588.

[92] Pan, M., Shan, C., Zhang, X., Zhang, Y., Zhu, C., Gao, G. and Pan, B. (2018). Environmentally Friendly in Situ Regeneration of Graphene Aerogel as a Model Conductive Adsorbent, *Environ. Sci. Technol.* 52: 739–746.

[93] Lei, Y., Chen, F., Luo, Y. and Zhang, L. (2014). Synthesis of three-dimensional graphene oxide foam for the removal of heavy metal ions, *Chem. Phys. Lett.* 593: 122–127.

[94] Shi, L., Chen, K., Du, R., Bachmatiuk, A., Rummeli, M. H., Xie, K., Huang, Y., Zhang, Y. and Liu, Z. (2016). Scalable seashell-based chemical vapor deposition growth of three-dimensional graphene foams for oil-water separation, *J. Am. Chem. Soc.* 138: 6360–6363.

[95] Huang, K., Yang, J., Dong, S., Feng, Q., Zhang, X., Ding, Y. and Hu, J. (2018). Anisotropy of graphene scaffolds assembled by three-dimensional printing, *Carbon* 130: 1–10.

[96] Shi, L., Li, Y., Rong, X., Wang, Y. and Ding, S. (2017). Facile fabrication of a novel 3D graphene framework/Bi nanoparticle film for ultrasensitive electrochemical assays of heavy metal ions, *Anal. Chim. Acta* 968: 21-29.

[97] Guo, X., Cui, R., Huang, H., Li, Y., Liu, B., Wang, J., Zhao, D., Dong, J. and Sun, B. (2020). Insights into the role of pyrrole doped in three-dimensional graphene aerogels for electrochemical sensing Cd(II), *Electroanal. Chem.,* 871: 114323.

[98] Lee, P. M., Chen, Z., Li, L. and Liu, E. (2015). Reduced graphene oxide decorated with tin nanoparticles through electrodeposition for simultaneous determination of trace heavy metals, *Electrochim. Acta* 174: 207–214.

[99] Arino, C., Serrano, N., Diaz-Cruz, J. M. and Esteban, M. (2017). Voltammetric determination of metal ions beyond mercury electrodes. A review, *Anal. Chim. Acta* 990 :11–53.

[100] Lin, X., Lu, Z., Dai, W., Liu, B., Zhang, Y., Li, J. and Ye, J. (2018). Laser engraved nitrogen-doped graphene sensor for the simultaneous determination of Cd(II) and Pb(II), *Electrochem. Chem*., 828: 41-49.

[101] Liu, F., Zhang, Y., Yin, W., Hou, C., Huo, D., He, B., Qian, L. and Fa, H. (2017). A high–selectivity electrochemical sensor for ultra-trace lead (II) detection based on a nanocomposite consisting of nitrogen-doped graphene/gold nanoparticles functionalized with ETBD and Fe_3O_4@TiO_2 core–shell nanoparticles, *Sens. Actuator* B 242: 889 -896.

[102] Manna, B. and Retna Ra, C. (2018). Nanostructured Sulfur-Doped Porous Reduced Graphene Oxide for the Ultrasensitive Electrochemical Detection and Efficient Removal of Hg(II), *CS Sustainable Chem. Eng.* 6, 5: 6175–6182.

[103] Chimezie, A. B., Hajian, R., Yusof, N. A., Woi, P. M. and Shams, N. (2017). Fabrication of reduced graphene oxide-magnetic nanocomposite (rGO-Fe_3O_4) as

an electrochemical sensor for trace determination of As(III) in water resources, *Electroanal. Chem.* 796: 33–42.

[104] Wei, Y., Gao, C., Meng, F. L., Li, H. H., Wang, L., Liu, J. H. and Huang X. J., (2012). SnO_2/reduced graphene oxide nanocomposite for the simultaneous electrochemical detection of cadmium(II), lead(II), copper(II), and mercury(II): an interesting favorable mutual interference. *J. Phys. Chem. C* 116(1): 1034–1041.

[105] Ramesha, G. K. and Sampath, S. (2011). In-situ formation of graphene lead oxide composite and its use in trace arsenic detection, *Sens. Actuators B* 160(1): 306–311.

[106] Karthik, R. and Thambidurai, S. (2017). Synthesis of cobalt doped ZnO/ reduced graphene oxide nanorods as active material for heavy metal ions sensor and antibacterial activity, *J Alloy Comp.* 715: 254–265.

[107] Zhou, H., Wang, X., Yu, P., Chen, X. and Mao, L. (2012). Sensitive and selective voltammetric measurement of Hg^{2+} by rational covalent functionalization of graphene oxide with cysteamine, *Analyst* 137(2): 305–308.

[108] Yuan, Y. H., Zhu, X. H., Wen, S. H., Liang, R. P., Zhang, L. and Qiu, J. D. (2018). Electrochemical assay for As (III) by combination of highly thiolrich trithiocyanuric acid and conductive reduced graphene oxide nanocomposites, *Electroanal. Chem.* 814: 97–103.

[109] Gupta, V. K., Yola, M. L., Atar, N., Ustundağ, Z. and Solak, A. O. (2013). A novel sensitive Cu(II) and Cd(II) nanosensor platform: graphene oxide terminated p-aminophenyl modified glassy carbon surface, *Electrochim. Acta* 112: 541–548.

[110] Gode, C., Yola, M. L., Yilmaz, A., Atar, N. and Wang, S. (2017). A novel electrochemical sensor based on calixarene functionalized reduced graphene oxide: application to simultaneous determination of Fe(III), Cd(II) and Pb(II) ions, *Coll. Inter. Sci.* 508: 525–531.

[111] Palanisamy, S., Thangavelu, K., Chen, S. M., Velusamy, V., Chang, M. H., Chen, T. W., Al-Hemaid, F. M. A., Ali, M. A. and Ramaraj, S. K. (2017). Synthesis and characterization of polypyrrole decorated graphene/β-cyclodextrin composite for low level electrochemical detection of mercury (II) in water, *Sens. Actuators B* 243: 888–894.

[112] Han, T., Jin, J., Wang, C., Sun, Y., Zhang, Y. and Li, Y. (2017). Ag Nanoparticles-Modified 3D Graphene Foam for Binder Free Electrodes of Electrochemical Sensors, *Nanomaterials* 7(2): 40.

[113] Lu, Z., Dai, W., Lin, X., Liu, B., Zhang, J., Ye, J. and Ye, J. (2018). Facile one-step fabrication of a novel 3D honeycomb-like bismuth nanoparticles decorated N-doped carbon nanosheet frameworks: Ultrasensitive electrochemical sensing of heavy metal ions, *Electrochem. Acta* 266: 94-102.

[114] Wu, J., Yang, L. N., Song, N. and Chen, J. R. (2014). Nanometer materials modified electrodes for detection of heavy metal ions by stripping voltammetry, *Appl. Mech. Mater.* 488-489:129-132.

[115] Wu, F., Lin, H. W., Yang, X. and Chen, D. Z. (2013). GS-Nafion-Au nanocomposite film modified SPCEs for simultaneous determination of trace Pb2þ and Cd2þ by DPSV, *Int. J. Electrochem. Sci.* 8: 7702-7712.

[116] Ping, J. F., Wang, Y. X., Wu, J. and Ying, Y. B. (2014). Development of an electrochemically reduced graphene oxide modified disposable bismuth film electrode and its application for stripping analysis of heavy metals in milk, *Food Chem.* 151: 65-71.

[117] Lu, Z. W., Zhang, J. J., Dai, W. L., Lin, X. N., Ye, J. P. and Ye, J. S. (2017). A screen-printed carbon electrode modified with a bismuth film and gold nanoparticles for simultaneous stripping voltammetric determination of Zn(II), Pb(II) and Cu(II), *Microchim. Acta* 184: 4731-4740.

[118] Podesva, P., Gablech, I. and Neuzil, P. (2018). Nanostructured Gold Microelectrode Array for Ultrasensitive Detection of Heavy Metal Contamination, *Anal. Chem.* 90, 2: 1161–1167.

[119] Xia, Y., Si, J. and Li, Z. (2016). Fabrication techniques for microfluidic paper-based analytical devices and their applications for biological testing: A review, *Biosens. Bioelectron.* 77: 774–789.

[120] Nie, Z., Nijhuis, C. A., Gong, J., Chen, X., Kumachev, A., Martinez, A. W., Narovlyansky, M. and Whitesides, G. M. (2010). Electrochemical sensing in paper-based microfluidic devices, *Lab Chip.* 10: 477 - 483.

[121] Shi, J., Tang, F., Xing, H., Zheng, H., Bi, L. and Wang, W. (2012). Electrochemical Detection of Pb and Cd in Paper-Based Microfluidic Devices, *J. Braz. Chem. Soc.,* 23: 1124-1130.

[122] Zou, Z., Jang, A., MacKnight, E., Wu, P.-M., Do, J., Bishop, P. L. and Ahn, C. H. (2008). Environmentally friendly disposable sensors with microfabricated on-chip planar bismuth electrode for in situ heavy metal ions measurement, *Sens. Actuators B* 134: 18–24.

[123] Ganjali, M. R., Rahmani, A. R., Shokoohi, R., Farmany, A., Khazaei, M. (2019). A highly sensitive and selective electrochemical mercury (II) sensor based on nanoparticles of Hg (II)-imprinted polymer and graphitic carbon nitride (g-C3N4). *Int. J. Electrochem. Sci.,* 14: 6420.

[124] Wu, S., Dai, X., Cheng, T. and Li, S. (2018). Highly sensitive and selective ion-imprinted polymers based on one-step electrodeposition of chitosan-graphene nanocomposites for the determination of Cr (VI). *Carbohydrate Polymers,* 195: 199–206.

[125] da Silva, W., Ghica, M. E. and Brett, C. M. (2020). Biotoxic trace metal ion detection by enzymatic inhibition of a glucose biosensor based on a poly (brilliant green)–deep eutectic solvent/carbon nanotube modified electrode, *Talanta*, 208: 120427.

[126] Abu-Ali, H., Nabok, A. and Smith, T. J. (2019). Development of novel and highly specific ssDNA-aptamer-based electrochemical biosensor for rapid detection of mercury (II) and lead (II) ions in water. *Chemosens.* 7: 27.

[127] Zhou, Q., Lin, Y., Lin, Y., Wei, Q., Chen, G. and Tang, D. (2016). Highly sensitive electrochemical sensing platform for lead ion based on synergetic catalysis of DNAzyme and Au-Pd porous bimetallic nanostructures, *Biosens. Bioelect.* 78: 236–243.

[128] Oroval, M., Coll, C., Bernardos, A., Marcos, M. D., Martinez-Manez, R., Shchukin, D. G. and Sancenon, F. (2017). Selective fluorogenic sensing of As(III) using aptamer-capped nanomaterials, *ACS Appl. Mater. Interfaces*, 9(13): 11332–11336.
[129] Nemiroski, A., Christodouleas, D. C., Hennek, J. W., Kumar, A. A., Maxwell, E. J., Fernandez-Abedul, M. T. and Whitesides, G. M. (2014) Universal mobile electrochemical detector designed for use in resource-limited applications, "Universal Mobile Electrochemical Detector Designed for Use in Resource-Limited Applications.", *Proc. Nat. Acad. Sci.* 111 (33): 11984–11989.
[130] Biyani, M., Biyani, R., Tsuchihashi, T., Takamura, Y., Ushijima, H., Tamiya, E. and Biyani, M. (2017). DEP-On-Go for Simultaneous Sensing of Multiple Heavy Metals Pollutants in Environmental Samples, *Sensors*, *17*(1): 45.
[131] MRC. *Portable Heavy metal analyzer.* In (Vol. 2020).

Chapter 11

A Brief Overview on Binary Pyridyl, Ternary Pyridyl and Triphenylphosphine Based Supramolecular Polymeric Complexes of Cu(I)

Goutam Kumar Patra*, Amit Kumar Manna, Meman Sahu and Kalyani Rout
Department of Chemistry, Guru GhasidasVishwavidyalaya, Bilaspur, Chhattisgarh, India

Abstract

The principal objective of crystal engineering is to understand the intermolecular covalent and non-covalent interactions in terms of self-assembly towards the design of novel functional materials for potential importance. By judicious selection of stereo-electronic preferences of the metal ions, spatial disposition of binding sites of the ligands and reaction conditions, the overall structure and consequently the functional properties of coordination compounds/polymers can be altered in a controlled manner. Studies on binary and ternary copper(I) complexes have been increasing because the flexibility of these ligands and the different steric characteristics of the co- ligands, can modify the compound geometry systematically. They are very important for their semiconducting, magnetic, photochromic and enriched photo-luminescence properties and their potential applications in solar energy conversion, light-emitting devices, luminescence-based sensors and probes of biological systems. There is a huge scope of research of these complexes both from the fundamental science and practical applications. In this chapter, we have reported crystal engineering pathways for

* Corresponding Author's Email: patra29in@yahoo.co.in.

In: Materials Science: Future Aspects
Editors: Kalpana Awasthi, Arti Srivastava and Mridula Tripathi
ISBN: 978-1-68507-843-0

developing binary pyridyl and ternary pyridyl and triphenylphosphine based supramolecular polymeric complexes of Cu(I).

Keywords: Cu(I) complexes, pyridyl ligands, binary complex, ternary complex, triphenylphosphine, coordination polymers, supramolecular architectures, crystal engineering, CuX (X = Cl, Br, I)

1. Introduction

1.1. A Brief Introduction to Supramolecular Chemistry

A supramolecule can be defined as a macromolecule or an array of small molecules and/or ions held by non-covalent interactions to result in an assembly having chemical and physical properties completely different from the starting motifs [1].

Atoms ⟶ Molecules ⟶ Supramolecule

According to F. Vogtle, "Supramolecular chemistry is defined as chemistry 'beyond the molecule', as the chemistry of tailored-shaped intermolecular-interaction. In 'supramolecules' information is stored in the form of structural peculiarities. Moreover, not only the combined action of molecules is called supramolecular, but also the combined action of characteristic parts of one and the same molecule."

According to J.-M. Lehn, "Supramolecule is an *oligo* molecular species that result from the intermolecular association of a few components based on the principles of molecular recognition." Nobel prize in chemistry in 1987 was awarded to Prof. J.-M. Lehn, C.J. Pedersen and D.J. Cram, for their breakthrough work in Supramolecular Chemistry. Supramolecular chemistry deals with the architecture and properties of polymeric arrays. The main driving forces in Supramolecular assemblies are

- Electrostatics (ion-ion, ion-dipole and dipole-dipole)
- Hydrogen bonding (4 – 120 kJmol-1)
- π - π stacking interactions (0 – 50 kJmol-1)
- Cation - π interaction (5 – 80 kJmol-1)
- Van der Waals forces (< 5 kJmol-1
- Hydrophobic effect

2. An Overview of Coordination Polymer

Polymeric metal-ligand frameworks are called coordination polymers. A coordination polymer is an entity constructed by a supramolecular approach through metal coordination and consisting of a backbone, which is held together by metal-ligand interactions. Similar supramolecular architectures are also called Metal-organic frameworks (MOFs) [2]. Coordination polymers are inorganic structures containing metal cation centres linked by ligands, extending in an array (Figure 1). They can also be described as polymers whose repeat units are coordination complexes, and coordination networks [3].

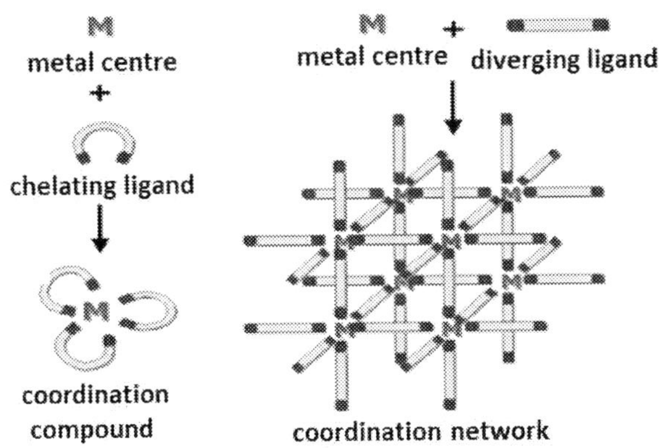

Figure 1. Formation of coordination network.

The structure of coordination polymers often times incorporates empty space in the form of pores or channels [4, 5]. This empty space is thermodynamically unfavourable. In order to stabilize the structure and prevent collapse, the pores or channels are often occupied by guest molecules. Guest molecules do not form bonds with the surrounding lattice but sometimes interact via intermolecular forces, such as hydrogen bonding or π-stacking. Most often, the guest molecule will be the solvent in which the coordination polymer was crystallized in, but can really be anything (other salts present, atmospheric gases such as oxygen, nitrogen, carbon dioxide, etc.). The presence of the guest molecule can sometimes influence the structure by supporting a pore or channel, where otherwise none would exist. Figure 2 shows different structural motifs and host-guest molecules.

3. Versatile Coordination Architectures of Cu(I)

Cu(I) ion is a well-suited cation because of its labile coordination modes with coordination numbers 2-5. It is well known that metal-organic polymeric compounds of Cu(I) exhibit photoluminescence properties and have potential applications in optoelectronic devices, fluorescent sensors and probes. Cu(I) salts catalyze many organic transformations like aryl guanidinylation, amination, photocycloaddition, oxidation of adrenaline, ferro-cytrochrome *c* oxidation etc. Copper complexes and clusters have attracted chemists because; most of these species play unique roles in both physical and biological research and application [6]. The photochemical and photophysical properties of copper(I) complexes have been studied in light of the d^{10} electronic configuration which affects their luminescent behaviour [7-9].

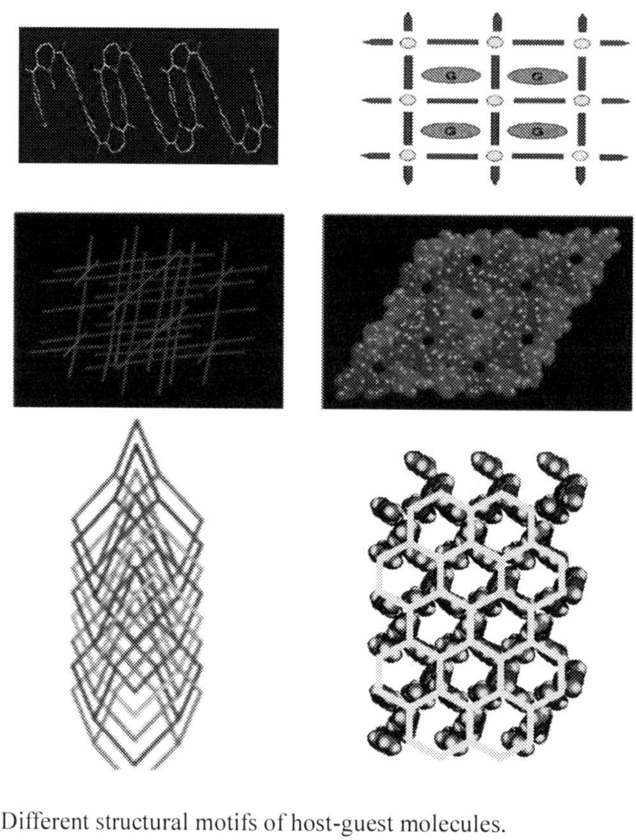

Figure 2. Different structural motifs of host-guest molecules.

Coordination chemistry of copper(I) complexes with polydentate ligands are particularly relevant to the search for model compounds that can mimic or even ideally duplicate some of the important physical and chemical properties of Cu(I) containing proteins [10, 11]. In 1961 Hathaway et al. [12] reported the preparation and properties of some tetrakis(methyl cyanide) copper(I) complexes. In 1983 J.M. Lehn and coworkers [13] reported a dinuclear Cu(I) complex $[Cu_2(pQP)_2](ClO_4)_2$ of a special quaterpyridine ligand pQP (Figure 3), in which two pQP molecules bind to two Cu(I) ions. Here Cu(I) adopts a distorted tetrahedral coordination geometry using a bipyridine subunit from each qaterpyridine chain.

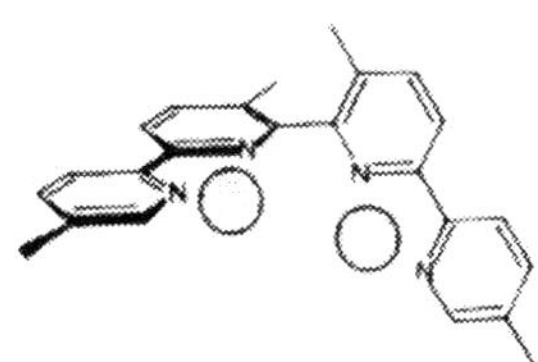

Figure 3. Structural formula for a single pQP unit.

In 1986 Lehn et al. [14] reported one dinuclear and another trinuclear Cu(I) complex derived from two oligo pyridine ligands containing two and three 2,2′- bipyridine subunits respectively, separated by 2- oxapropylene bridges. The copper(I) complexes of the above two ligands have been obtained by treating chloroform solution of the ligand with acetonitrile solution containing a slight excess of Cu(I) (perchlorate or trifluoroborate). These complexes can also be prepared by adding an aqueous solution of Cu(II) salt to a suspension of the ligand (BP_2 or BP_3) in acetonitrile followed by reduction with aqueous hydrazine (Scheme 1). Both the complexes are quite stable in air. The two Cu(I) complexes contain two ligand molecules and two or three Cu(I) ions. They are chiral and have double-helical structures in which two molecular strands are wrapped around two or three Cu(I) ions (Figure 4). The most important feature of this work is, the characteristic properties of the trinuclear complex resemble the DNA double helix.

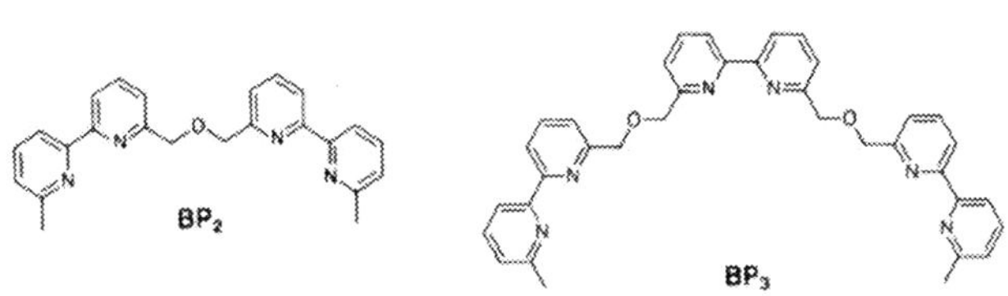

Scheme 1. Chemical Structure of the two ligands used in the study of J.-M. Lehn.

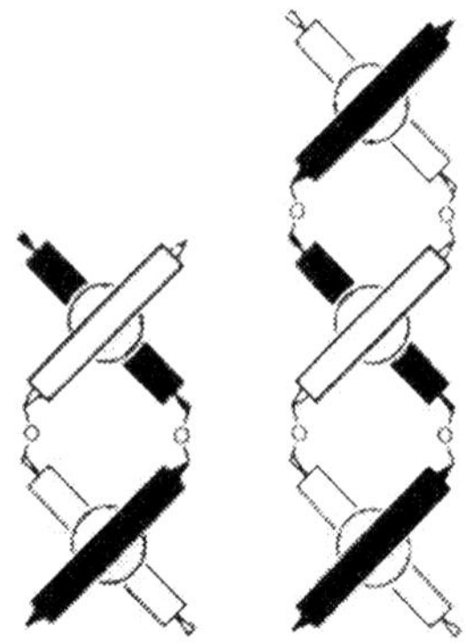

Figure 4. Schematic representation of the double helicates $[Cu_2(BP_2)_2]^{2+}$ and $[Cu_3(BP_3)_2]^{3+}$.

In 1998, Fleming et al. [15] synthesized a tetradentate ligand **L**, containing two bidentate pyridyl-pyrazole moieties linked by a flexible o-$CH_2C_6H_5CH_2$ spacer (Scheme 2). The coordination behaviour of the ligand towards Cu(I) and Cu(II) was studied. With Cu(I), $[Cu(MeCN)_4]$ $[PF_6]$, the ligand forms a dinuclear double helicate $[Cu_2\mathbf{L}_2]$ $[PF_6]_2$ by bridging two metal ions, but while reacting with Cu(II) the ligand acts as tetradentate chelate to a single metal centre to form a Cu(II) complex [Cu**L**Cl] $[BF_4]$ (Figure 5).

In 2006, Marie Hutin and coworkers [16] reported tricopper double helicate (**1**) synthesized from a reaction between 8-aminoquinoline, 1,10-phenanthroline -2,9-dicarbaldehyde and copper(I) tetrafluoroborate (Scheme 3). The dynamic covalent imine (C = N) bonds facilitate the participation of the complexes in two such reactions which were unknown in helicate chemistry earlier.

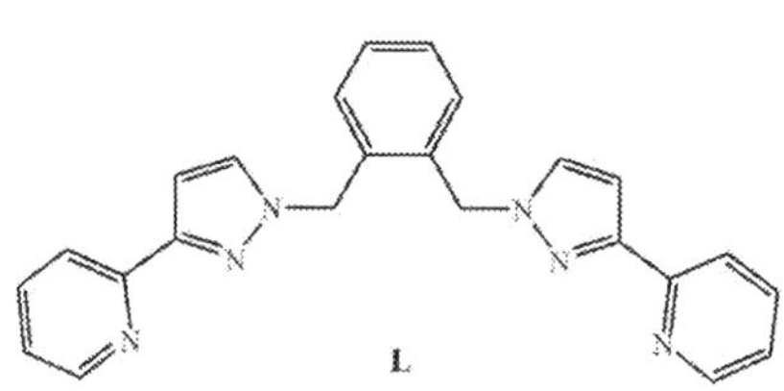

Scheme 2. Structure of the ligand used in this study.

Constable, Potts and Thummel [17-19] also reported similar helicate Cu(I) complexes containing three Cu(I) centres which were surrounded by two ligands each having six sp^2 nitrogen donor atoms. R.N. Mukherjee's group [20] reported a copper(I) complex $[(L^2)$ $Cu(MeCN)_2][ClO_4]$ synthesized from a new bidentate N-donor ligand L^2, 1-benzyl-[3-(2-pyridyl)] pyrazole. K. Saito et al. [21] prepared a series of Cu(I) mixed-ligand complexes containing dmp (2,9-dimethyl-1,10-phenanthroline) and one of the simple diphosphine ligands $[Ph_2P(CH_2)_nPPh_2]$. The two complexes $[Cu(dppp)(dmp)]PF_6$ and $[Cu_2(dppb)_2(dmp)_2](PF_6)_2$ were characterized by X-ray structure analysis. [dppp = 1,3-bis(diphenylphosphino) propane] and [dppb = 1,4-bis(diphenylphosphino) butane]. R.P. Houser and coworkers [22] synthesized four Cu(I) complexes from a family of pyridylmethylamide ligands. A. Caselli et al. [23] have prepared and characterized two new Cu(I) complexes with chiral pyridine containing macrocyclic ligands. Their use in catalytic asymmetric cyclo-propanation reactions has been investigated.

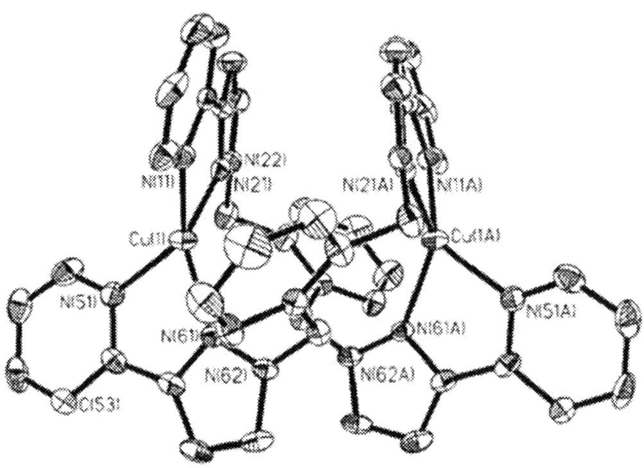

Figure 5. View of the crystal structure of$[Cu_2L_2][PF_6]$.

4

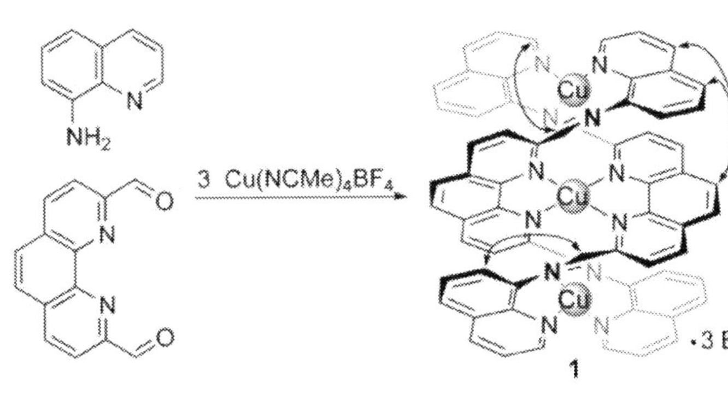

(1)

Scheme 3. Synthesis of the Cu(I) helical complex.

S. Kitagawa and coworkers [24] reported three novel Cu(I)–C_2H_4 adducts from 4-(2-pyridyl) pyrimidine ligand. They synthesized the three complexes by using $[Cu(C_2H_4)_n]NO_3$ and $[Cu(MeCN)_4]BF_4$ as metallating agents. Changing of solvent from Me_2CO to MeOH gave a different product. Y. Wang et al. [25] reported the first optical oxygen sensor, based on Cu(I) complex-polystyrene composite nanofibrous membranes showing high sensitivity (I_0/I_{100} = 15.56), good linear Stern-Volmer characteristics (R^2 = 0.9966) and short response/recovery time, these results represent the best values reported for oxygen sensors based Cu(I) complexes. Generally, the copper(I) halide aggregates, ($Cu_xX_yL_z$) where X = Cl, Br or I; L = N, S or P donating organic ligand, could exist as discrete compounds or as CuX-based building blocks in coordination polymers. The investigation of copper(I) halide aggregates probably started a hundred years ago, when the technique of X-ray diffraction crystallography was unexplored in chemical research.

In 1970s, the first and so far most prominent member in the copper(I) halide family, $Cu_4I_4py_4$, was reported by Hardt and Pierre. They discovered the significant luminescent property of the $Cu_4I_4py_4$ cluster which they termed "luminescence thermochromism" [26]. Hardt and co-workers also reported the existence of at least three different complexes with the empirical formula of $CuIL_z$ (z = 1, 2 or 3), regulated by reaction stoichiometry [27]. They suggested the structural configuration of the $Cu_4I_4py_4$ cluster mostly on the basis of indirect spectroscopic characterization since X-ray diffraction crystallography was unknown till then. In1976, the first crystallographic evidence for the $Cu_4I_4py_4$ tetramer revealed its tetrahedral cubane-like configuration [28]. Ford et al. [29] did an extensive study on the luminescent properties of $Cu_4I_4Py_4$ and its derivatives both experimentally and theoretically [30].

In recent years inorganic clusters or aggregates can be incorporated in metal complexes as "connectors" in place of single metal ions. Champness and coworkers did the preliminary research and understood that copper halide aggregates could be incorporated in coordination polymers [31]. Moreover, using bridging ligand (connecting the copper halide aggregates) instead of monodentate or bidentate chelating ligands could give rise to more complicated copper(I) halide architectures. Thus, the crystal engineering of copper(I) halides extended from only discrete oligomers to one-dimensional (1D) or even two-dimensional(2D) and three-dimensional (3D) architectures.

A group of Swedish and Australian chemists composed a review on anionic copper halides (which they termed "halocuprate complexes") based on their structural diversity [32]. They considered the crystal engineering of copper halides as a "difficult system." Holt et al. [33] obtained three CuI crown ether-based crystal structures–a discrete compound, namely $[Cu_4I_6][Cu_8I_{13}]K_7(12\text{-crown-}4)_6$, containing two anionic copper(I) iodide clusters, and two 1D coordination polymers, namely $[Cu_4I_6]K_2(15\text{-crown-}5)_2$ and $[Cu_3I_4]K(\text{dibenzo-}24\text{-crown-}8)$ – from the traditional solution reaction of CuI and different crown ethers in the presence of aqueous KI.

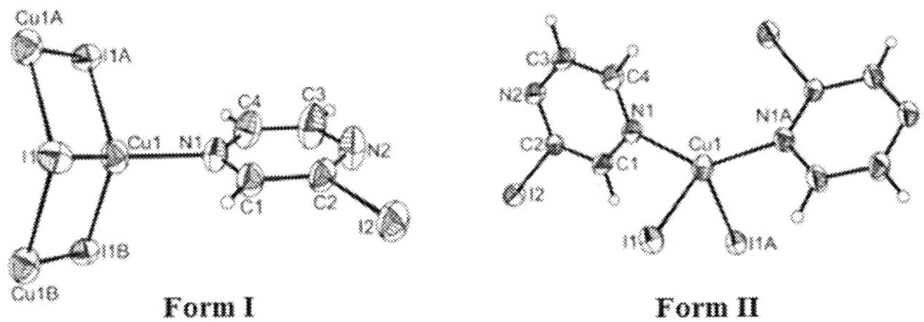

Figure 6. Structures of Form I & II of CuI(2-iodopyrazine).

In 2003 Nather et al. [34] reported two modifications of the new copper(I) coordination polymer CuI(2-iodopyrazine), obtained by the reaction of CuI and 2-iodopyrazine in acetonitrile. During this reaction, bright yellow crystals of form **I** appeared first which transformed to bright red coloured crystals of form **II** within a few minutes. On heating both forms decomposed to copper(I) iodide (Figure 6). Crystal structure of catena[CuI(2-iodopyrazine-*N*)] and crystal structure of poly[CuI(*í*-2-iodopyrazine-*N*,*N*′)] have been shown in Figures 7 and 8.

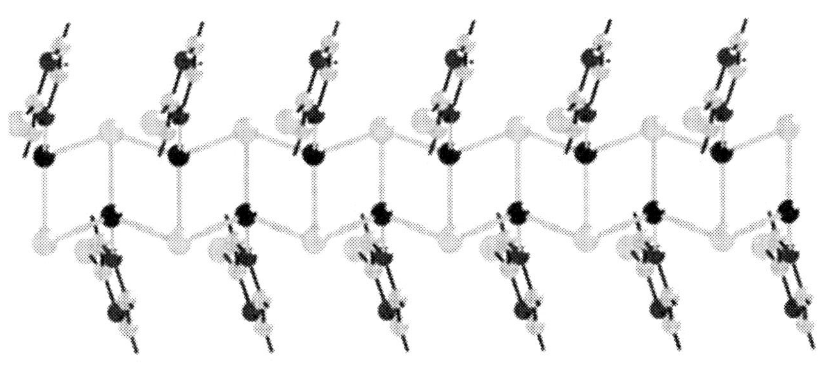

Figure 7. Crystal structure of catena [CuI(2-iodopyrazine-*N*)] with view along the crystallographic *b*-axis.

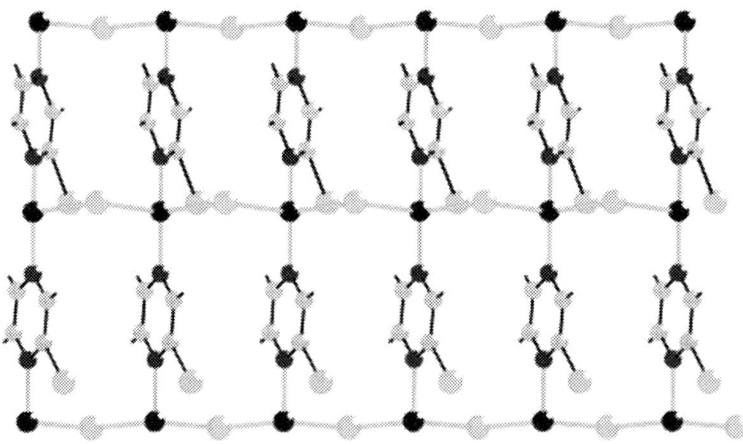

Figure 8. Crystal structure of poly[CuI(*i*-2-iodopyrazine-*N*,*N*/)] with view along the crystallographic *b*-axis.

In 2004, Nather et al. [35] reported a ligand rich 1:1 compound poly-[CuCl(μ_2 -pyrimidine – N,N /) **(2)** by reaction of copper chloride with pyrimidine in acetonitrile which upon heating loses half of its ligands and transformed into the new 12:1 coordination polymer poly – [$(CuCl)_2$ (μ_2 - pyrimidine – N,N /) **(3).** If copper(I) iodide is used in the synthesis instead of CuCl the new 1:1 coordination polymer poly [CuI(μ_2 -pyrimidine – N,N /) **(4)** forms which on heating transformed into the new 2:1 compound poly [$(CuI)_2$(μ_2 -pyrimidine – N,N /) **(5)** (Figure 9).

Elena Cariati and coworkers [36] described the structural, emissive and nonlinear optical properties of new CuI adducts with para-substituted trans-stilbazolic and pyridine ligands (Scheme 4). Single crystal X-ray diffraction results indicate that the para substituent on the organic ligand greatly influences the structural motif by its steric (tert butyl), electronic/steric (dimethylamino) or bridging donor (cyno) properties. Two absolutely new structural motifs, polymeric and oligomeric, are found when trans stilbazole and pyridine carry a dimethylamino group in para position. A surprising photoemission behaviour is observed. They synthesized CuI adducts by reacting CuI with the following ligand systems.

Dan Li et al. [37] reported linear chain polymers $[Cu(pymtH)X]_n$ (X = Cl, Br), obtained from the reaction of copper(I) chloride or bromide with neutral pyrimidine -2-thione ligand (pymtH) which can act as a bridging N,S donor. In contrast copper(I) iodide under same conditions gave the dimeric complex $[Cu(pymtH)_2I]_2$ with the (pymtH) ligand (Scheme 5).

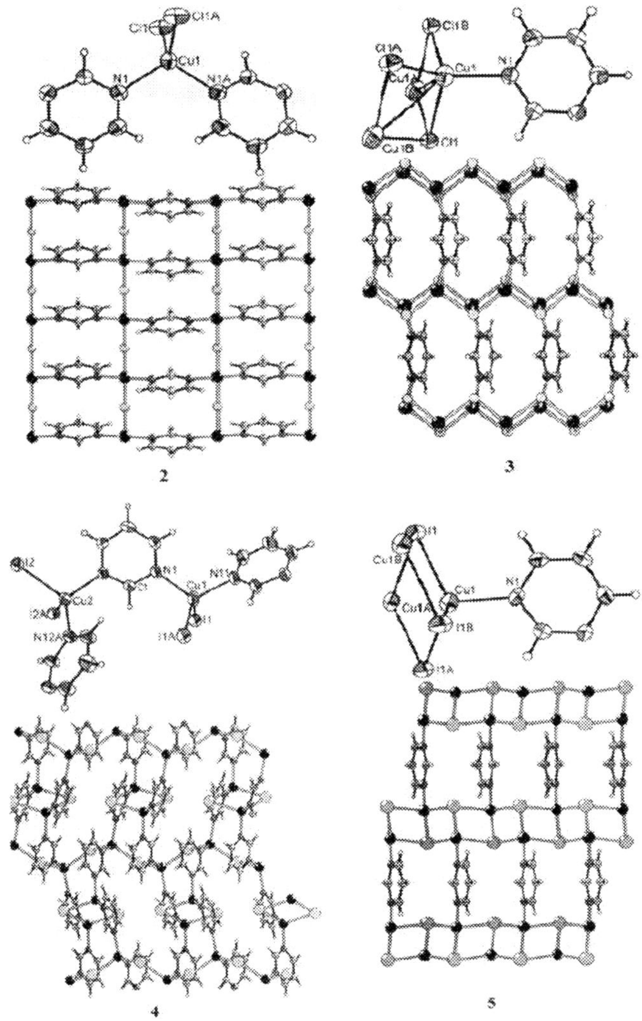

Figure 9. Crystal structures of **2**, **3**, **4** and **5**.

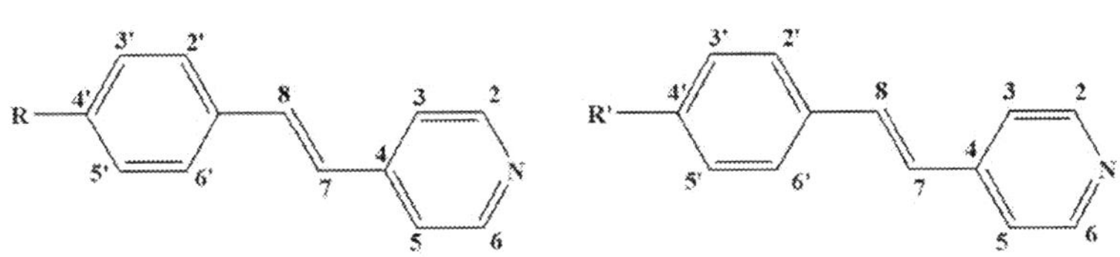

R = H, NMe_2, ^tBu and R' = NMe_2

Scheme 4. Ligand systems used in this study Elena Cariati and coworkers[36].

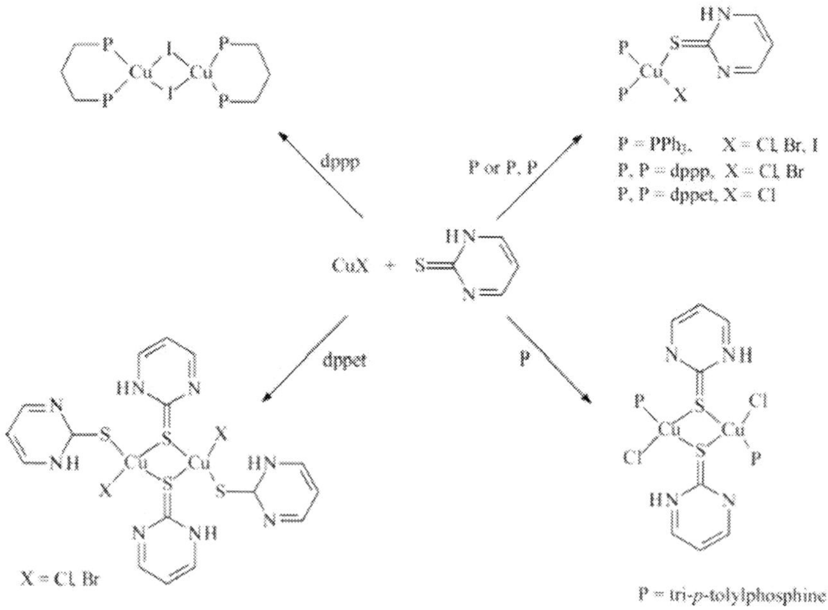

Scheme 5. Synthesis of the ternary Cu(I) complexes.

Thebault et al. [38] prepared and characterised a family of coordination polymers [CuI(dpt)]$_n$ (**6**) [dpt = 2,4-bis(4-pyridyl)-1,3,5-triazine], [CuI (dpb)]$_n$(**7**)[dpb = 1,4-bis-(4-pyridyl)-benzene], [(CuI)$_3$(dpypy)$_2$]$_n$ (**8**) [CuI (dpypy)]$_n$(**9**)[dpypy = 3,5-bis(4-pyridyl)- pyridine], and [Cu$_3$I$_3$(pypm)]$_n$(**10**) [pypm = 5-(4-pyridyl)pyrimidine] by reacting CuI with a range of angular bidentate or tridentate N-donor ligands (Figure 10). The architectures of the coordination polymers were controlled by ligand design (Scheme 6).

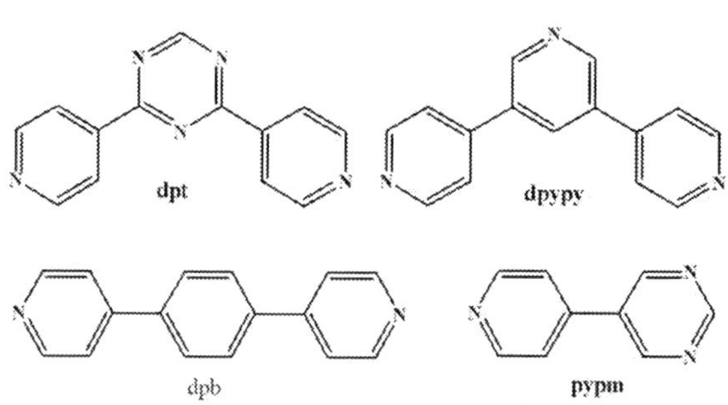

Scheme 6. Ligand systems used in this study of Thebaultet al.[38].

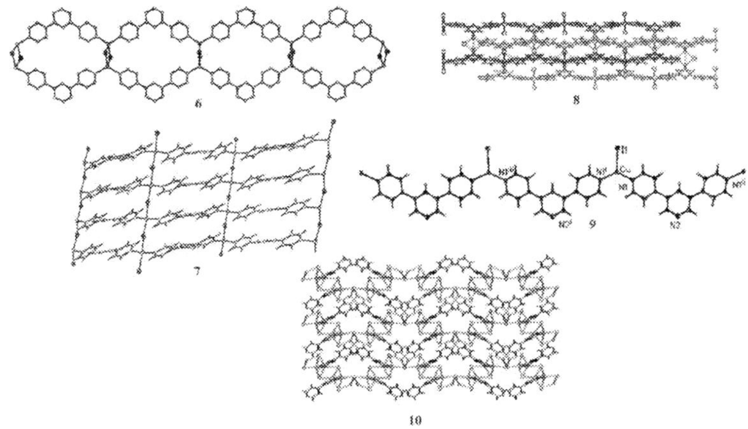

Figure 10. Crystal structures of **6**, **7**, **8**, **9** and **10**.

T. H. Kim et al. [39] reported two coordination polymers $\{[Cu_4I_4(L_3)_2](CH_3CN)_{0.5}\}_n$ (**11**) and $[CuI(L_4)]_n$ (**12**), prepared from reaction of copper(I) iodide with two asymmetric dithioether ligands with cyclohexyl (L_3) and phenyl (L_4) end-groups. L_3 with CuI gave a 1D channel-type coordination polymer **(11)** interconnected by cubane-type tetranuclear Cu_4I_4 cluster units, whereas, a 2D brick-wall-type coordination polymer **(12)** with rhomboid dinuclear $Cu–I_2–Cu$ nodes were isolated from the reaction of L_4 with copper(I) iodide (Scheme 7).

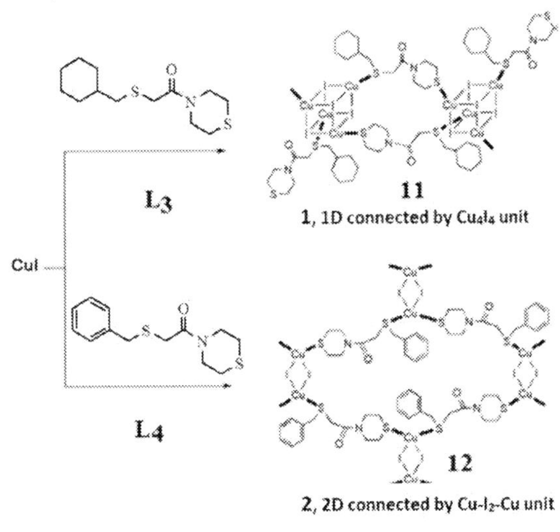

Scheme 7. Reaction scheme for the syntheses of **11** and **12**.

Wu et al. [40] reported five new copper(I) iodide coordination polymers, $(Cu_{10}I_{10})$ $(Cu_6I_6)[(Cu(Bta)_2]_3^-$ $(DMBta)_4^+\cdot(I^-)\cdot x(I_2)$ ($x \approx 1.5$) (**13**), $[Cu_2I_2(MBta)]$ (**14**), $[Cu_2I_2(MTa)]$ (**15**), $(Cu_5I_5)[Cu(Ta)_2]^- \cdot (DETa)^+$(**16**), and $[Cu_3I(Bta)_2]$ (**17**) [(Bta) benzotriazole, DMBta = 1,3-*N*-dimethyl-benzotriazolium, MBta = 1-*N*-methylbenzotriazole, MTa = 1-*N*-methyl-triazole, Ta = 1,2,3-triazole, DETa = 1,3-*N*-diethyltriazolium)], synthesized by the solvo (hydro) thermal reactions of CuI, KI, and benzotriazole (or 1,2,3-triazole) in methanol (or ethanol or water) and structurally characterized. Complex (**13**) shows an unusual 2D→3D interpenetrated anionic network containing two new, discrete $Cu_{10}I_{10}/Cu_6I_6$ cluster units. Complex (**14**) and (**15**) show similar 1D double-stranded stairlike ribbon structures. Complex (**16**) exhibits 2D anionic layers which consist of Cu_5I_5 cluster units and complex (**17**) shows a 2D simple sheet structure (Figure 11).

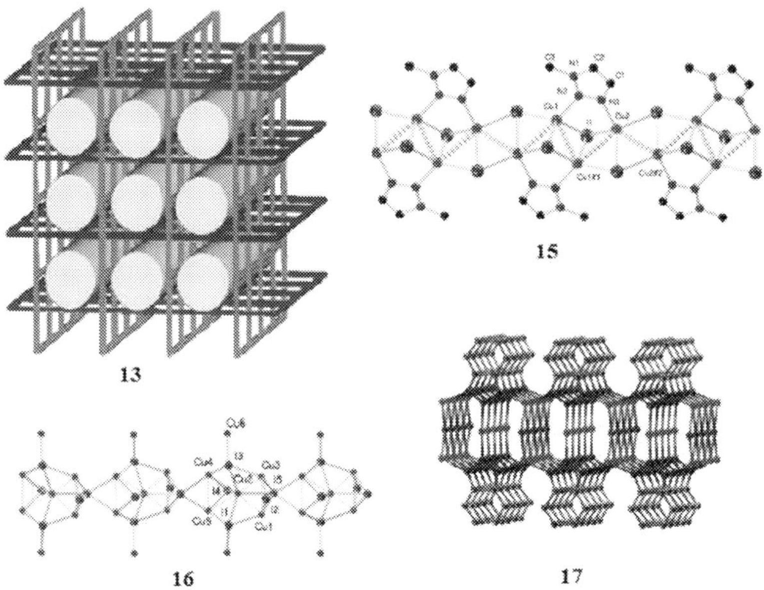

Figure 11. Bridged structures of **13**,**15**,**16** and **17**.

Hao and Zhang [41] reported three phosphorescent acentric cuprous halide coordination polymers, namely, $[Cu_2(pyt)Cl]$ (**18**), $[Cu_2(pyt)Br]$ (**19**), and $[Cu_5(pytH)_3Cl_5]$ (**20**), synthesized through the treatment of copper halides with achiral ligand pyridine-4-thiol (pyHt)by the hydro(solvo) thermal method (Figure 12).

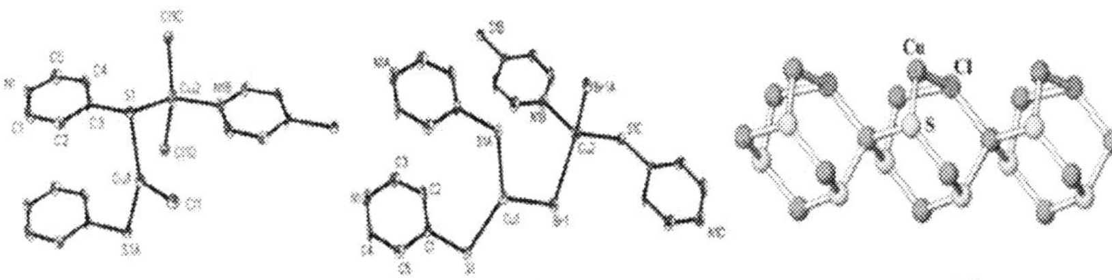

18 19 20

Figure 12. The Coordination environments of copper sites in **18, 19** and **20.**

T. H. Kim et al. [42] reported one Copper (I) coordination polymer $[Cu_3I_3L_5]_n$ **(21)** synthesized from the reaction between CuI and 1,4-bis((cyclohexylthio)acetyl)piperazine (L_5). Structural and photophysical studies of **21** in solid state were reported. The bonding of Cu_3I_3 produces staircase polymers interconnected by the sulfur atoms of L_5 (Figure 13).

Yang Chen and coworkers [43] prepared Four $[Cu_2I_2]$- or $[Cu_4I_4]$-based coordination polymers, $[CuI(bpp)]_n$(**22**), $\{[Cu_3I_3(bpp)_3]\cdot 2aniline \cdot MeCN\}_n$ (**23**), $\{[Cu_2I_2\text{-}(bpp)_2]\cdot 2aniline\}_n$ (**24**), and $[Cu_4I_4(bpp)_2]_n$ (**25**), by solvo thermal reactions of CuI or $Cu_2(OH)_2CO_3$ with 1,3-bis(4-pyridyl)propane (bpp) in different solvent systems (Figure 14). Solvent effects on the assembly of coordination polymers along with the luminescent properties were studied.

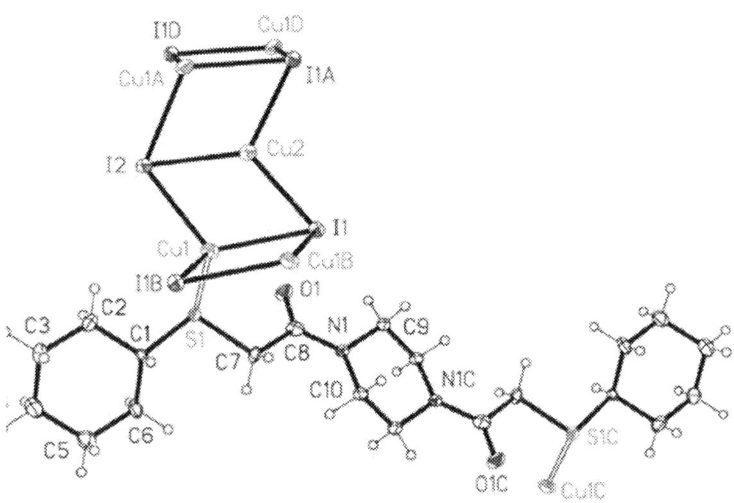

Figure 13. ORTEP drawing of **21** showing 50% probability thermal ellipsoid.

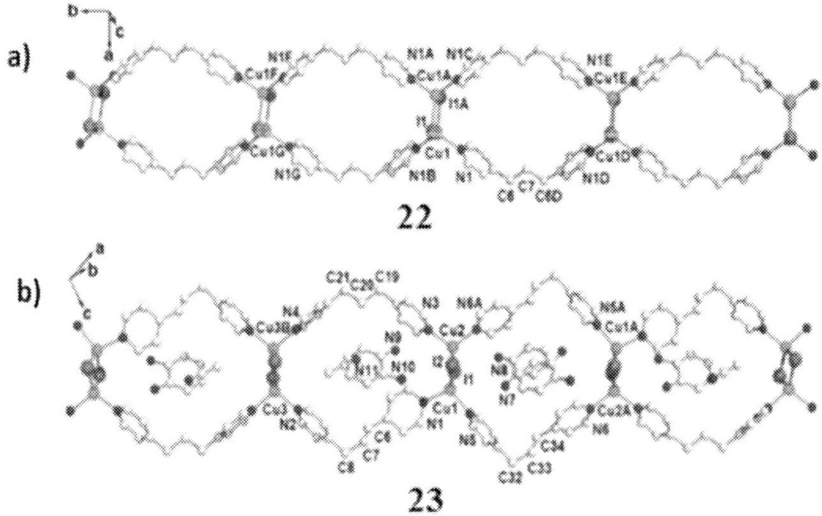

Figure 14. (a) View of a section of the 1D double chain of **22** (b) View of a portion of the 1D double chain of **23**.

Mian Li and coworkers [44] reported three anionic-tetrazolate-based coordination polymers, namely $[Cu_2(\mu_3\text{-I})(\mu_5\text{-Cpta})]_n$**(1)**, $[Cu_5(\mu_4\text{-I})(\mu_4\text{-Mtta})_3(CN)]_n$**(24)** (Figure 15) and $[Cu_5(\mu_6\text{-I})(\mu_2\text{-I})(\mu_4\text{-Mtta})_3]_n$**(25)** [Cpta = 5-(3-cyanophenyl)tetrazolate, Mtta = 5-methyltetrazolate], which exhibit structure-related green, cyan and blue luminescence, respectively.

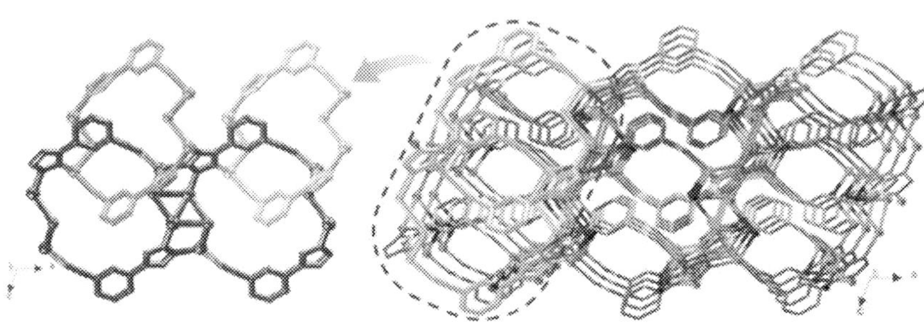

Figure 15. Complex **24**; the overall 3D framework.

Wei and coworkers [45] synthesized and characterized an intriguing supramolecular tetrahedral aggregate $[Cu_4I(ptt)_3(Hptt)_3]_4$ **(26)**; (Hptt = 1-phenyl-1*H*-tetrazole-5-thiol) based on novel copper(I) iodide thiolate clusters in the solid-state (Figure 16). This complex exhibits temperature-dependent green photoluminescence.

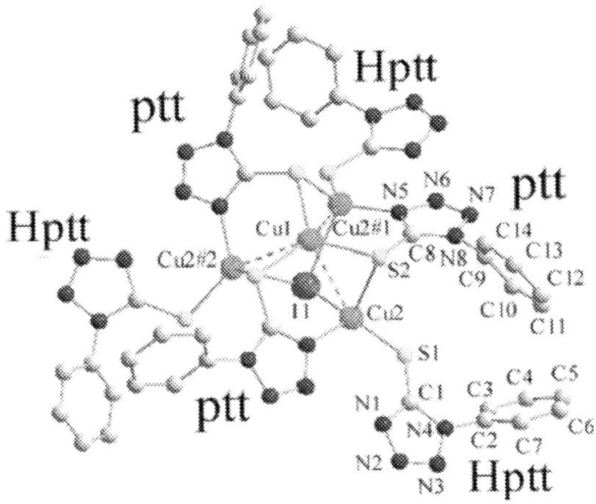

Figure 16. View of the metal-cluster molecule with C_3 symmetry in **26.**

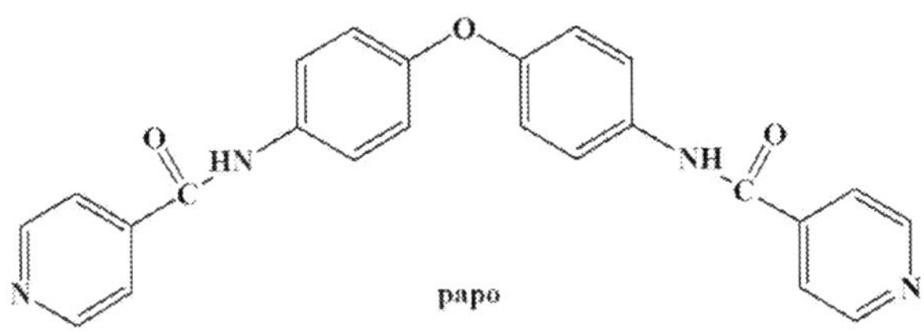

Scheme 8. Ligand used in the study of Biing-ChiaoTzeng and Tsung-Yi Chang [46].

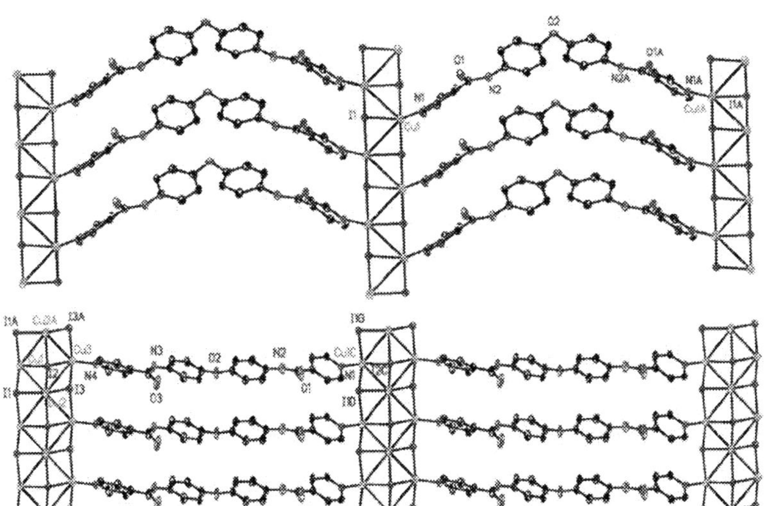

Figure 17. The 2-D ORTEP diagram of **27** and **28.**

Biing-ChiaoTzeng and Tsung-Yi Chang [46] reported four new metal-organic frameworks derived from copper(I) iodide and N,N$^{/}$-bis (pyridylcarbonyl)-4,4$^{/}$-diaminodiphenyl ether (papo)(Scheme 8). In the absence of CH_3CN the reaction gave [CuI(papo)]$_n$ (**27**) and [Cu_3I_3(papo)]$_n$**(28)** (Figure 17), whereas in the presence of CH_3CN complex [Cu_3I_3(papo)]$_n$(**29**) and [Cu_8I_4(dmtrz)$_4$]$_n$**(30)**;(dmtrz = 3,5-dimethyl-1,2,4-triazolate) were obtained. In the same pot in the presence of CH_3CN, a red and mononuclear Cu(II) complex, [Cu(dmtap)$_2$]I_2**(31)**, (dmtap = 2,4-dimethyl-1,3,5-triazapentadiene), was also isolated.

Li Qin et al. [47] reported the synthesis and characterization of two novel mixed-ligand CuI complexes [Cu(NP)(DPEphos)]$^+$, where NP = 8-dipheny-lphosphanylquinoline (dppq) (**31**) and 2-methyl-8-diphenyl-phosphanyl-quinoline (mdppq) (**32**) (Figure 18). They exhibit photoluminescence properties.

Ying Zhang and coworkers [48] reported three 3-D copper iodide frameworks, (Cu_4I_4)(DABCO)$_2$(solvent)$_x$ (**33**), (Cu_6I_6)(Cu_8I_8)(DABCO)$_5$-(imidazole) (**34**), and (Cu_2I_2)$_{1/2}$(Cu_5I_5)(DABCO)$_2$ (**35**), constructed with DABCO ligands (DABCO = 1,4-diazabicyclo-[2.2.2]-octane)and various neutral CuI clusters. 1, 2, and 3 display an interpenetrating diamond-type net, a 5-connected interpenetrating net, and a (4,6)-connected net, respectively (Figure 19). Compound **35** displays chiral symmetry, while compound **34** shows photoluminescence at 298 K.

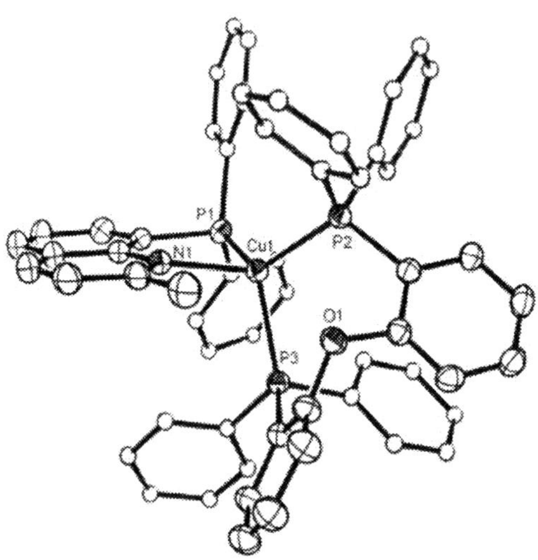

Figure 18. ORTEP diagram of the complex cation in **32**.

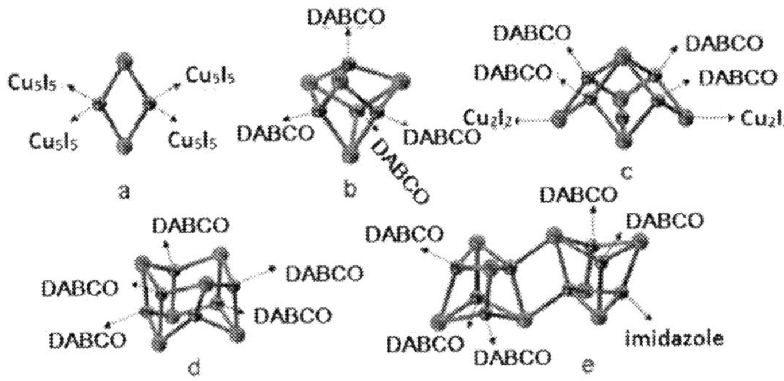

Figure 19. Various neutral CuI clusters and their connectivity: (a) Cu_2I_2; (b) Cu_4I_4; (c) Cu_5I_5; (d) Cu_6I_6; (e) Cu_8I_8.

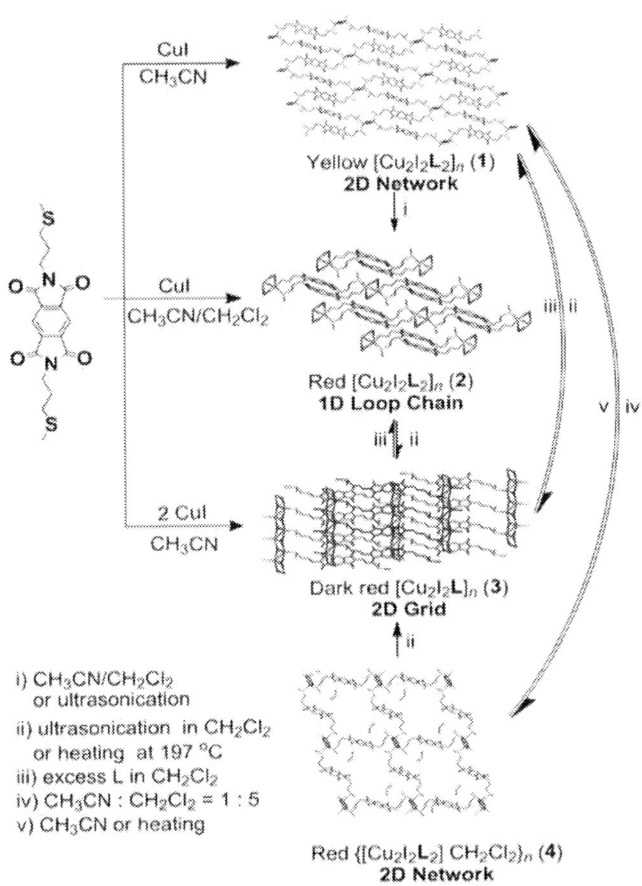

Scheme 9. Preparation of coordination polymers **36-39.**

Park et al. [49] synthesized and structurally characterized four copper(I) coordination polymers $[Cu_2I_2L''_2]_n$**(36)**, $[Cu_2I_2L''_2]_n$**(37),** $[Cu_2I_2 L'']_n$ **(38),** and $\{[Cu_2I_2L''_2]_3\ CH_2Cl_2\}_n$ (**39**), reacting ligand L" = N,N$^{/}$-bis[3-(methylthio) propyl]pyromelliticdiimid, with copper iodide in acetonitrile and in dichloromethane-acetonitrile solution. Structural transformations between the polymers were controlled by the appropriate solvent composition, mole ratio, or temperature (Scheme 9).

Jinfang Zhang [50] reported one copper(I) complex $\{[Cu_2\ I_2(C_{12}H_8N_6)(C_{18}H_{15}\ P)_2].2CH_3\ CN\}_n$ derived from centrosymmetric dinuclear $Cu_2\ I_2(PPh_3)_2$ complex unit and 3,6-di-pyridyl- 1,2,4,5- tetrazine ligands (Scheme 10).

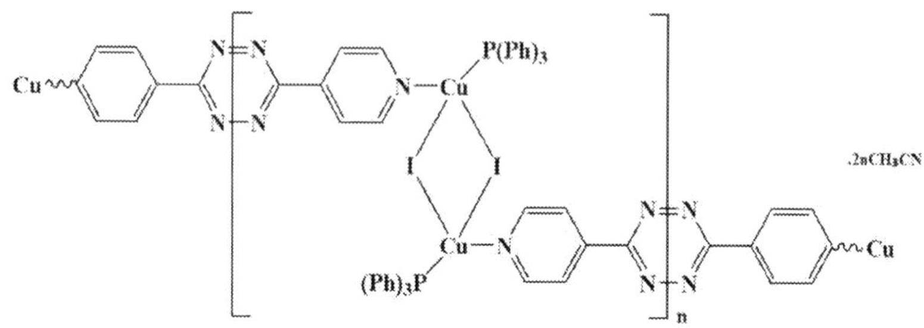

Scheme 10. Chemical structure of the complex.

Bhattacharya et al. [51] has synthesized two novel 3-D coordination polymers of the ligands N, N Bis-pyridin-4-yl-methylene hydrazine (L_1") and N, N /-bis-(1-pyridin-4-yl-ethethylidene)- hydrazine (**L_2"**). Ligands **L_1"&** **L_2"** are shown in Scheme 11.

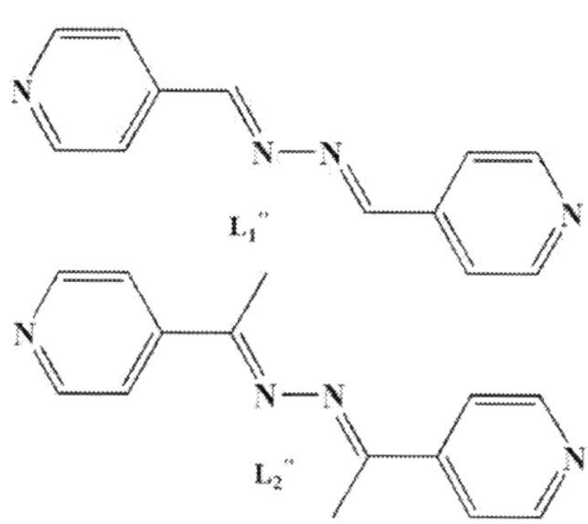

Scheme 11. Chemical structures of polydentateimino-pyridyl ligands.

The complex **40** crystallizes in the triclinic P $\bar{1}$ space group having one formula unit in the asymmetric unit, complex **41** crystallizes in the tetragonal P $\bar{4}$b2 space group with half formula unit included in the asymmetric unit. The Cu(I) atom in both the complexes adopts a slightly distorted tetrahedral geometry generated by the binding of one of the pyridyl nitrogens of the ligands (Figure 20). The average Cu-N bond distances around the Cu(I) ion in **40** and **41** are 2.042 and 2.038 Å and are consistent. The N-Cu-N angles lie within the range of 102.6(4)° to 113.3(4)° in **40**, while in **41** it ranges between 100.8(5)° - 119.8(6)° indicating distortion from the ideal tetrahedral system.

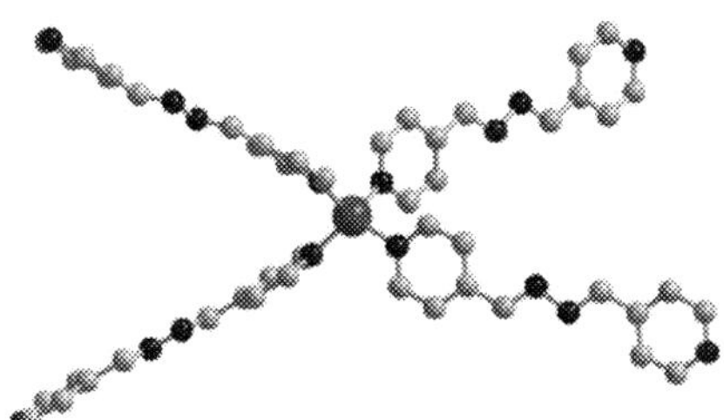
40

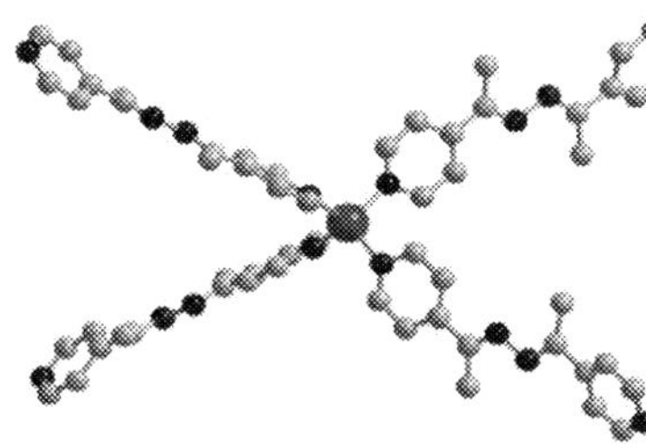
41

Figure 20. Molecular building blocks of **40** and **41**.

Mondal et al. [52, 53] synthesized six new copper (I) coordination polymers $[Cu_2(\mu\text{-}L_3'')(\mu\text{-}I)_2)(PPh_3)_2]_n$(**42**), $[Cu_2(\mu\text{-} L_3'')(\mu\text{-}Br)_2(PPh_3)_2]_n$ (**43**) and $[Cu_2(\mu\text{-} L_3'')(\mu\text{-}Cl)_2(PPh_3)_2.CH_3CN]_n$ (**44**), $[Cu_2(L_4'')_2(\mu\text{-}I)_2)(PPh_3)_2]_2$(**45**), $[Cu_2(\mu\text{-} L_4'')(\mu\text{-}Br)_2(PPh_3)_2.CH_2Cl_2]_\infty$ (**46**) and $[Cu_2(\mu\text{-} L_4'')(\mu\text{-}Cl)_2(PPh_3)_2]_\infty$ (**47**). Where L_3'' and L_4'' are (1E,2E)-1,2-diphenyl-1,2-bis((1-(pyridine-4-yl)ethylidene)hydrazineylidene)ethane and (1E,2E)-1,2-diphenyl-1,2-bis ((pyridin-4-ylmethylene)hydrazineylidene)ethane respectively(Scheme 12 and Scheme 13). The crystal structures of the complexes **42** and **44** have been determined by single-crystal X-ray analysis and it has been found that these are 1-D coordination polymers containing bridged halide ions as well as the azino-pyridyl ligand L_3''. Matching of X-ray powder patterns and theoretical studies reveal that the complex **43** is also a 1-D coordination polymer. The crystal structures of the complexes **44** and **45** have been determined by single-crystal X-ray analysis and it has been found that complex **42** is a dimer with -{Cu^I-(μ-I_2)-Cu^I}- rhomboid core whereas complex **46** is a 1-D coordination polymer containing bridged Br^{-1} ion as well as the ligand L_4''. Matching of X-ray powder pattern of **47** with the simulated powder data obtained from the single crystal of **46**, and DFT studies reveal that complex **47** is also a 1-D coordination polymer like complex **46.** The ligands **L**^{1}and **L**^{2} is non-emissive at room temperature in dichloromethane, whereas the complexes **42-47** are photoluminescent. The $E_{1/2}$ values of the Cu^{I}-Cu^{II} couple for **42**, **43** and **44** are within 0.75-0.8 (vs. Ag/AgCl in 1 M KCl, scan rate 100 mV s^{-1}). The $E_{1/2}$ values of the Cu^{I}-Cu^{II} couple of **45**, **46** and **47** are 1.06 V, 0.85 V and 0.81 V (vs. Ag/AgCl in 1 M KCl, scan rate 100 mV s^{-1}) respectively.

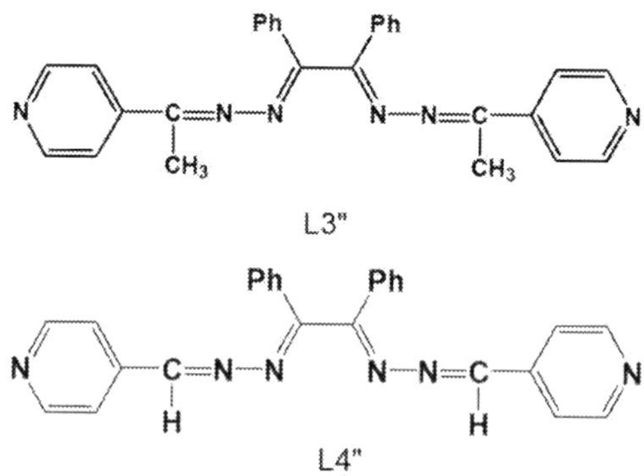

Scheme 12. Ligands system used in the study of Mondal et al. [52].

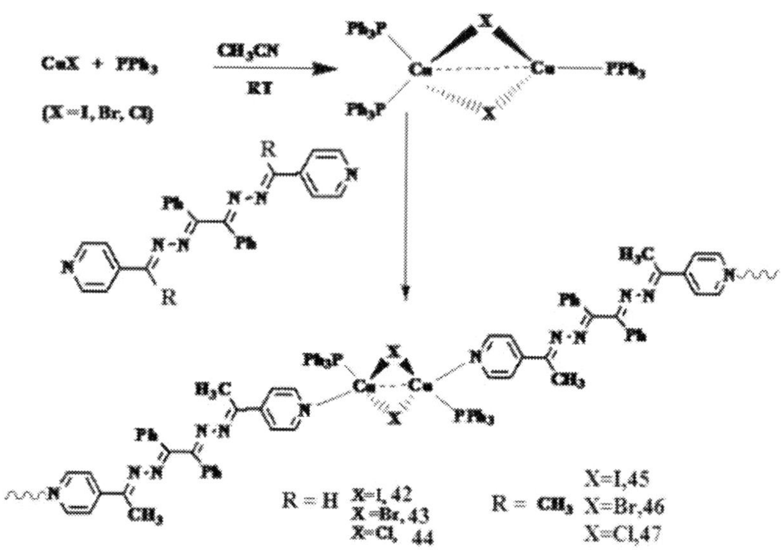

Scheme 13. Synthetic procedure for the preparation of complexes **42-47**.

The Cu(I) complexes 42-47 have been synthesized in good yields by reacting the azio-pyridyl ligands with CuX (X = I, Br and Cl) and PPh_3 in 1:1:1 proportion in acetonitrile solvent at room temperature. The syntheses of the complexes are summarized in Scheme 13 These complexes are stable in solid state for about 2–3 weeks in air.

Conclusion

In the present chapter, we have discussed the frontier of crystal engineering strategies through the binary pyridyl and ternary pyridyl and triphenylphosphine based inorganic–organic hybrid coordination oligomers/polymers of Cu(I). The electronic and structural characteristics of the N-heterocyclic ligands play an important role in determining the overall structure of the polynuclear complexes and consequently their semi-conducting, magnetic, photochromic and enriched photoluminescence properties. This study demonstrates the versatile modes by which organic ligands and metal ion auxiliaries can self-assemble into robust hybrid organic–inorganic arrays. It also illustrates several facile synthetic procedures that can be utilized in the supramolecular construction of such coordination polymers.

Acknowledgment

The authors would like to thank Guru Ghasidas Vishwavidyalaya, Bilaspur for various help.

References

[1] Steed, J.W. and Atwood J.L. (2003). *Supramolecular Chemistry.* United Kingdom: John Wiley and Sons.

[2] Lucia, C., Ciani, G. and Proserpio, D.M. (1999). *Coordination polymers and supramolecular architectures.* http://dcssi.istm.cnr.it/1995-1999/coordination_polymers.htm. Retrieved 26 March 2010.

[3] Ye, B.H., Tong, M.L. and Chen, X.M. (2005). Metal-organic molecular architectures with 2,2′-bipyridyl-like and carboxylate ligands. *Coord. Chem. Rev.* 249 : 545-565.

[4] Gao, E.Q., Cheng, A.L., Xu, Y.X., He, M.Y. and Yan, C.H. 2005. From low-dimensional manganese(II) azido motifs to higher-dimensional materials: Structure and magnetic properties. *Inorg. Chem.*44 : 8822-8835.

[5] Bu, X.H., Tong, M.L., Chang, H.C., Kitgawa, S. and Baten. S.R. (2003). A neutral 3D copper coordination polymer showing 1D open channels and the first interpenetrating NbO-type network. *Angew. Chem. Int. Ed.* 43 : 192-195.

[6] Armaroli, N., Accorsi, G., Cardinali, F. and Listorti, A. (2007). Photochemistry and photophysics of coordination compounds: Copper. *Top.Curr. Chem.* 280 : 69-115.

[7] Ford, P.C., Cariati, E. and Bourassa, J. (1999). Photoluminescence properties of multinuclear copper(I) compounds. *Chem. Rev.*99 : 3625-3647.

[8] Yam, V.W.W. and Lo, K.K.W. (1999). Luminescent polynuclear d10 metal complexes. *Chem. Soc. Rev.*28 : 323-334.

[9] Barbieri, A., Accorsi, G. and Armaroli, N. (2008). Luminescent complexes beyond the platinum group: the d10 avenue. *Chem. Commun.* 19: 2185-2193.

[10] (a) Zhang, Y., Wang, S., Enright, G.D. and Breeze, S.R. (1998). Tetraacetylethanedianion (Tae) as a bridging ligand for molecular square complexes: CoII4(Tae)4(Dpa)4, Dpa = Di-2-pyridylamine, a chiral molecular square in the solid state. *J. Am. Chem. Soc.*120 : 9398-9399; (b) Cotton, F.A., Lin, C. and Murillo, C. (2001). Supramolecular arrays based on dimetal building units. *Acc. Chem. Res.*34 : 759-771.

[11] (a) Duan, C.Y., Liu, Z.H., You, X.Z, Xue, F. and Mak, T.C.W. (1997). A self-assembled macrocyclictetra nuclear molecular square [Co(HL)]44+ [H2L = tetra(2-pyridyl)thiocarbazone]. *J. Chem. Soc., Chem. Commun.*4 : 381-382; (b) Fitchett, C.M. and Stell, P.J. (2000). Chiral heterocyclic ligands. Part 9. Homo-configurational coordination polymers based on a C2-symmetric, linear-bridging ligand. *New J. Chem.* 24 : 945-947.

[12] Hathaway, B.J., Holah, D.G., Postlethwalte, J.D., (1961). The preparation and properties of some tetrakis(methylcyanide)copper(I) complexes. *J. Chem. Soc.* 630 : 3215-3218.

[13] Lehn, J.M., Sauvage, J.P., Simon, J., Ziesse, R., Leopardi, C.P., Germain, G., Declercq, J.P., Meersche, M.V. (1983). Synthesis and metal complexes of a conformationally restricted quaterpydine. Crystal structure of its dimeric dinuclear Cu(I) complex, $(Cu_2(PQP)_2)^{2+}$. *Nouv. J. Chim.* 7 : 413-420.

[14] Lehn, J.M., Rigault, A., Siege, J., Harrowfield, J., Chevrien, B. and Moras, D., (1987). Spontaneous assembly of double-stranded helicates from oligobipyridine ligands and copper(I) cations: structure of an inorganic double helix. *Proc. Nat. Acad. Sci.* USA 84 : 2565-2569.

[15] Fleming, J.S., Mann, K.L.V., Coachman, S.M., Jeffery, J.C., McCleverty, J.A. and Ward M.D. (1998). Double-helical dinuclear copper(I) and mononuclear copper(II) complexes of a compartmental tetradentate bridging ligand: crystal structures and spectroscopic properties. *J. Chem. Soc., Dalton Trans.*12 :2047-2052.

[16] Hutin, M., Frantz, R., Nitschke, J.R. (2006). A dynamic tricopper double helicate. *Chem. Eur. J.* 12 : 4077-4082.

[17] Constable, E.C., Ward, M.D. and Tocher, D.A. (1991). Molecular helicity in inorganic complexes; bi- and tri-nuclear complexes of 2,2′:6′,2″: 6″,2‴:

6''',2'''':6'''',2'''''-sexipyridine and the crystal and molecular structure of bis(μ-2,2':6',2'':6'',2''':6''',2'''':6'''',2'''''-sexipyridine-κ3N,N', N'':κ3N''', N'''',N''''') dicadmiumhexafluorophosphate–acetonitrile (1/4). *J. Chem. Soc., Dalton Trans.*7 : 1675 –1683.

[18] Potts, K.T., Keshavarz. –K.M., Tham, F.S., Abruna, H.D. and Arana, C. (1993). Metal ion-induced self-assembly of functionalized 2,6-oligopyridines. 3. Metal-metal interaction and redox state-induced transformations in double-stranded helicates derived from functionalized quinquepyridine and sexipyridine. *Inorg. Chem.*32 : 4436-4449.

[19] Zong, R. and Thummel, R.P. (2005). 2,2':9',2' '-Ter[1,10]phenanthroline. *Inorg. Chem.* 44 : 5984-5986.

[20] Mukherjee, J. and Mukherjee, R.N. (2006). Reaction with dioxygen of a Cu(I) complex of 2-[3-(2'- pyridyl)pyrazol-1-ylmethyl]benzene triggers ethyl acetate hydrolysis: acetato-/pyrazolato-, dihydroxo-and diacetato-bridged Cu(II) complexes. *Dalton Trans.* 13: 1611-1621.

[21] Saito, K., Arai, T., Takahashi, N., Tsukuda, T. and Tsubomura, T. (2006). A series of luminescent Cu(I) mixed-ligand complexes containing 2,9-dimethyl-1,10-phenanthroline and simple diphosphine ligands. *Dalton Trans.*37: 4444-4448.

[22] Yang, L., Powell, D.R. and Houser, R.P. (2007). Structural variation in copper(I) complexes with pyridylmethylamide ligands: structural analysis with a new four-coordinate geometry index, τ4. *Dalton Trans.* 9 : 955-964.

[23] Caselli, A., Cesana, F., Gallo, E., Casati, N., Macchi, P., Sisti, M., Celentano, G. and Cenini, S. (2008). Designing new ligands: asymmetric cyclopropanation by Cu(I) complexes based on functionalised pyridine-containing macrocyclic ligands. *Dalton Trans.* 32 : 4202-4205.

[24] Maekawa, M., Nabei, A., Tominaga, T., Sugimoto, K., Minematsu, T., Okubo, T., Kurodasowa, T., Munakata, M. and Kitagawa, S. (2009). A unique chair-shaped hexanuclear Cu(I) metallamacrocyclic C2H4 adduct encapsulating a BF4− anion. *Dalton Trans.*3 : 415-417.

[25] Wang, Y., Li, B., Liu, Y., Zhang, L., Zuo, Q., Shiab, L. and Su, Z. (2009). Highly sensitive oxygen sensors based on Cu(I) complex–polystyrene composite nanofibrous membranes prepared by electrospinning. *Chem. Commun.* 39 : 5868-5870.

[26] Hardt, H.D. and Pierre, A. (1973). Fluorescence thermochromism of pyridine copper iodides and copper iodide. *Z Anorg. Allg. Chem.*402 : 107-112.

[27] (a) Ahna, H.D.D. and Hardt, H.D. (1972). Fluoreszenz und fluoreszenz-thermochromiebei [Fluorescence and fluorescence thermochromism in] Alkyl-pyridinokupfer(I)-halogeniden. *Z Anorg. Allg. Chem.*387 : 61-71;

(b) Hardt, H.D. and Gechnizdjani, H. (1973). Fluoreszenz-thermochrome Verbindungen von Kupfer(I)-jodidmitÄthyl- und Aminopyridinen [Fluorescence thermochromic compounds of copper(I) iodide with ethyl and aminopyridines]. *Z Anorg. Allg. Chem.*397 : 23-30;

(c) Hardt, H.D. and Stoll, H.J. (1981). NeueErgebnissezu den Fluoreszenzeigen-schaften von Sesqui-, Bis- und [New results on the fluorescence properties of

sesqui-, bis- and] Tris(triphenylphosphan)kupfer(I)-Halogeniden. *Z Anorg. Allg. Chem.* 480 : 193-198;
(d) Hardt, H.D. and Stoll, H.J. (1981). Wirkung von Substituenten und Anionen auf die Fluoreszen zeigenschaften gemischt ligandiger [Effect of substituents and anions on fluorescence properties of mixed ligands] (Amin-triphenyl-phosphan) kupfer(I)-Halogenide. *Z Anorg. Allg. Chem.* 480: 199-204.

[28] Raston, C.L. and White, A.H. (1976). Crystal structure of the copper(I) iodide–pyridine (1/1) tetramer. *J. Chem. Soc., Dalton Trans.* 21 : 2153-2156.

[29] (a) Ford, P.C. and Vogler, A. (1993). Photochemical and photophysical properties of tetranuclear and hexanuclear clusters of metals with d10 and s2 electronic configurations. *Acc. Chem. Res.* 26 : 220-226; (b) Vitale, M. and Ford, P.C. (2001). Luminescent mixed ligand copper(I) clusters (CuI)n(L)m (L = pyridine, piperidine): thermodynamic control of molecular and supramolecular species. *Coord. Chem. Rev.* 219–221 : 3-16.

[30] Angelis, F.D., Fantacci, S., Sgamellotti, A., Cariati, E., Ugo, R., and Ford, P.C. (2006). Electronic transitions involved in the absorption spectrum and dual luminescence of tetranuclearcubane [Cu4I4(pyridine)4]cluster: A density functional theory/time-Ddependent density functional theory investigation. *Inorg. Chem.*45 : 10576-10584.

[31] Blake, A.J., Brooks, N.R., Champnes, N.R., Hanton, L.R., Hubberstey, P. and Schröder, M. (1998). Copper(1) halide supramolecular networks linked by N-heterocyclic donor bridging ligands. *Pure Appl. Chem.* 70 : 2351-2357.

[32] Arnby, C.H., Jagner, S. and Dance, I. (2004). Questions for crystal engineering of halocuprate complexes: concepts for a difficult system. *Cryst. Eng. Comm.* 6 : 257-275.

[33] Rath, N.P. and Holt, E.M. (1985). Copper(I) iodide complexes of novel structure:[Cu4I6][Cu8I13]K7(12-crown-4)6, [Cu4I6]K2(15-crown-5)2, and [Cu3I4]K(dibenzo-24-crown-8). *J. Chem. Soc., Chem. Commun.*10 : 665-667.

[34] Nather, C., Wriedt, M. and Jess, I. (2003). Dimorphism of a new CuI coordination polymer: Synthesis, crystal structures and properties of catena[CuI(2-Iodopyrazine-N)] and Poly[CuI(μ2-2-Iodopyrazine-N,N‘)]. *Inorg. Chem.* 42 : 2391-2397.

[35] Nather, C. and Je, B.I. (2004). Synthesis, Crystal Structure and Thermal Reactivity of New Copper(I) Halide Pyrimidine-Containing Coordination Polymers. *Eur. J. Inorg. Chem.*14 : 2868-2876.

[36] Cariati, E., Roberto, D., Ugo, R., Fond, P.C., Gally, S. and Siron, A. (2005). New structural motifs, unusual quenching of the emission, and second harmonic generation of copper(I) iodide polymeric or oligomeric adducts with para-substituted pyridines or trans-stilbazoles. *Inorg. Chem.*44 : 4077-4085.

[37] Li, D., Shi, W.J. and Hou, L. (2005). Coordination polymers of copper(I) halides and neutral heterocyclic thiones with new coordination modes. *Inorg. Chem.* 44: 3907-3913.

[38] Thebault, F., Barnett, S.A., Blake, A.J., Wilson, C., Champness, N.R. and Schroder, M. (2006). Control of Copper(I) Iodide Architectures by Ligand Design: Angular versus Linear Bridging Ligands. *Inorg. Chem.* 45 : 6179-6187.

[39] Kim, T.H., Shin, Y.W., Lee, S.S. and Kim, J. (2007). Luminescent staircase copper(I) coordination polymer based on planar Cu3I3. *Inorg. Chem. Commun.* 10 : 717-719.

[40] Wu, T., Li, M., Li, D. and Huang, X.C. (2008). Anionic cunIn cluster-based architectures induced by in situ generated N-alkylated cationic triazoliumsalts. *Crystal Growth & Design* 8 : 568-574.

[41] Hao, Z.M. and Zhang, X.M. (2008). Phosphorescent acentric cuprous halide coordination polymers for nolinear optical materials. *Crystal Growth & Design* 8(7): 2359-2363.

[42] Kim, T.H., Yong, W.S., Park, K.-M. and Kim, J. (2009). Phyllo-poly[[μ2- 1,4-bis(cyclohexylsulfanylmethyl)benzene-κ2 S:S']-(μ2-nitrato-κ2O:O')silver(I)]. *Acta Cryst.* E65: m385.

[43] Chen, Y., Li, H.X., Liu, D., Lu, L., Li, N.Y., Ye, H.Y., Zhang, Y. and Lang, J.P. (2008). Solvent effects on the assembly of [Cu2I2]- or [Cu4I4]-based coordination polymers: Isolation, structures and luminescent properties. *Crystal Growth & Design* 8: 3810-3816.

[44] Li, M., Li, Z. and Li, D. (2008). Unprecedented cationic copper(I)–iodide aggregates trapped in "click" formation of anionic-tetrazolate-based coordination polymers. *Chem. Commun.* 29 : 3390-3392.

[45] Wei, W., Wu, M., Gao, Q., Zhang, Q., Huang, Y., Jiang, F. and Hong, M. (2009). A novel supramolecular tetrahedron assembled from tetranuclearcopper(I) cluster molecules via aryl embrace interactions. *Inorg. Chem.* 48 : 420-422.

[46] Tzeng, B.C. and Chang, T.Y. (2009). Toward copper(I)-iodide-based coordination architectures via N,N′-Bis(pyridylcarbonyl)-4,4′-diaminodiphenyl ether with different solvent compositions.*Crystal Growth & Design* 9: 5343-5350.

[47] Qin, L., Zhang, S.W, Wang, J., Lu, C., Cheng, Y. and Wang, L. (2009). Novel luminescent iminephosphine complex of copper(I) with high photochemical and electrochemical stability. *Dalton Trans.* 43 : 9388-9391.

[48] Zhang, Y, Wu, T., Liu, R., Dou, T., Bu, X. and Feng, P. (2010).Three-dimensional photoluminescent frameworks constructed from size tunable CuI clusters. *Crystal Growth & Design Comm.* 10 : 2047-2049.

[49] Park, G., Yang, H., Kim, T.H. and Kim, J. (2011). Copper(I) coordination polymers of N,N′-Bis[3-(methylthio)propyl]pyromelliticdiimide: Crystal transformation and solvatochromism by halogen−π interactions. *Inorg. Chem.* 50 : 961-968.

[50] Zhang, J. (2010). Catena-Poly[[[(triphenylphosphane)copper(I)]-di-μ-iodido-[(triphenylphosphane)copper(I)]-μ-[3,6-bis(4-pyridyl)-1,2,4,5-tetrazine]] acetonitrile disolvate]. *ActaCryst.* E 66 : m 1480.

[51] Mukherjee, A., Chakrabarty, R., Patra, G.K. (2009). Novel 3-dimensional sixfold interpenetrating diamondoid networks of copper(I) coordination polymers of polypyridyl ligands – Syntheses, characterization and crystal structures. *Inorg. Chem. Commun.*12 : 227-1230.

[52] Mondal, J., Mukherjee, A., Patra, G.K. (2017). CuX (X = I, Br and Cl) based coordination polymers of azino-pyridyl ligand and PPh3: Structural, spectral, electro-chemical, and DFT studies. *Inorg. Chim. Acta* 463 : 44-53.

[53] Mondal, J., Pal, P.K., Mukherjee, A., Patra, G.K. (2017). Synthesis, structural variation, photophysical, electrochemical and DFT studies of CuX (X = I, Br and Cl) mixed ligand complexes of an azino-pyridyl ligand and triphenylphosphine. *Inorg. Chim. Acta* 466 : 274-284.

Chapter 12

A Literature Review on Asymmetric Membranes Based on Poly (Ethylene-Co-Vinyl Alcohol) Polymer

Isha Gupta*
Department of Chemistry, Guru Ghasidas Vishwavidyalaya, Bilaspur, Chhattisgarh, India

Abstract

Membranes are now important components in a variety of fields. In fact, applications in oil and petrochemical, biopharmaceutical, power, and energy-related industries are gradually expanding from traditional fields such as water desalination and purification and food processing. Asymmetric membranes are made up of a number of layers, each with its own structure and permeability, and can be thought of as hierarchically structured systems with layers that are well-designed to overcome transport constraints. A variety of poly (ethylene-co-vinyl alcohol) blended with other polymers has been used for the development of asymmetric membranes by implementing numerous approaches such as phase inversion, electrospinning, solvent casting, and others. Certain polymer properties, such as solubility in non-toxic solvents, mechanical and thermal stability, non-toxicity, solvent resistance, and separation capabilities, are highly desired depending on the application. EVAL (poly (ethylene-co-vinyl alcohol)) is a polymer that combines the above-mentioned properties with excellent film-forming capabilities, chemical and mechanical stability, and tuned hydrophilicity, making it a viable candidate for membrane preparation. This review covers the main

* Corresponding Author's Email: ishagupta10299@gmail.com.

In: Materials Science: Future Aspects
Editors: Kalpana Awasthi, Arti Srivastava and Mridula Tripathi
ISBN: 978-1-68507-843-0

preparation methods and applications of porous EVAL-based asymmetric membranes, along with introductory material.

Keywords: asymmetric membranes, EVAL (poly (ethylene-co-vinyl alcohol)), phase inversion, electrospinning, solvent casting, excellent film-forming capabilities, solvent resistance

1. Introduction

Asymmetric or anisotropic porous membranes have a gradient in pore morphology and permeability, and a top dense layer, or "skin" also known as the permselective layer [1, 2], of 0.1–5μm thickness is usually supported on top of a porous, 50–200μm thick layer. These barriers andsupporting layers can be made of the same material or different materials; a less porous top or dense layer serves as a discriminating barrier, while the underlying macroporous layer serves as mechanical support. An integrally asymmetric membrane is one in which the dense top layer is made of the same materials as the dense bottom layer [1]. The surface layer determines the separation properties and permeation rates, and the substructureserves only as mechanical support with little separating function. The thin surface layer determines in large part or entirely the resistance to mass transfer.

Thickened membranes are capable of absorbing the compressive forces used in separation. The thin film is always on the high-pressure side of the membrane, i.e., the feed side, since this allows the support layer to be used to its full potential in stabilizing the thin film. Thesemembranes had the advantage of higher fluxes, and almost all commercial processes use such membranes [2]. Asymmetric porous membranes have widely been utilized in municipal and industrial water purification systems for the removal or recovery of contaminants [3, 4] and in biomedical applications like protein filtration [5], plasma separation [6], hemodialysis [7], and blood oxygenation [8]. A wide range of applications uses membranes fabricated or modified with biocompatible polymers, such as artificial organs, drug delivery systems, tissue engineering, and biosensors [9-11]. Kuraray introduced the EVAL™ brand of EVAL [poly (ethylene-co-vinyl alcohol)] to themarket in 1972. Kuraray began commercial production of EVAL after obtaining the patent in 1971. Later, Nippon Gohsei and Chang Chun began commercial EVAL production as well [12, 13]. Due to continued market growth, they have more than doubled their global production capacity from 45,000 tons in 2001 to 92,000 tons in 2016, with another 11,000 tons planned for 2018 [14-16].

EVAL is a non-biodegradable [17], non-toxic, hydrophilic, biocompatible [18], semi-crystalline random copolymer [19, 21] with vinyl alcohol and ethylene monomer units [20] (ethylenecontent ranging from 27 to 47 mol. %) [17, 21]. The hydrophilic/hydrophobic balance, in particular, varies depending on the polymer's composition [19].

The synthesis of EVAL is a two-step procedure [23, 24]. The first stage is a copolymerization process between ethylene and vinyl acetate, which produces the random copolymer of ethylene vinyl acetate. Vinyl alcohol is unstable and can't be separated; hence, the first step is a copolymerization reaction between ethylene and vinyl acetate. In the second phase, the trans-esterification of polyethylene vinyl acetate with methanol produces EVAL, along with methyl acetate as a byproduct [25].

EVAL crystallizes over the entire range of ethylene compositions and has excellent thermal stability and chemical resistance [26]. Because of their good blood compatibility and hydrophilicity, EVAL membranes have recently attracted a lot of research interest in the fields of biomedical science and water treatment [27, 28]. Its high impermeability to gases and hydrocarbons, its high tensile strength, its ability to form films, and its safety has made it an ideal packaging material. [29] According to recent reports, EVAL can be used in hollow fiber dialyzers, wound dressing materials, neural repair scaffolds, and periodontal ligament regeneration [30-33].

The following characteristics make EVAL film desirable.

- *Antistatic properties:* EVAL resin prevents dust from accumulating on the package because it is a highly antistatic polymer.
- *Luster and transparency:* EVAL resins have a high gloss and low haze, resulting in excellent clarity. The use of EVAL resin on a package's outer surface provides excellent sparkle and improves the package's appearance.
- *Printability:* In its molecule chains, EVAL resin contains a -OH group that makes it easy to print.
- *Resistance to oil and organic solvents:* Oils and organic solvents are resistant to EVAL resins, making them ideal for packaging oily foods, edible oils, mineral oils, agricultural pesticides, and organic solvents.
- *Weather resistance:* The weatherability of EVAL resins is excellent. The polymerretains its color and does not yellow or become opaque

when exposed to outdoor conditions. Mechanical property changes are minor, indicating a high level of resistanceto weathering.

- *Permeability:* EVAL resins have excellent gas barrier properties (oxygen, carbon dioxide, nitrogen, and helium) and maintain their barrier properties over a wide range of humidity. EVAL's oxygen barrier properties depend on its ethylene content. . EVAL resins are capable of retaining fragrances and preserving thearoma of package contents. At the same time, unwanted odors cannot enter or leave the package simultaneously [22].

In this review, we will study some of the preparatory methods for asymmetric membranes based on EVAL.

2. Preparation of Asymmetric Membranes Based on EVAL

2.1. Solvent Casting

Solvent casting is a technique that creates porous polymers by evaporating solvent from a suspension containing solid particles called porogen and polymeric solutions, then removing the porogen from the polymer matrix. Porogens found in this domain include sodium chloride, ammonium bicarbonate, and glucose with various crystal sizes [34-38]. Abdulaziz et al. prepared a poly(ethylene-co-vinyl alcohol)/poly(ε-caprolactone) Blend for Bioscaffolding in which the goal was to (i) improve cell adhesion to PCL and improve its mechanical properties by blending it with EVAL, and (ii) create a bioscaffold made up of an interconnected porous construct. As part of this experiment, we applied differential scanning calorimetry (DSC), Fourier-transform infrared spectroscopy (FTIR), X-ray diffraction (XRD), and scanning electron microscope (SEM) techniques to examine the miscibility of a series of EVAL/PCL blends prepared by solvent casting. The wettability of the surfaces was determined using a dynamic mechanical analysis, and the mechanical properties were determined using static contact angle measurements. Cell adhesion was determined using the Kueng method of colorimetric detection of bound cells and confirmed using the Yamada and Kennedy method of phase-contrast microscopy to measure the flattening of adherent cells [39]. Using microparticles of naphthalene, the porogen created interconnected pores. A study was conducted to determine how interconnected the pores of miscible materials are. PCL and EVAL were dissolved in DMF at 80°C in a 50 mL Erlenmeyer flask under constant stirring until completely dissolved and a viscous solution was formed. Under continuous stirring, the two prepared solutions were mixed together to make an EVAL/PCL/DMF ternary solution. The solution was poured into a Teflon Petri dish and then deposited horizontally on a square Styrofoam plate floating on the surface of a crystallizer filled with water and dried at ambient temperature for 48 hours to obtain a film sample of perfectly homogeneous thickness. The Teflon plate containing polymeric material was then dried in a vacuum oven at 60°C for 24 hours to extract the residual solvent crusted in the obtained film. The same method was used to make a series of EVAL/PCL blends containing 10, 25, 50, 75, and 90 wt. %.

Table 1. Preparation of PCL/EVAL blends in dimethylformamide [40]

Sample	EVAL/PCL (wt. %)	EVAL(g)	PCL(g)
PCL	100:0	0.0	10.0
EVAL/PCL90	90:10	1.0	9.0
EVAL/PCL75	75:25	2.5	7.5
EVAL/PCL50	50:50	5.0	5.5
EVAL/PCL25	25:75	7.5	2.5
EVAL/PCL10	10:90	9.0	1.0
EVAL	0:100	10.0	0.0

EVAL, with the experimental conditions listed in Table 1.

2.2. Phase Inversion

2.2.1. Non-Solvent Induced Phase Separation

A homogeneous polymer solution is immersed in a non-solvent bath in the NIPS process, and the non-solvent penetrates the polymer solution, causing phase separation [41–45]. The Loeb–Sourirajan phase inversion method is widely used to produce asymmetric membranes [46]. In this technique, a cast film of a polymer solution in a suitable solvent is immersed in a coagulation bath containing a non-solvent of the polymer. The solvent and non-solvent diffuse into each other during this process, causing thermodynamic instability in different parts of the system. Because of the phase separation of the polymer and the solvents, a polymer lean phase is formed by the nucleation of non-solvent, resulting in the formation of pores. Later on, the polymer-rich phase yields crystalline membrane structures. Riyasudheen et al. prepared poly (ethylene-co-vinyl alcohol)/poly (vinyl pyrrolidone) blend membranes in which they described how they used the NIPS method to develop an asymmetric EVAL/PVP blend polymer microfiltration membrane with a DMSO/water solvent/non-solvent system. In the phase inversion of the EVAL/PVP/DMSO/water quaternary system, the importance of thermodynamic and rheological properties was investigated. The membranes' morphology and permeation characteristics, such as flux and protein rejection, were investigated. The impact of PVP on hydrophilicity and proteinadsorption was also investigated. IR spectroscopy was used to determine the amount of residual PVP after membrane formation. Where dialysis membranes are used, the prepared membranes can be considered ideal candidates. EVAL was dissolved in the DMSO/PVP mixture at 60 °C with stirring to make casting solutions. Table 2 lists the various compositions (the weight ratios of DMSO to PVP were changed). To remove air bubbles, the casting solutions were sonicated for 1½ hours. At 28°C, the viscosities of thecasting solutions were measured using a Brookfield digital rheometer. The solutions were cast on a glass plate with a 150μm clearance using a doctor's blade. After 10 seconds of air contact, the nascent membrane was immersed in a distilled water coagulation bath. The remaining solvent was then removed by immersing the membranes in distilled water for 24 hours. To prevent the microbial attack, the prepared membranes were kept in 0.01 M formalin.

Table 2. Composition in weight ratios and viscosity data of membrane casting solutions are shown below [47]

	Casting solution composition			
Samples	EVAL (Wt. %)	PVP (Wt. %)	DMSO (Wt. %)	Viscosity
EP0	15	0	85	525.61
EP2	15	2	83	822.10
EP4	15	4	81	894.54
EP6	15	6	79	1288.86
EP8	15	8	77	1884.54
EP10	15	10	75	3653.52

2.2.2. Immersion-Precipitation

Immersion precipitation occurs when a thin film of a concentrated polymer solution is immersed in a non-solvent bath. The polymer solution becomes unstable due to the exchange of solvent from the polymer solution and non-solvent from the coagulation bath.A polymer lean phase and a polymer-rich phase result from the separation of liquid-liquid phases. The polymer lean phase creates pores within the membrane's matrix, created by the polymer-rich phase [48].

Cheng et al. prepared membranes from non-solvent–solvent–crystalline polymer systems in which light transmission experiments were used to investigate the precipitation rate of the EVAL solution in 2-propanol and water, and a ternary mass transfer model was used to analyze the diffusion trajectories in the phase diagram. The effect of non-solvent on membrane structures was discussed using both thermodynamic and kinetic properties for a ternary system containing a crystalline polymer. The direct immersion–precipitation method was used to make the membranes. To make a 20 wt. % homogeneous solution, an appropriate amount of EVAL was dissolved in DMSO. To form membranes, this solution was uniformly dispersed on a glass plate (ca. 100 mm) and then immersed in the non-solvent bath. To obtain the membrane structure, freeze-dried samples of the membranes were examined using a scanning electron microscope (SEM) [49].

With microscopic to mesoscopic inhomogeneity, the film spontaneously de-mixes into a polymer-lean phase and a polymer-rich phase. The quench deepens after phase separation until the polymer-rich phase becomes concentrated enough to vitrify or crystallize, arresting the network structure that forms the membrane pores [50].

2.2.3. Thermally Induced Phase Separation (TIPS)

One of the most widely used methods for producing microporous membranes is the TIPS process [51, 52]. In general, a homogeneous polymer solution is made by blending the polymer with a high-boiling-point diluent at a high temperature, then the solution is cooled to induce phase separation, and finally, a microporous structure is formed after the diluent is extracted with a volatile solvent [53]. Lv et al. prepared EVOH/PVP membranes via thermally induced phase separation. PVP would not be lost during membrane formation because phase separation is thermally induced rather than non-solvent–solvent exchange induced in the phase inversion process. It is simple to investigate the distribution of PVP. To the best of our knowledge, no research in this area has been published. According to the results of ATR-FT-IR and DSC measurements, the compatibility of EVOH and PVP was discussed. The impact of PVP content on the phase diagram and membrane morphology was also investigated. Finally, the membranes' hydrophilicity and antifouling properties were assessed.

To obtain a homogeneous mixture, certain amounts of EVOH44, PVP, and PEG300 were added to a tube, heated to 150–180°C, and maintained for

4 hours. The homogeneous solution was then immersed in liquid nitrogen, which caused phase separation and solidification. The samples were taken and stored in the refrigerator for later use. A small sample was placed between a pair of microscope coverslips. To prevent diluent loss due to evaporation, a 200μm Teflon film with a circular opening in the center was inserted between the coverslips. To achieve uniformity, each sample was heated on a hot stage at 150–180°C for 5 minutes. The samples were then cooled in 15°C air. A thermocouple buried in the sample measured a cooling rate of 23°C/min. After thesamples had solidified, the membranes were obtained. To extract PEG300 and precipitate PVP, the membranes were immersed in acetone for 24 hours. After the acetone evaporated at room temperature, the porous membrane was formed.

All of the samples' EVOH, PVP, and PEG300 weight ratios are listed in Table 3.

Table 3. The samples' chemical composition is shown below [20]

Sample No.	EVAL/PEG300/PVP(w/w/w)
S1	1/4/0
S2	1/4/0.125
S3	1/4/0.25
S4	1/4/0.5

2.2.4. Dry Cast Phase Inversion

The dry cast process is one of the methods for changing the thermodynamic state of a polymer solution to achieve phase inversion. Evaporation of non-solvent and solvent from an initially homogeneous single-phase solution characterizes it. External effects cause the homogeneous polymer solution to become thermodynamically unstable at first, resulting in the formation of a two-phase solution due to evaporation. As a result, the phase dividesinto polymer-poor and polymer-rich phases. When the polymer-rich phase solidifies, it forms a solid matrix that envelops the polymer lean phase, which is rich in solvent and non-solvent and fills the pores. This method avoids the complications that come withusing coagulation baths in the wet cast method. As the solvent evaporates from the solution, the polymer concentration is increased to reduce the solvation character of the polymer-solvent. In both the dry and wet cast phase inversion processes, the evaporation step has a significant impact on the final membrane morphology [54]. Tylkowski et al. prepared Asymmetrical Poly(ethylene-co-vinyl alcohol) MembranesBlended with Azobenzene Polymers in which a new dry-cast method for making EVAL asymmetrical composite membranes with azobenzene polymers, in which rough structures on the substrate surface and the presence of a photosensitive moiety result in a large change in wettability after UV irradiation. To ensure that the azo moieties were in E configurations, all azo-containing solutions andazo-containing membranes were prepared under red light and kept in dark bottles or dark boxes. The host material in our experiment was EVAL, with a 1:1 weight ratio between the azo-P1 or azo-P2 solution and the EVAL solution. The membranes' final compositions are listed in Table 4:

- membrane A: 100 wt.% EVAL;
- membrane B: 99.94 wt.% EVAL+ azo-P1 0.06 wt.%;
- membrane C: 99.94 wt. % EVAL + azo-P2 0.06 wt. %;

Membranes were made in the ratio mentioned above. At 70°C, EVAL was dissolved in DMSO to form a 20 wt. % polymer solution. At room temperature, the solution was stirred for 24 hours. To make a 0.0112 wt, each azo polymer was separately dissolved in THF at room temperature. The azo-P1 and azo-P2 solutions were then mixed separately with a 20% EVAL solution, stirred at room temperature for about 2 hours, and then degassed under vacuum for 30 minutes. After that, the solutions were cast on uniformly thick Petri dishes to make films. For 48 hours, the solvents were evaporated in a furnace. The dry films were removed from the Petri dishes after evaporation and stored in a dark box. In the case of membrane A, a 20 wt. % solution of EVAL was mixed in a 1:1 ratio with pure THF and the film was made as described previously. They used a well-known oven principle to obtain asymmetrical cross-sections of the membranes: the temperature of the panel heater in the furnace set to temperature x is a few degrees higher than the temperature at its center. Temperature readings taken in the furnace at various heights from the panel heater confirmed our suspicions: When the furnace temperature was set to 40°C, the temperature on the surface of the panel heater of the furnace was 46°C, which was 8°C higher than the temperature in the furnace's center.

By preparing an additional membrane (D) using the same procedure as membrane A but evaporating the solvent at a set temperature of 65°C, the effect of temperature on crystallinity, and the morphology of EVAL membranes was verified.

Table 4. Chemical composition, thickness, and contact angles of membranes A-C [55]

Composition				Contact Angle			
Membrane	EVAL (wt. %)	Guest (wt. %)*	Membrane thickness	CA of the top porous surfaces before irradiation	CA of the top porous surface after irradiation	CA of the bottom dense surface before irradiation	CA of the bottom dense surface after irradiation
A	100	-	274 ± 5	48.1 ± 1.2	48.1 ± 0.7	44.4 ± 0.9	44.1 ± 1.0
B	99.94	0.06	254 ± 5	62.0 ± 0.7	36.1 ± 0.8	56.9 ± 0.5	54.4 ± 0.6
C	99.94	0.06	256 ± 5	62.1 ± 0.9	36.1 ± 0.4	56.2 ± 1.0	54.6 ± 0.3

* Membrane B: azo-P1; membrane C: azo-P2.

2.2.5. Electrospinning

Electrospinning is a voltage-driven process that produces fibers and particles from a polymer solution and is governed by electrohydrodynamic phenomena. A solution contained in a reservoir (typically a syringe) and tipped with a blunt needle (for needle-based electrospraying), a pump, a high-voltage power source, and a collector are the most basic components of this technique.

The spinning process begins when the solution is pumped at a constant flow rate and a specific voltage is applied to create an electric field between the needle tip and the collector. At the liquid's surface, a charge builds up. When electrostatic repulsion exceeds surface tension, the liquid meniscus is deformed into a conically shaped structure known as the Taylor cone. The charged liquid jet is ejected towards the collector after the Taylor cone has been formed. Collectors include stationary flat plates, rotating drums, mandrels, and discs, among others. Solid fibers will form as the solvent evaporates from the whippingmotion that occurs during its flight time from the Taylor cone to the collector, depending on the solution viscosity. A non-woven fiber mat is deposited on the collectoras a result. A non-woven fiber mat is deposited on the collector as a result [56]. Mayuri et al. prepared electrospun poly (ethylene-co-vinyl alcohol) membranes in whichthe effect of EVAL membrane structure on leukodepletion from whole blood was evaluated, including the role of fiber diameter, pore size, and other factors.

To tune the resulting membrane properties, EVAL was electrospun from both 2- propanol: water mixture and HFIP. To begin, the required weight of EVAL pellets was dissolved in a 70:30 mixture. 2-propanol: water mixture, vigorously stirred overnight at 60-65°C to a concentration of 9 w/v%. This solution was electro spun twice with an 18-gauge needle with an internal diameter of 1.2 mm. To be more specific, the flow rate was varied from 1 mL/min to 10 mL/h, and the applied voltage was likewise varied, as shown in Table 4. EVAL was electrospun from HFIP once more because the flow rate could not be reduced any lower than 10 mL/h using the 2-propanol: water solvent mixture. A sufficient amount of EVAL pellets were added to a proportional amount of HFIP and kept at 27 ± 2°C overnight for agitation. The EVAL was used at aconcentration of 9 w/v% in HFIP with a flow rate of 1 mL/h through a 21-gauge needle. To stabilize the fiber jet, a voltage of 10–11 kV (Gamma High Voltage, USA) was required once more. The electrospun fibers were collected onto a mandrel set at 500RPM (tangential velocity ¼ 2.62m/s) and 1500 RPM (tangential velocity ¼ 7.85m/s) at a distance of 13 cm from the spinneret each time (Table 4). The electrospinning was continued until the membranes were approximately 0.56± 0.02 mm thick. All membranes were washed 3-4 times in distilled water using ultrasound sonication and dried to a constant weight at 50 °C under vacuum. Prior to the filtration study, the membranes were sterilized with ethylene oxide gas.

Table 5. The electrospinning parameters taken are shown below [57]

Solvent	Concentration	Flow rate	Distance	Applied voltage	Collection speed (RPM)
2-propanol: water (70:30)	9w/v %	10mL/h 1mL/min	13cm	11-12kV 14-15kV	500 and1500 (tangential
HFIP		1mL/h		10-11kV	velocities = 2.62m/sand7.85m/s, respectively

Conclusion

Asymmetric membranes have the advantage of higher fluxes, and almost all commercial processes use such membranes. This review provides an up-to-date account of recent developments in the field of asymmetric membranes based on EVAL [Poly (ethylene-co-vinyl alcohol)]. Different methods for the preparation of asymmetric membranes based on EVAL have been reviewed with schematic diagrams and required parameters. Despite the various applications and advantages of asymmetric membranes based on EVAL, several issues related to their use still remain to be addressed.

References

[1] Gohil Jaydevsinh M., Choudhury Rikarani R., 2019. Introduction to nanostructured and Nano-enhanced polymeric membranes: preparation, function, and application for water purification, *Nanoscale Materials in Water Purification*: 25-57.

[2] *Membrane types – Asymmetric*: http://www.separationprocesses.com/Membrane/MT_Chp02b.htm.

[3] Baker R. W., 2000. *Membrane Technology and Applications*, McGraw- Hill, New York.

[4] Majewska-Nowak K., Kowalska I., Kabsch-Korbutowicz M., 2005. Ultrafiltration of SDS solutionsusing polymeric membranes, *Desalination* 184:415–422.

[5] Ye S. H., Watanabe J., Iwasaki Y., Ishihara K., 2003. Antifouling blood purification membrane composed of cellulose acetate and phospholipid polymer, *Biomaterials* 24: 4143–4152.

[6] Ran F., Nie S., Zhao W., Li J., Su B., Sun S., 2011. Biocompatibility of modified polyethersulfone membranes by blending an amphiphilic triblock copolymer of poly (vinyl pyrrolidone)-b-poly (methyl methacrylate)-b-poly (vinyl pyrrolidone), *Acta Biomater.* 7(9): 3370–3381.

[7] Yang Q., Chung T. S., Weber M., 2009. Microscopic behavior of poly (vinyl pyrrolidone) hydrophilizing agents on phase inversion polyethersulfone hollow fiber membranes for hemofiltration, *J. Membr. Sci.* 326(20): 322–331.

[8] Su B., Fu P., Li Q., Tao Y., Li Z., Zao H., 2008. Evaluation of polyethersulfone high flux hemodialysis membrane in vitro and in vivo, *J. Mater. Sci. Mater. Med.* 19(7): 745–751.

[9] Alvaro de Queiroz A. A., Abraham G. A., Higa O. Z., 2006. Controlled release of 5-fluorouridine from radiation-cross linked poly (ethylene-co-vinyl acetate) films, *Acta Biomater*. 2: 641– 650.

[10] Uchiyama T., Watanabe J., Ishihara K., 2002. Biocompatible polymer alloy membrane for implantable artificial pancreas, *J. Membr. Sci*. 208(6): 39–48.

[11] Shim J., Woo J. J., Moon S. H., Kim G. Y., 2009. A preparation of a single-layered enzymemembrane using asymmetric pBPPO base film for development of pesticide detecting biosensor, *J. Membr. Sci*. 330(1-2): 341–348.

[12] Wagner Jr. J. R., 2016. *Multilayer Flexible Packaging*; Elsevier Science: Oxford *2*:384/410.

[13] Finch, C. A., 1992. *Polyvinyl Alcohol: Developments*; John Wiley & Sons Ltd.: Hoboken 2:700.

[14] ICIS Chemical Business. *Eval to Build EVOH Plant to Meet Global Demand*: https://www.icis.com/resources/news/2001/06/25/141585/eval-to-build-evoh-plant-to-meet-global-demand/(last accessed 13 March 2017).

[15] Kuraray America Inc. *Kuraray Americas EVAL Plant to Increase Production by More than 17 Million Pounds*: http://www.kuraray.us.com/kuraray-americas-eval-plant-to-increase-production-by-more-than-17-million-pounds/ (last accessed 13 March 2017).

[16] Kuraray Co. Ltd. *Kuraray – EVAL™*: http://www.evalevoh.com (last accessed 28 February 2017).

[17] Mayuri P. V., Bhatt Anugya, Joseph Roy, P. Ramesh, 2016. Effect of photo grafting 2-hydroxyethyl acrylate on the hemocompatibility of electrospun poly (ethylene-co-vinyl alcohol) fibro porous mats, *Materials Science and Engineering:* C (60):19-29.

[18] Merck, *Poly (vinyl alcohol-co-ethylene)*: https://www.sigmaaldrich.com/IN/en/product/aldrich/414093.

[19] Gene Medlock, 2015. *Barrier Resins | Properties, Processing & Handling of EVOH, Pt. 1 – Paper, Film, and Foil Converter*: https://www.pffc- online.com/flexpack/12690-barrier-resins-properties-processing-handling-of-evoh-pt-1.

[20] Rui, Lv, Zhou Jing, Du Qiangguo, Wang Haitao, Zhong Wei, 2006. Preparation and characterization of EVOH/PVP membranes via thermally induced phase separation, *Journal of Membrane Science* 281(1-2): 700–706.

[21] Chang Hsu-Hsien, Beltsios Konstantinos, and Chen Yih-Hang, Lin Dar-Jong, Cheng Liao-Ping, 2014. Effects of Cooling Temperature and Aging Treatment on the Morphology of Nano- and Micro- Porous Poly (ethylene-co-vinyl alcohol) Membranes by Thermal Induced Phase Separation Method, *J. Appl. Polym. Sci.* 131: 40374.

[22] Mckeen Laurence W., 2017. A volume in Plastics Design Library, Chapter 10-Polyvinyls and Acrylics-10.2-Ethylene-Vinyl Alcohol Copolymer, *Film Properties of Plastics and Elastomers* 4: 273-316.

[23] Iwanami T.; Hirai Y., 1983. Ethylene-vinyl alcohol Resins for Gas-Barrier Material. *Tappi J.* 66: 85–90.

[24] Mokwena K. K., Tang J., 2012. Ethylene-vinyl alcohol: A Review of Barrier Properties for Packaging Shelf Stable Foods. *Crit. Rev. Food Sci. Nutr.* 52: 640–65.

[25] Maes Caroline, Luyten Wout, Herremans Geert, Peeters Roos, Carleer Robert, Buntinx Mieke, 2018. Recent Updates on the Barrier Properties of Ethylene Vinyl Alcohol Copolymer (EVOH): A Review, *Polymer Reviews* 58(2): 209–246.

[26] Okaya T., Ikari K., in: C. A. Finch (Ed.), 1992. *Polyvinyl Alcohol Developments*, Wiley, New York (Chapter 8).

[27] Guerra N., Barbani L., Lazzeri L., Lelli L., Palla M., Rizzo C., 1995. The activation of human plasma prekallikrein as a hemocompatibility test for biomaterials. II. Contact activation byEVAL and EVAL–SMA copolymers, *J. Biomater. Sci.* 4665.

[28] Young T. H., Lai J. Y., You W. M., Cheng L. P., 1997. Equilibrium phase behavior of the membrane-forming water–DMSO–EVAL system, *J. Membr. Sci.* 128: 55–65.

[29] Pemberton L., Jaeger R. De, Gengembre L., 1998. Surface grafting of a poly (organophosphazene)onto poly (vinyl alcohol) and poly (ethylene-co-vinyl alcohol) using triethoxysilane as a coupling agent, *J. Appl. Polym. Sci.*, 69(10): 1965–1974.

[30] Xu, C., F. Xu, B. Wang, T. Lu, 2011. Electrospinning of poly (ethylene-co-vinyl alcohol) nanofibres encapsulated with Ag nanoparticles for skin wound healing, *J. Nanomater.* 2011:1-7.

[31] Young T. H., Lin C. W., Cheng L. P., Hsieh C. C., 2001. Preparation of EVAL membranes with smooth and particulate morphologies for neuronal culture, *Biomaterials* 22(13): 1771– 1777.

[32] Peng C., Tsutsumi S., Matsumura K., Nakajima N., Hyon S. H., 2001. Morphologic study and syntheses of type I collagen and fibronectin of human periodontal

ligament cells cultured on poly (ethylene-co-vinyl alcohol) (EVA) with collagen immobilization, *J. Biomed. Mater. Res.* 54(2): 241–246.

[33] Sakurada Y., Sueoka A., Kawahashi M., 1987. Blood purification device using membranes derived from poly (vinyl alcohol), and copolymer of ethylene and vinyl alcohol, *Polym. J.*, 19 (5),501–513.

[34] Pattison M. A., Wurster S., Webster T. J., Haberstroh K. M., 2005. Three-dimensional, Nano-structured PLGA scaffolds for bladder tissue replacement applications. *Biomaterials* 26: 2491–2500.

[35] Kim H. J., Kim U. J., Leisk G. G., Bayan C., Georgakoudi I., Kaplan D. L., 2007. Bone regeneration on macroporous aqueous-derived silk 3-D scaffolds. *Macromol. Biosci* 7: 643–655.

[36] Sokolsky-Papkov M., Agashi K., Olaye A., Shakesheff K., Domb A. J., 2007. Polymer carriers for drug delivery in tissue engineering. *Adv. Drug. Deliv. Rev.* 59: 187–206.

[37] Shin M., Abukawa H., Troulis M. J., Vacanti J. P., 2008. Development of a biodegradablescaffold with interconnected pores by heat fusion and its application to bone tissue engineering. *J. Biomed. Mater. Res. A84*: 702–709.

[38] Yoon J. J., Song S. H., Lee D. S., Park T. G., 2004. Immobilization of cell adhesive RGD peptide onto the surface of highly porous biodegradable polymer scaffolds fabricated by a gas foaming/salt leaching method. *Biomaterials* 25: 5613–5620.

[39] Yamada K. M., Kennedy D. W., 1984. Dualistic nature of adhesive protein function: Fibronectin and its biologically active peptide fragments can auto inhibit fibronectin function. *J. Cell Biol.* 99: 29–36.

[40] Ali Alghamdi Abdulaziz, Alattas Hussain, Saeed Waseem Sharaf, Al-Odayni Abdel-Basit, Alrahlah Ali, Aouak Taieb, 2020. Preparation and Characterization of Poly (ethylene-co-vinyl alcohol)/poly (ε-caprolactone) blend for bio scaffolding Applications, *Int. J. Mol. Sci.* 21: 588

[41] Shang M. X., Matsuyama H., Teramoto M., 2003. Preparation and membrane performance of poly (ethylene-co-vinyl alcohol) hollow fiber membrane via thermally induced phase separation, *Polymer* 44: 7441–7447.

[42] Fu S. S., Matsuyama H., Teramoto M., Lloyd D. R., 2003. Preparation of microporous polypropylene membrane via thermally induced phase separation as support of liquid membranes used for metal ion recovery, *J. Chem. Eng. Jpn.* 36: 1397–1404.

[43] Shang M. X., Matsuyama H., Maki T., Teramoto M., Lloyd D. R., 2003. Effect of crystallization and liquid-liquid phase separation on phase-separation kinetics in poly (ethylene-co-vinyl alcohol)/glycerol solution, *J. Polym. Sci. Polym. Phys.* 41: 194–201.

[44] Matsuyama H., Iwatani T., Kiitamura Y., Tearamoto M., Sugoh N., 2001. Formation of porous poly (ethylene-co-vinyl alcohol) membrane via thermally induced phase separation, *J. Appl.Polym. Sci.* 79: 2449–2455.

[45] Matsuyama H., Iwatani T., Kitamura Y., 2001. Solute rejection by poly (ethylene-co-vinyl alcohol)membrane prepared by thermally induced phase separation, *J. Appl. Polym. Sci.* 79: 2456–2463.

[46] Loeb G. S., Sourirajan S., 1963. Seawater demineralization by means of an osmotic membrane, *Adv. Chem. Ser*. 38: 117–132.

[47] Riyasudheen N., Sujith A., 2012. Formation behavior and performance studies of poly (ethylene-co-vinyl alcohol)/ poly (vinyl pyrrolidone) blend membranes prepared by non-solvent induced phase inversion method, *Desalination* 294: 17–24.

[48] Boom R. M., 1992. *Membrane formation by immersion precipitation: The role of a polymeric additive*. University of Twente, 230.

[49] Cheng Liao-Ping, Young Tai-Horng, Chuang Wen-Yuan, Chen Li-Yen, Chen Leo-Wang, 2001. The formation mechanism of membranes prepared from the nonsolvent–solvent–crystalline polymer systems, *Polymer* 42: 443–45.

[50] Ismail A. F., Norida R., Sunarti A. R., 2002. Latest development on the membrane formation for gas separation *Songklanakarin J. Sci. Technol*., 24(Suppl.): 1025-1043.

[51] Mulder M., 1996. *Basic Principles of Membrane Technology*, Kluwer Academic Publishers, Dordrecht, 74 pp.

[52] Witte P. Van de, Dijkstra P. J., Berg J. W. A. Van den, Feijen J., 1996. Phase separation processesin polymer solutions in relation to membrane formation, *J. Membr. Sci.* 117: 1–31.

[53] Lloyd D. R., Kim S. S., Kinzer K. E., 1991. Microporous membrane formation via thermally induced phase separation. II. L–L phase separation, *J. Membr. Sci*. 64: 1–11.

[54] Arya R. K., 2013. Drying Induced Phase Separation, *Journal of Chemical Engineering*, 27 (1): 12–20.

[55] Tylkowski Bartosz, Peris Sergio, Giamberini Marta, Garcia-Valls Ricard, Reina Jose A., Ronda Joan C., 2010. Light-Induced Switching of the Wettability of Novel Asymmetrical Poly (vinyl alcohol)-co-ethylene Membranes Blended with Azobenzene Polymers, *Langmuir* 26(18): 14821–14829.

[56] *Nanoscience Instruments–Electrospinning*: https://www.nanoscience.com/techniques/electrospin/.

[57] Mayuri P. V., Bhatt Anugya, Sabareeswaran A., Ramesh P., 2021. Effect of membrane parameters and filter structure on the efficiency of leukocyte removal by electrospun poly (ethylene-co-vinyl alcohol) membranes, *Journal of Biomaterials Science*, Polymer Edition 32(5): 595-612.

Editors' Contact Information

Dr. Kalpana Awasthi
Assistant Professor,
Department of Physics, K. N. Govt. P. G. College,
Gyanpur, Bhadohi, India
Email: awasthi.k@gmail.com.

Dr. Arti Srivastava
Assistant Professor,
Department of Chemistry, Guru Ghasidas Vishwavidyalaya,
Bilaspur, India
Email: artifeb@gmail.com.

Dr. Mridula Tripathi
Associate Professor,
Department of Chemistry, CMP Degree College,
Prayagraj, India
Email: mtcmpau@gmail.com.

Index

F

G

H